BIOMECHANICS
— OF —
SKELETAL MUSCLES

Vladimir M. Zatsiorsky, PhD
Pennsylvania State University

Boris I. Prilutsky, PhD
Georgia Institute of Technology

Human Kinetics

Library of Congress Cataloging-in-Publication Data

Zatsiorsky, Vladimir M., 1932-
  Biomechanics of skeletal muscles / Vladimir M. Zatsiorsky, Boris I. Prilutsky.
    p. ; cm.
  Includes bibliographical references and index.
  ISBN-13: 978-0-7360-8020-0 (hardcover)
  ISBN-10: 0-7360-8020-1 (hardcover)
  I. Prilutsky, Boris I., 1957- II. Title.
  [DNLM: 1. Muscle, Skeletal--physiology. 2. Biomechanics--physiology. 3. Movement--physiology. WE 500]
  612.74045--dc23

2011039905

ISBN-10: 0-7360-8020-1 (print)
ISBN-13: 978-0-7360-8020-0 (print)

The web addresses cited in this text were current as of August 2011, unless otherwise noted.

Acquisitions Editor: Loarn D. Robertson, PhD; **Managing Editor:** Bethany J. Bentley; **Assistant Editors:** Steven Calderwood and Derek Campbell; **Copyeditor:** Julie Anderson; **Indexer:** Susan Danzi Hernandez; **Permissions Manager:** Martha Gullo; **Graphic Designer:** Fred Starbird; **Graphic Artist:** Yvonne Griffith; **Cover Designer:** Keith Blomberg; **Art Manager:** Kelly Hendren; **Associate Art Manager:** Alan L. Wilborn; **Illustrations:** Courtesy of Vladimir M. Zatsiorsky, Boris I. Prilutsky, and Julia Kleyman unless otherwise noted; **Printer:** Thomson-Shore, Inc.

Printed in the United States of America   10 9 8 7 6 5 4 3 2 1

The paper in this book is certified under a sustainable forestry program.

Human Kinetics
Website: www.HumanKinetics.com

United States: Human Kinetics
P.O. Box 5076
Champaign, IL 61825-5076
800-747-4457
e-mail: humank@hkusa.com

Canada: Human Kinetics
475 Devonshire Road Unit 100
Windsor, ON N8Y 2L5
800-465-7301 (in Canada only)
e-mail: info@hkcanada.com

Europe: Human Kinetics
107 Bradford Road, Stanningley
Leeds LS28 6AT, United Kingdom
+44 (0) 113 255 5665
e-mail: hk@hkeurope.com

Australia: Human Kinetics
57A Price Avenue
Lower Mitcham, South Australia 5062
08 8372 0999
e-mail: info@hkaustralia.com

New Zealand: Human Kinetics
P.O. Box 80
Torrens Park, South Australia 5062
800 222 062
e-mail: info@hknewzealand.com

E4720

To Rita—for 54 years of love and 51 years of caring.

—V.Z.

To my daughter, Anya—for inspiration.

—B.P.

CONTENTS

PREFACE

This book is the third volume of a three-volume series on the biomechanics of human motion. The first two volumes, *Kinematics of Human Motion* (1998) and *Kinetics of Human Motion* (2002), and this book share a common format. The books have been prepared to be studied in sequence, each book as a one-semester course, for example.

All volumes are designed for readers interested in technical interpretations of human movement, specifically for graduate students in human movement science and neighboring fields such as exercise and sport science, biomechanics, motor control, movement physiology, biomedical engineering, ergonomics, and physical and occupational therapy. Despite a rather detailed exposition, for graduate students this should still be an introductory-level text. Students interested in specific areas of muscle biomechanics are advised to also consult research monographs and review papers.

At this time a great deal of knowledge on skeletal muscle biomechanics has been accumulated, but it is scattered among the literature and is not yet fully organized. Although several basic textbooks are available, there is a substantial gap between the content of these elementary courses and the cutting edge of research. This has made it difficult for some students to understand the related scientific literature. The objective of the book is to begin filling that void.

Skeletal muscles are biological motors that exert force and produce mechanical work. When considered together with the tendons, muscles also act as force transmitters and shock absorbers. There are numerous other, less obvious muscle functions. For instance, muscles produce heat and assist in thermoregulation. Muscle receptors contribute to the kinesthesia, the sense of one's body posture and mechanical state. This book concentrates, however, solely on muscles' biomechanical functions. Its scope is from sarcomere to human motion. Molecular mechanisms of muscle force generation are not covered, as there already exist many excellent books on this subject.

The science of biomechanics is based on biological experiments and mechanical models. The text of the book reflects this duality. An attempt has been made to maintain a proper balance between mathematical sophistication and experimental evidence. Although we wanted to make the book understandable to a wide audience, we also wanted to prepare readers for independent understanding of the scientific literature. Hence, we constantly struggled with too hard (complex), too soft (simple), and just right. Given that students have different educational backgrounds, in particular different levels of mathematical proficiency, it was impossible to find one level of complexity that is appropriate for everyone. The texts that are just right for some students, for instance, the students with limited mathematical schooling, may look trivial for students with engineering diplomas. As compared with the previous volumes, the present book is more biological: it is mainly based on experimental facts collected in biomechanical research. For readers with a strong mathematical background, the book includes sections of advanced material. These sections—marked by 《 at the beginning and 》 at the end of a section—could be bypassed by those readers who would prefer to have a conceptual overview of the issues. Following is a brief sketch of the chapter contents.

The book consists of two parts. Part I addresses the mechanical behavior of single muscles—from the sarcomere to the entire muscle level. Chapter 1 describes muscle architecture—the internal anatomy of the muscles. Until recently this information was obtained from cadavers, but recent developments in experimental techniques such as magnetic resonance imaging and ultrasound recording have allowed for in vivo muscle architecture studies. Muscle architecture is not completely identical across individuals (everyone has his or her anatomy), yet the traditional approach has been to average anatomical data. With newer imaging methods, individual differences in muscle architecture can be addressed and potentially used to explain interindividual variations in motor task performance. Evidence suggests that interindividual differences are larger than was previously thought. Still, in the present textbook they are mainly neglected, in order to keep the book volume in reasonable limits. Chapter 2 deals with the mechanical properties of tendons and passive muscles. Passive muscle properties are a benchmark for comparison with active muscle properties. Chapter 3 addresses the biomechanics of active muscles. Muscles work not only as force and power generators but also as force transmitters and shock absorbers. Chapter 4 is intended to familiarize the students with the force transmission and shock absorption aspects of muscle action, that is, eccentric action.

Part II is concerned with in vivo muscle behavior. It addresses the various issues of muscle functioning in human motion. Chapter 5 deals with the transformation from muscle force to joint moments. Chapter 6 discusses the function of two-joint muscles in human movements. Chapter 7 is devoted to eccentric muscle action in human motion. Chapter 8 considers biomechanical aspects of muscle coordination, in particular muscle redundancy and optimization.

Besides the main text, Refreshers and From the Literature boxed sections are included. Refreshers, as their name implies, are brief reminders of what the readers are assumed to already know. Those who are not familiar with the highlighted concepts should consult special literature. From the Literature texts are practical examples that illustrate the concepts under discussion. Section numbers are boldfaced. By necessity, in-text citations and reference lists are limited. We apologize to our colleagues if we have unintentionally been unable to cite your work.

While writing the book the authors tackled a challenge of limiting the book size. Obesity is a serious problem not only for people and nations but also for books. Large, fatty books are difficult to read. The line should be drawn somewhere. The initial plan was to write a textbook for graduate students and other newcomers to the field of biomechanics of human motion. After 12 years of work and the three volumes published, the first author admits that a comprehensive book on biomechanics of human motion has not been written. Even more, such a book that would include an extensive range of topics, from muscle forces and intra-abdominal pressure to posture and locomotion, has not even been started. [I earnestly hope for others to craft such a book. Now when I am approaching the ninth decade of my life, the tender age of infirmity, and work with two eyeglasses interchangeably, alas this task is not for me. I have to look for new interests in my life—wish me good luck. V.Z.]

We are greatly indebted to the readers of the first two volumes who have sent comments and reported errors. We humbly ask the readers of the present volume to do likewise.

Vladimir M. Zatsiorsky
Boris I. Prilutsky

Acknowledgments

The authors are indebted to many colleagues whose ideas and experimental results contributed to the development of biomechanics of skeletal muscles, a fast-growing field of scientific knowledge. We are under a special debt of gratitude to Dr. Mark L. Latash (Penn State University, USA), Dr. Todd C. Pataky (Shinshu University, Japan), and Dr. Thomas J. Burkholder (Georgia Institute of Technology, USA) for reading the entire text and making suggestions. We thank Drs. Huub Mass (Vrije Universiteit Amsterdam, the Netherlands) and Dilson J.E. Rassier (McGill University, Canada) for reviewing selected chapters of the book and Drs. Robert J. Gregor (Georgia Institute of Technology and University of Southern California, USA) and T. Richard Nichols (Georgia Institute of Technology, USA) for insightful discussions of muscle function.

I

MUSCLE ARCHITECTURE AND MECHANICS

Part I addresses mechanical behavior of a single muscle—from a sarcomere to the entire muscle. The following terminology is adopted. The term *muscle architecture* is used to describe the internal anatomy of the muscles, that is, how the muscles are built. The term *muscle morphometry* denotes the anatomical characteristics of the muscles in the body, such as location of their origin and insertion and moment arms.

Essentially, skeletal muscles are biological motors that exert force and produce mechanical work. In their coupling with the tendons, muscles also act as force transmitters and shock absorbers. Above and beyond these kinetic roles, muscles have numerous other functions. For instance, they produce heat and assist in thermoregulation. Muscle receptors contribute to kinesthesia, or body sense. However, this book concentrates solely on muscles' kinetic functions.

Muscles can exert forces while

- maintaining a constant length (static or isometric action),
- shortening (concentric or miometric action), or
- lengthening (eccentric or plyometric action).

In the preceding expressions, the suffix *metric* means length, the prefix *iso* means same or constant, *mio* means less, and *plio* (*pleo*) means more. In the United States sport literature, *plyometrics* has become a common spelling, with *pliometrics* as an alternative. The term *concentric* means "coming to the center" and *eccentric* means "away from the center." These terms are not perfect, and such terms as *shortening contraction* and *lengthening contraction* have been suggested instead. This is a matter of personal taste, however. For some users, but not for others, the word *contraction* means shortening. For such users the term *shortening contraction* is a pleonasm and the term *lengthening contraction* is an oxymoron.

Research on Muscle Mechanics

Muscle mechanics have been studied *in vivo* (within a living body, as it is), *in situ* (in the original biological location but with partial isolation), and in *vitro* (isolated from a living body and artificially maintained, literally "in glass"). Given evident technical limitations and ethical restraints, not all measurements can be performed on human muscles. Experiments on the individual muscles of nonhuman mammals have also their own limits. It is difficult to maintain muscles' biological life when they are isolated from the body for long periods of time. In contrast, skeletal muscles of amphibians when placed in physiological solution keep their properties for several hours. Therefore, many experiments on muscle mechanics have been performed on the muscles of frogs and toads, with the frog sartorius muscle being the most popular object of research. Although some of the main mechanical properties of the muscles of amphibians, nonhuman mammals, and humans are similar, this similarity is limited. Without direct measurements on live human muscles, similarity and dissimilarity cannot be established. This book describes the properties of human skeletal muscles unless stated otherwise.

Chapter 1 discusses the muscle architecture. Biomechanical properties of tendons and passive muscles are addressed in chapter 2. Biomechanics of active muscles are discussed in chapter 3, whereas chapter 4 deals with the eccentric muscle action.

1

MUSCLE ARCHITECTURE

This chapter addresses the main architectural characteristics of the skeletal muscles. The main mechanically relevant feature of muscle architecture is the arrangement of muscle fibers relative to the axis of force generation. Until recently, this information for human muscles could only be obtained from cadavers. The bodies were dissected, abundant connective and fat tissues were removed to reveal the muscle fibers, and muscles were then studied as separate entities. The interconnections among the muscles were sometimes noted, but their functional importance was commonly overlooked. Only basic anatomy was studied, and the adjustments occurring during force production and movement, for instance, fiber length changes, could not be measured. The state of affairs started changing in the 1990s. Developments in experimental techniques such as magnetic resonance imaging (MRI) and ultrasound recording have allowed study of muscle architecture in living people. Muscle imaging is a booming area of research with a large potential that is still at its beginning. At present, the detailed architecture of the majority of human muscles (>600!) is not investigated and remains unknown.

Muscle architecture is not completely identical across individuals. The traditional approach is to average anatomical data. New imaging methods can address individual differences in muscle architecture, which potentially can be used to explain interindividual differences in performance of mechanical tasks. These differences undoubtedly exist, and evidence suggests that they are larger than were previously thought. However, in this book these differences are mainly neglected to keep the length of the book within reasonable limits.

Chapter 1 addresses geometry of the muscles and their constituents. The basic methods of kinematic geometry that are explained in the first two chapters of *Kinematics of Human Motion* are applicable for the description of muscle architecture. They are not sufficient, however; the geometry of some muscles is too intricate to be described by those global methods.

The chapter starts with the description of muscle fascicles and their arrangements (section **1.1**). Parallel-fibered and fusiform muscles (section **1.1.1**), pennate muscles (section **1.1.2**), and convergent and circular muscles (section **1.1.3**) are discussed. Muscle fascicle curvature is considered in section **1.2**. The Frenet frames used to describe geometry of curved lines in three-dimensional space are also explained here. Section **1.3** addresses the fiber architecture in the fascicles. The spanning and nonspanning fibers are described, and the concepts of myotendinous force transmission and myofascial force transmission are introduced. In section **1.4**, muscles are analyzed as fiber-reinforced composites. In the composites, the fibers possessing high tensile strength are immersed in another material (called *matrix*), which binds the fibers together and transfers internal loads. Section **1.5** discusses fiber, fascicle, and muscle length (section **1.5.1**) as well as the various length–length ratios (section **1.5.2**). Section **1.6** is concerned with muscle path and muscle centroids. The following issues are described in successive subsections: **1.6.1**—straight-line representation of muscle path, **1.6.2**—centroid model of muscle path, **1.6.3**—curved and wrapping muscles, **1.6.4**—twisted muscles, and **1.6.5**—muscles attached to more than two bones. Section **1.7** concentrates on cross-sectional areas of the muscles, both physiological and anatomical areas. Finally, section **1.8** deals with muscle attachment geometry. As with all other chapters in the book, chapter 1 ends with a Summary (section **1.9**), Questions for Review (section **1.10**), and Literature List (section **1.11**).

• • • ANATOMY REFRESHER • • •

Muscles

Skeletal muscles are made up of bundles of fibers (muscle cells) running alongside each other (figure 1.1). Individual fibers are enclosed in a tubular cellular membrane, *sarcolemma*, and surrounded by a sheath of collagenous tissue known as the *endomysium* (literally, "within muscle"). The endomysium is connected to the sarcolemma via the basal lamina, which is believed to play a special role in force transmission from the fiber. The basal lamina covers the muscle fibers and interfaces with the endomysium. Each fiber is bound to adjacent fibers to form bundles (called *fascicles* or in Latin *fasciculi*). The

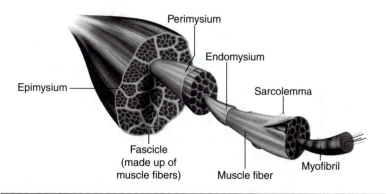

Figure 1.1 Main structural elements of a muscle.
© Human Kinetics.

fascicles are enclosed within the sheaths of connective tissue called *perimysium* (literally, "around muscle"). The perimysium bounds adjacent fascicles. The entire muscle is surrounded by a sheath of connective tissue called *epimysium* (literally, "on top of muscle"). The epimysium is continuous with *fascia*, a fibrous envelopment that binds adjacent muscles together.

Muscles are generally attached to bones by tendons. Flattened wide tendons are known as aponeuroses. An aponeurosis is often found in series with a tendon. A muscle together with the attached tendons is called the *muscle–tendon unit* or the *muscle–tendon complex*. A muscle itself, without the attached tendons, is often referred as the *muscle belly*. The muscle belly length equals the distance from the proximal to distal tendon. The proximal attachment of a muscle to a bone or other tissue is known as its *origin* and the distal attachment as the *insertion*. Muscles of a body segment are organized in groups, muscle compartments (*compartments* for short). Each compartment is coated by fascia (septum). The muscles in each compartment are served by the same nerve and have a common blood supply. For instance, the forearm has two compartments, anterior (supplied by the median nerve) and posterior (supplied by the radial nerve).

A muscle fiber contains threadlike *myofibrils*. Each myofibril consists of serially connected elements, *sarcomeres*, the smallest contractile units of the skeletal muscle separated one from another by the *Z-membranes* (*Z-lines*, *Z-discs*; from the German *Zwischenscheibe*,

(continued)

"the band in between"). Sarcomeres are composed of three major myofilaments, or simply filaments, which are long chains of protein subunits: *myosin*, in the middle; *actin*, at either side of the sarcomere; and *titin* (figure 1.2). The tension in the sarcomere is produced by the heads of myosin molecules that connect to the actin filaments (*cross-bridges*), rotate, and thereby pull on the actin filaments. Muscle fibers shorten when thin actin filaments slide alongside thick myosin filaments. Myosin and actin are therefore named the *contractile proteins*. The highly extensible titin filaments connect the thick filaments to the Z-membranes that are the terminal borders of the sarcomere. The myofibril extensibility is mainly attributable to the extensibility of titin

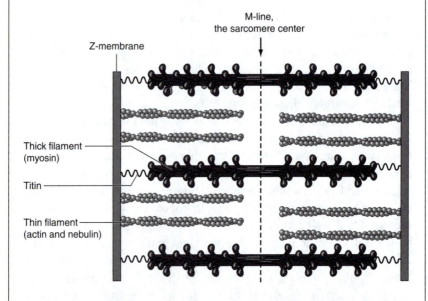

Figure 1.2 A schematic of a sarcomere. The thin filaments are composed of actin molecules arranged along the nonextensible protein nebulin, which makes the thin filaments inelastic. The thin filaments are attached to Z-membranes. The position of each myosin (thick) filament in the center of the sarcomere is maintained by titin filaments that connect the thick filaments to the Z-membranes. The sarcomere architecture is greatly oversimplified in this schematic.

© Human Kinetics.

filaments. For the whole muscle also, the parallel connective tissues are important. Titin exists in different structural variants (isoforms) that exhibit different extensibility.

Table 1.1 summarizes the data on the dimensions and number of various muscle elements.

Table 1.1   Muscle Elements in Numbers

Property	Dimension
Actin (thin) filament thickness	50-60 Å (angstroms, 1 Å = 10^{-10} m)
Actin filament length	2.0 µm (micrometer, 1 µm = 10^{-6} m)
Myosin (thick) filament thickness	110 Å
Myosin filament length	1.65 µm
Myofibril thickness	1-2 µm
Sarcomere length in humans, range	2.5-4.5 µm
Number of crossbridges at each end of a myosin filament	≈100
Maximum force that one crossbridge can exert	5.3 pN (pico Newton, pico—the factor 10^{-12})
The total maximum force at the center of the myosin filament	530 pN
Number of sarcomeres in a human muscle myofibril	
Tibialis	8,000-15,000
Soleus	14,000
Medial gastrocnemius	15,000-16,000
Thickness of a basic lamina	30-50 nm (nanometer, nano—the factor 10^{-9})
Thickness (diameter) of a muscle fiber	10-100 µm
Number of myofibrils in one muscle fiber	Up to 8,000
Number of fibers in human muscles (rounded to 1,000)	
First lumbrical	10,000
First dorsal interosseum	40,000
Brachioradialis	129,000
Tibialis anterior	271,000
Medial gastrocnemius	1,033,000
Density of muscle	1.054-1.060 g/cm³

Sources: Huxley 1966; Lieber 2009; Matthews 2002

1.1 MUSCLE FASCICLES
AND THEIR ARRANGEMENTS

Representative publications: Gans 1982; Otten 1988

Muscles exist in a variety of forms. In a single fascicle the fibers are parallel, but the arrangement of fascicles in the entire muscle can vary. Muscle fibers exert forces along their longitudinal axes, whereas muscles exert forces along tendons. In various muscles, fascicles are connected to tendons at different angles, and hence the direction of an individual muscle fiber force and the muscle force direction can be different.

All muscles are three-dimensional and should be studied as such. As early as 1667, Nicholas Steno (1638-1686) performed a geometric analysis of muscle contraction in three dimensions. He represented a muscle—we would call such a muscle unipennate, discussed subsequently—as a parallelepipedon (figure 1.3). However, for centuries muscle architecture has been described and analyzed in a plane. This planar representation greatly simplifies mathematical analysis of muscle performance but is not adequate for many muscles.

Skeletal muscles are commonly classified as *parallel-fibered, fusiform, pennate, convergent,* or *circular.* With the exception of circular muscles, this

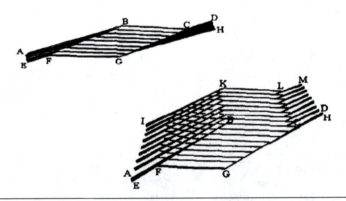

Figure 1.3 Modeling pennate muscles as simple geometrical objects. Upper panel, a planar model. Bottom panel, modeling in 3-D. A muscle as a parallelepipedon. In both models, the individual muscle fibers are of the same length and parallel. Closer to the bone connection, the tendons become thicker. The schematic is attributed to N. Stensen (1667).

From *Niels Stensen's Elements of Myology,* 1667. Reprinted from T. Kardell, 1994, *Steno on muscles: Introduction, texts, translation (Transactions of the American Philosophical Society),* vol. 84, part I (Darby, PA: Diane Publishing), p. 226.

Methods of Research

Cadaver Dissection

This is a classic method of muscle architecture research that has been used for centuries. The method is limited, however. Muscle architecture parameters such as fiber lengths and pennation angles change during muscle contraction: fibers shrink and angles change during tissue fixation. To study muscle architecture and morphometry in living people, several new methods have been developed.

Ultrasound Methods

These methods use high-frequency sound waves. Ultrasound devices operate at frequencies much larger than the limit of human hearing (approximately 20 kHz), usually in the range of 2 to 18 MHz. Because of wavelength properties, lower frequencies produce poorer resolution but image deeper into the body. In muscle research, ultrasound methods exist in two forms: *ultrasonography*, in which ultrasound waves bounce off tissues and are converted into an image, and *sonomicrometry*, which measures muscle fiber length based on the time for ultrasound propagation. The latter method is invasive—it requires inserting piezoelectric crystals into the muscle—and currently is not used in humans.

Magnetic Resonance Imaging (MRI)

The method uses a powerful magnetic field to align the nuclear magnetization of hydrogen atoms in the tissue. The orientation of the magnetization is then systematically changed by radio waves, causing the hydrogen atoms to produce a rotating magnetic field. The method yields good-quality images of the body in any plane. MRI exists in several variants, such as functional MRI, cine phase contrast MRI, and diffusion MRI imaging (reviewed in Blemker et al. 2007).

classification scheme is not very exact. These muscle types are not separate entities but rather the extreme cases of adaptable muscle architecture. Fiber arrangement in every muscle is unique, and many muscles do not fit a simple schema. All muscles except for circular muscles, when active, generate and transmit tension via tendons to bones. Forces generated by circular muscles are not transmitted to tendons; instead, tension is directed toward the center of the circle.

1.1.1 Parallel Fibered and Fusiform Muscles

In parallel-fibered and fusiform muscles, muscle fascicles are arranged along the line of muscle force action. Some muscles are flat bands with the fibers running in parallel. Such straplike or sheetlike muscles are known as parallel fibered. Some of these muscles are short (e.g., the intercostals interni), whereas others (e.g., the pronator quadratus, the rhomboids, and quadratus lumborum) are longer but otherwise no different. At the extreme of length are the sartorius, gracilis, pectoralis minor, rectus abdominis, obliquus externus abdominis, and semitendinosus. Some of the parallel-fibered muscles (e.g., the rectus abdominis and semitendinosus) are traversed by tendinous intersections (for the rectus abdominis, three in number), which serve as the place of insertion for fibers.

The fusimotor muscles are spindle-shaped; these muscles narrow down at the end. In the spindle-shaped muscles the fibers located closer to the outer muscle border are longer and curved; they surround a shorter straight fiber at the muscle core. The biceps brachii, biceps femoris, psoas, and teres muscles are examples of such a muscle. The centerline and peripheral fascicles in these muscles may have different functional properties (discussed in more detail in chapter 3).

In some parallel-fibered and fusiform muscles, especially in long ones, fibers do not run the whole length of their fascicles but rather end somewhere in the middle. As a result, fiber length can be much smaller than the fascicle length. A question to inspire your curiosity: How can fiber force be transmitted to the tendon in a muscle where the fibers do not run from one end of the fascicle to another? This is discussed later in sections **1.3** and **3.1.5**.

1.1.2 Pennate Muscles

In pennate (or pinnate) muscles, fascicles are attached to the tendon at an angle (figure 1.4). Pennate muscles resemble feathers; their name comes from the Latin *penna* for "feather." In such muscles, fibers lie at an angle to the line of action of the entire muscle, and the directions of fiber shortening and tendon movement are different.

As compared with parallel and fusiform muscles, pennate muscles provide some mechanical advantages as well as disadvantages. The main advantage is that more fibers can be filled into a given volume of muscle. Hence, the total force of all the muscle fibers can be greater than in parallel-fibered muscles of the same volume. However, given the oblique orientation of the muscle fibers with respect to the muscle line of action, only a component of fiber force contributes to the muscle force. Furthermore, fibers in pennate muscles are shorter and have fewer sarcomeres in series than in parallel muscles of equal volume, so their maximum displacements and velocities are smaller.

▪ ▪ ▪ *FROM THE LITERATURE* ▪ ▪ ▪

Why Do We Have Pennate Muscles?

Source: Leijnse, J.N. 1997. A generic morphological model of a muscle group—application to the muscles of the forearm. *Acta Anat (Basel)* 160(2):100-111

The main factors driving pennate muscle evolution may be purely geometric: the total area of the skeleton is too small to provide the required origin area for all muscle fibers of the body. Muscle fibers should find their attachment at the insufficient surface area of the skeleton. This is achieved by enlarging the skeletal surface via muscle origin aponeuroses. Aponeuroses provide flexible surfaces of origin that permits muscle shape alteration to fit the space available in the body. The aponeurotic sheets must be thin enough to find their origins on the small available skeletal areas yet strong enough to transmit muscle forces to the bone. For the purpose of illustration, consider a bone with no surface area at all, like a stick diagram. If a thin aponeurosis is attached to such a stick, it occupies only a line, but it can provide a place for attachments of many muscle fibers (figure 1.4). With such an arrangement, one-dimensional structures (like "stick" bones) are transformed into two-dimensional ones.

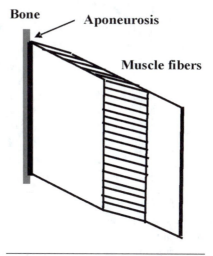

Figure 1.4   Unipennate muscle model attached to a one-dimensional skeletal segment through the aponeurosis.

1.1.2.1 Planar Models of Pennate Muscles

Depending on the number of fiber directions, the pennate muscles are classified as unipennate, bipennate (or herringbone), and multipennate. The extensor digitorum muscle, an extrinsic hand muscle that extends the finger joints, and the peroneus longus, the muscle that performs ankle plantar flexion and eversion, are unipennate; the rectus femoris is bipennate; and the deltoid and serratus anterior are multipennate. Some muscles (e.g., the extensor carpi radialis brevis and the flexor carpi radialis) have partly unipennate and partly bipennate fiber arrangements. In the pennate muscles, the tendons have two parts: internal and external (figure 1.5, *b* and *c*). The internal tendons are also addressed as the *aponeuroses* or *tendinous sheets*.

In pennated muscles, individual fibers exert force on the aponeurosis in parallel, whereas along the aponeurosis the forces are transmitted in series with each fiber force being added to the fiber forces that precede it. As a result, the parts of the aponeurosis (internal tendon) that are located closer to the external muscle tendon transmit larger forces. This explains

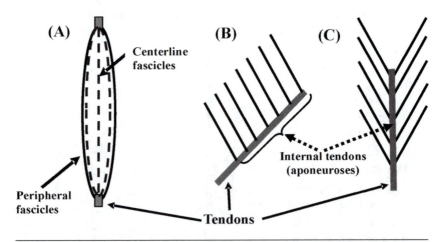

Figure 1.5 Muscle fiber arrangements in the muscles. *(a)* Fusimotor (spindle-shaped) muscle, *(b)* unipennate, and *(c)* bipennate. In *b* and *c*, the external and internal tendons (aponeuroses) are shown. In the spindle-shaped muscles *(a)*, the centerline fascicles and the fascicles on the muscle periphery have different length. Depending on the fascicle location in the muscle—closer to or farther from the centerline—the fascicles also have different curvature.

why the aponeurosis usually becomes thicker from the muscle center to the muscle edge.

Unipennate muscles come in two main varieties: with one or two internal tendons (figure 1.6). The one-tendon unipennate muscles (figure 1.6*a*) have the second attachment with the bone. Such muscles generate at the origin both a force and a force couple (they pull the bone and exert a moment of force on it). At the insertion, they have internal and external tendons in the same direction. In the two-tendon muscles the internal and external tendons are in different directions (figure 1.6, *b* and *c*). Such muscles do not generate a couple, only pulling force. These muscles exist in two forms: (a) with parallel fibers of equal length (e.g., semimembranosus) and (b) with fibers of different lengths and different pennation angles (oblique muscles; e.g., within the clavicular head of pectoralis major muscle, the fiber lengths increase from superior to inferior).

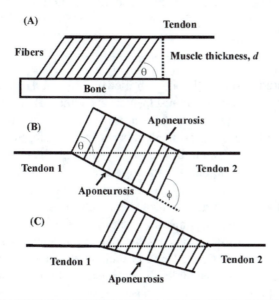

Figure 1.6 Unipennate muscles. *(a)* One-tendon muscle; *d* is muscle thickness. Such a muscle can generate both a force and a moment of force on the bone. *(b)* Two-tendon muscle with parallel fibers, of equal length. θ is the angle between the line of muscle force action and muscle fiber, and φ is the angle between aponeurosis and muscle fiber. *(c)* Two-tendon muscle with the fibers of unequal length (the oblique unipennate muscle).

■ ■ ■ *FROM THE LITERATURE* ■ ■ ■

Pectoralis Major Consists of a Uniform Clavicular Head and a Segmented Sternal Head With Six or Seven Segments (Bundles)

Source: Fung, L., B. Wong, K. Ravichandiran, A. Agur, T. Rindlisbacher, and A. Elmaraghy. 2009. Three-dimensional study of pectoralis major muscle and tendon architecture. *Clin Anat* 22(4):500-508

Eleven cadaveric specimens were examined. The muscle belly consisted of an architecturally uniform clavicular head and a segmented sternal head with six or seven segments. Within the clavicular head, the fiber bundle lengths increased from superior to inferior, whereas the mean fiber bundle lengths of sternal head were greatest in segments three through five, found centrally. The mean lateral pennation angle was greater in the clavicular head (29.4° ± 6.9°) than in the sternal head (20.6° ± 2.7°).

The magnitude of pennation is quantified by *pennation angle*. This is the angle between the direction of muscle fibers and either (a) the line of muscle force action (external portion of the tendon) or (b) the aponeurosis (internal tendon) (figure 1.6*b*). The first method is generally preferred. However, in the muscles of complex architecture, such as the soleus, because of technical difficulties the angle often is measured relative to the aponeurosis. If not mentioned otherwise, in this book pennation angle refers to the fiber direction relative to the line of muscle force action.

In different muscles, the pennation angle varies from 0° (parallel-fibered muscles) to 30°. In many muscles the pennation angle is not uniform throughout the muscle (figure 1.6*c*); in such muscles the individual fibers pull the tendons at different angles. For instance, in the soleus some muscle fascicles are close to being horizontal whereas others are almost vertical. The architectural data on some muscles are presented in table 1.2.

The pennation angle is subject to change. The angle increases with muscle hypertrophy induced by strength training or anabolic administration as well as during growth. The angle decreases when muscle size decreases, for example, with advanced age or disuse atrophy. The angles of pennation are larger in males than in females. This difference reflects the difference in muscle thickness. Because strength training increases muscle thickness and pennation angle, pennate muscles are generally associated with a positive correlation between muscle force and pennation angle.

Table 1.2   Architecture of Lower Extremity Muscles, Examples

Muscle and the number of studied specimen	Muscle length, cm	Fiber length, cm	Lf coefficient of variation, %	Sarcomere length, μm	Pennation angle, °	Physiological cross-sectional area, cm²
Psoas, $n = 19$	24.25 ± 4.75	11.69 ± 1.66	12.4 ± 5.9	3.11 ± 0.0.28	10.6 ± 3.2	7.7 ± 2.3
Gluteus medius, $n = 16$	19.99 ± 6.42	7.33 ± 1.57	20.3 ± 11.8	2.40 ± 0.18	20.5 ± 17.3	33.8 ± 14.4
Sartorius, $n = 20$	44.81 ± 4.19	40.30 ± 4.63	6.4 ± 4.2	3.11 ± 0.19	1.3 ± 1.8	1.9 ± 0.7
Rectus femoris, $n = 21$	36.28 ± 4.73	7.59 ± 1.28	9.7 ± 4.6	2.42 ± 0.30	13.9 ± 3.5	13.5 ± 5.0
Vastus lateralis, $n = 19$	27.34 ± 4.62	9.94 ± 1.76	9.1 ± 6.1	2.14 ± 0.29	18.4 ± 6.8	35.1 ± 16.1
Vastus medialis, $n = 19$	43.90 ± 9.85	9.68 ± 2.30	10.7 ± 5.7	2.24 ± 0.46	29.6 ± 6.9	20.6 ± 7.2
Tibialis anterior, $n = 21$	25.98 ± 3.25	6.83 ± 0.79	6.6 ± 4.0	3.14 ± 0.16	9.6 ± 3.1	10.9 ± 3.0
Gastrocnemius medial head, $n = 20$	26.94 ± 4.65	5.10 ± 0.98	13.4 ± 7.0	2.59 ± 0.26	9.9 ± 4.4	21.1 ± 5.7
Gastrocnemius lateral head, $n = 20$	22.35 ± 3.70	5.88 ± 0.95	15.8 ± 11.2	2.71 ± 0.24	12.0 ± 3.1	9.7 ± 3.3
Soleus, $n = 19$	40.54 ± 8.32	4.40 ± 0.99	16.7 ± 6.9	2.12 ± 0.24	28.3 ± 10.1	51.8 ± 14.9

The authors studied 27 muscles from 21 human lower extremities (height of the specimen 168.5 ± 9.3 cm, mass 82.7 ± 15.3 kg). Values in the table are expressed as mean ± standard deviation. Lf is the fiber length normalized to a sarcomere length of 2.7 μm. More comprehensive data on architectural properties of human upper and lower extremity muscles are presented in Lieber (2009).

The data are from Ward et al. 2009.

When a joint angle changes, the distance from the muscle–tendon origin to insertion *(muscle length)* also changes. What is happening with the pennation angle in this case is not well established. Some authors have reported that in relaxed muscles, pennation angles depend on the joint position, whereas other authors have not been able to confirm this relation.

During muscle contraction, with or without joint movement, pennation angle increases; that is, the fibers rotate. Rotation of the fibers about their origin as they shorten is considered the main functional feature distinguishing pennate muscles from parallel-fibered and fusiform muscles. The fibers of some parallel fibered muscles (e.g., sartorius) and fusiform muscles (e.g., biceps brachii and brachioradialis) insert at an angle into the tendons, but the fibers move purely in translation. In contrast, the seemingly parallel fibers of the medial gastrocnemius rotate as they shorten, and hence the muscle is considered pennate fibered. Analysis of fiber shortening and rotation in pennate muscles is presented later in the text.

■ ■ ■ *FROM THE LITERATURE* ■ ■ ■

Pennation Angle From Newborn to the Elderly

Source: Binzoni, T., S. Bianchi, S. Hanquinet, A. Kaelin, Y. Sayegh, M. Dumont, and S. Jequier. 2001. Human gastrocnemius medialis pennation angle as a function of age: from newborn to the elderly. *J Physiol Anthropol Appl Hum Sci* 20(5):293-298

The aim was to quantify changes in human skeletal muscle pennation angle values during growth and adult life. The human gastrocnemius medialis muscle of 162 subjects (96 males and 66 females) in the age range 0 to 70 years was scanned with ultrasonography. The pennation angle increased monotonically starting from birth (0 years) and reached a stable value after the adolescent growth spurt. The present findings indicate that pennation angle evolves as a function of age.

Pennation Angles Are Greater in Bodybuilders

Source: Kawakami, Y., T. Abe, and T. Fukunaga. 1993. Muscle-fiber pennation angles are greater in hypertrophied than in normal muscles. *J Appl Physiol* 74(6):2740-2744

Muscle-fiber pennation angles were measured with the use of ultrasonography in the long and medial heads of the triceps brachii in 32 male subjects (from untrained subjects to highly trained bodybuilders). The pennation angles were determined as the angles between the fibers and an aponeurosis (angle ϕ in figure 1.6). The pennation angles were in the range of 15° and 53° for the long head and 9° to 26° for the medial head. Significant differences were observed between normal subjects and bodybuilders in muscle thickness and pennation angles ($p < .01$), and there were significant correlations between muscle thickness and pennation angles for both long ($r = .884$) and medial ($r = .833$) heads of triceps, suggesting that muscle hypertrophy involves an increase in fiber pennation angles.

1.1.2.2 Pennation in Three Dimensions

Representative publications: Scott et al. 1993; Benard et al. 2009

The data cited previously are obtained from planar measurements, commonly performed with a simple protractor. Usually only average pennation angles of superficially located fibers are measured. This approach is sufficiently accurate for the flat, wide muscles. However, many muscles have complex three-dimensional architecture and should be studied in 3-D. The angle of pennation in 3-D is usually defined as the angle between the tangent vector of the fascicle and the tangent vector of the aponeurosis at the point of contact (i.e., the angle φ not θ is determined). The angle is measured in the plane formed by these two vectors, the pennation plane. If both the fascicles and the aponeurosis are the straight-line segments, the pennation plane can be determined from three-dimensional muscle images. The orientation of the pennation plane is defined with respect to a skeleton. Unfortunately, when some imaging techniques (e.g., ultrasonography) are used, this plane is not known before the examination and if it is not determined properly, only the projection of the pennation angle on the plane of measurement (visualization plane, image plane) is recorded. Evidently, the actual pennation angle is larger than its projections on any plane (figure 1.7).

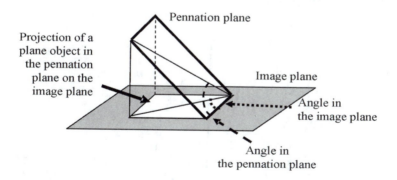

Figure 1.7 The angles measured in the pennation plane and in the image plane (visualization plane). Imaging is performed in the top-down direction. The angles measured in the actual plane of pennation and in the visualization plane (image plane) may differ from each other.

■ ■ ■ *FROM THE LITERATURE* ■ ■ ■

Ultrasound Probe Orientation Affects the Results of Muscle Architecture Measurements

Source: Klimstra, M., J. Dowling, J.L. Durkin, and M. MacDonald. 2007. The effect of ultrasound probe orientation on muscle architecture measurement. *J Electromyogr Kinesiol* 17(4):504-514

To determine the effect of probe orientation on the measurement of muscle architecture parameters, the human tibialis anterior muscle was imaged from nine different probe orientations during concentric contractions at four joint angles. Probe orientations were changed ±5° in two directions. The measurements included pennation angle, fascicle length, and muscle thickness. Statistically significant differences were found between probe orientations.

To avoid distortion and compute muscle architecture correctly, the imaging should be done in at least two projections. The three-dimensional orientation of a fascicle within a muscle can be determined from its projections on two orthogonal planes, for example, planes *XOZ* and *YOZ* (see figure 1.8). The orientation of a fascicle with length *l* is described by two angles: θ, the fascicle angle relative to the horizontal plane, and φ, the angle relative to the *YOZ* plane.

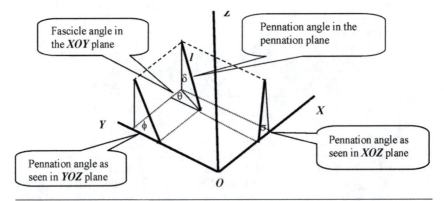

Figure 1.8   The pennation angle in 3-D and its derivation from two orthogonal projections.

Adapted from S.H. Scott, C.M. Engstrom, and G.E. Loeb, 1993, "Morphometry of human thigh muscles: Determination of fascicle architecture by magnetic resonance imaging," *Journal of Anatomy* 182(Pt 2): 249-257.

If the projection angles of the unknown pennation angle δ onto the planes *XOZ* and *YOZ* are σ and φ, respectively, the angle δ can be computed as

$$\delta = \tan^{-1}\left(\frac{\tan^2 \sigma \tan^2 \phi}{\tan^2 \sigma + \tan^2 \phi} \right) \qquad [1.1]$$

and the angle θ as

$$\theta = \sin^{-1}\left(\frac{\tan \delta}{\tan \sigma} \right). \qquad [1.2]$$

The derivation is based on elementary geometry and is left to the readers.

■ ■ ■ FROM THE LITERATURE ■ ■ ■

Fascicle Architecture From Images in Two Projections

Source: Scott, S.H., C.M. Engstrom, and G.E. Loeb. 1993. Morphometry of human thigh muscles. Determination of fascicle architecture by magnetic resonance imaging. *J Anat* 182(Pt 2):249-257

The striation patterns spanning individual muscles in longitudinally oriented magnetic resonance (MR) images were used to measure the orientation of muscle fascicles. The angles of striations were measured at several positions within vastus medialis and semimembranosus from sagittal and frontal plane MR images. Mathematical techniques described previously were used to infer the 3-D orientation of fascicles based on these striation angles. The pennation angle, defined as the angle between the fascicles and the line of action of the muscle, predicted from the MR images, was similar to directly measured values. The pennation angle varied along the length of the muscle; in vastus medialis, pennation angle ranged from 5° to 50° in a proximodistal direction. Procedures were developed and validated to compute fascicle length by projection of fascicle orientation across the three-dimensional shape of the muscles.

When pennation angle in a single image plane is used to calculate muscle force, only a projection of the force exerted by the fascicles onto this plane is determined whereas other force components are disregarded (for help understanding this, peruse figure 1.2 in *Kinetics of Human Motion*, p. 7).

1.1.3 Convergent and Circular Muscles

Convergent muscles, for instance, pectoralis major, infraspinatus (the muscle rotates arm laterally, figure 1.9), subscapularis (rotates arm medially), serratus

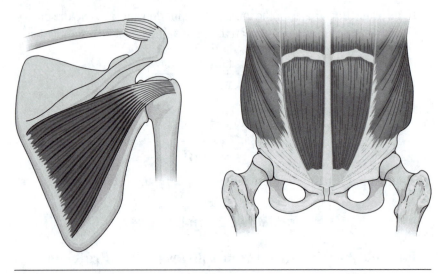

Figure 1.9 Examples of convergent muscles: infraspinatus, posterior view, and the inferior portion of the rectus abdominis.

anterior, deltoid, temporalis, obturator externus, and the inferior portion of the rectus abdominis, are approximately triangular in shape. They are broad at the origin and converge to a small attachment area at the insertion, giving to the muscle a fanlike appearance. Because of that they are also called the *fanned muscles* or *triangular muscles*. Muscles of this type are mainly located at the trunk. The individual fibers of a convergent muscle pull in dissimilar directions and hence their action at the joint can be different. The action may also depend on the joint position. The convergent muscles can also be seen as multipennate.

In a circular muscle, the fibers are oriented concentrically around an opening. When the muscle contracts, the opening decreases. An example is the orbicularis oris muscle of the mouth.

1.2 MUSCLE FASCICLE CURVATURE: FRENET FRAMES

Representative publications: Muramatsu et al. 2002; Blemker et al. 2005

In section **1.1**, muscle fibers and fascicles were considered straight-line segments, whereas the description was mainly based on the planar approach. However, many muscles are curved and have complex three-dimensional geometry. The curvature, if it exists, depends on the force exerted by the muscle and its current length (i.e., on the joint angle).

■ ■ ■ **FROM THE LITERATURE** ■ ■ ■

Curvature of Muscle Fascicles Depends on the Exerted Force and Joint Position

Source: Muramatsu, T., T. Muraoka, Y. Kawakami, A. Shibayama, and T. Fukunaga. 2002. *In vivo* determination of fascicle curvature in contracting human skeletal muscles. *J Appl Physiol* 92(1):129-134

Fascicle curvature of human medial gastrocnemius muscle was determined in vivo by ultrasonography during isometric contractions at three (distal, central, and proximal) locations and at three ankle angles. The curvature significantly increased from rest to maximum voluntary contraction (0.4-5.2 m^{-1}). In addition, the curvature at maximum voluntary contraction became larger in the order dorsiflexed, neutral, plantar flexed (i.e., the curvature increased with the decrease in muscle length). Thus, both contraction levels and muscle length affected the curvature. Fascicle length estimated from the pennation angle and muscle thickness, under the assumption that the fascicle was straight, was underestimated by ~6%.

A comment from the book authors: With increased force, the curvature of an isolated fascicle or fiber is expected to decrease. In contrast, in the present study the curvature increased with force. This can be due to increased pressure from the neighboring fascicles and muscles or to the fact that with increasing force, the muscle belly shortens more because of tendon stretch.

The elementary geometric models of the fascicles are not always sufficient to describe their architecture and performance. Only simplified intuitive explanation of the mathematical methods is provided here. The description starts with replacing a muscle fiber (or fascicle, muscle line of action) with an imaginary flexible band, a curved line in space. In experimental research, the band coordinates are usually measured at selected points, discretely. Hence, the first task is to represent the band as a smooth curve (a continuous function). This is achieved by using spline functions (*splines* for short). Mathematicians borrowed the term from shipbuilders: In craftsmanship, a spline is a long, pliable strip of metal or wood used to make smooth curves. In numerical analysis, the spline is a curve composed of pieces that pass through or near the connecting points while keeping the entire curve as smooth as possible. Various formulas can be used for splines, but the cubic splines whose pieces are polynomial functions of the third degree are used most often. The smoothness of cubic splines is achieved by making the polynomial pieces as well as their first and second derivatives continuous at the connecting points.

■ ■ ■ FROM THE LITERATURE ■ ■ ■

Muscle Architecture in Three Dimensions

Source: Blemker, S.S., and S.L. Delp. 2005. Three-dimensional representation of complex muscle architectures and geometries. *Ann Biomed Eng* 33(5):661-673

Magnetic resonance images of the psoas, iliacus, gluteus maximus, and gluteus medius muscles were obtained from a single female subject at two different leg positions. The boundaries of the muscles and bones of interest were outlined manually. Muscle images taken from 4 to 7 mm thick slices distributed 1 to 3 mm apart were digitized. The four template fiber geometries describing the trajectory of muscle fibers in a given part of the muscle were introduced: parallel simple fibers, pennate fibers, parallel curved fibers, and fanned fibers. The templates consisted of interpolated spline curves. The architecture of specific muscles was determined (figure 1.10) by establishing a correspon-

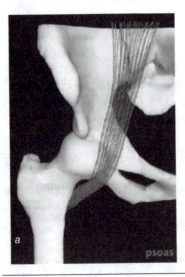 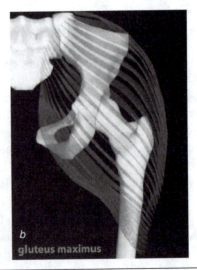

Figure 1.10   Fiber geometries mapped to the *(a)* psoas, *(b)* gluteus maximus.

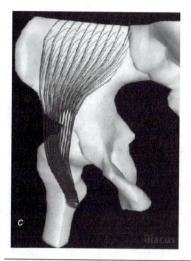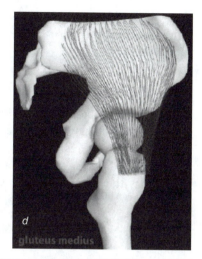

Figure 1.10 *(continued)*   *(c)* iliacus, and *(d)* gluteus medius.

With kind permission from Springer Science+Business Media: *Annals of Biomedical Engineering,* "Three-dimensional representation of complex muscle architectures and geometries," 33(5), 2005, 661-673, S.S. Blemker and S.L. Delp, fig. 5.

dence between the templates and the muscle images such that each material coordinate in the template has a corresponding coordinate in the muscle. It is evident from the figure that modeling the analyzed muscles as planar structures consisting of straight-line muscle fibers would be an oversimplification.

For the description of curved lines in 3-D, the method of choice is *Frenet frames* (after French mathematician Jean Frédéric Frenet, 1816-1900). This method has an advantage of line curvature computation (inverse of the radius). If a force is transmitted along the band, the method allows for computing the lateral forces acting normally from the band on the neighboring tissues (provided that the force acting along the band is known).

The Frenet frames describe the curved line in space locally, at a given point on the line. Each frame consists of a point on the line and three orthogonal vectors—the vectors that are perpendicular to one another and have unit length (figure 1.11). The frame can be seen as moving along the line. There is a clear analogy between describing a line in space and describing the trajectory of a moving point. Any of these can be described parametrically by three coordinates $X, Y, Z = f(s)$, where s is the distance along the curve from the origin or another parameter such as time. For easy visualization, we assume that a curve represents the trajectory of a point in time. This will allow us to use such intuitively simple terms as velocity and acceleration.

The three orthogonal vectors of a Frenet frame are as follows:

1. The tangent vector **T** is a unit vector in the direction of the derivative at the point of interest. If the curve *u* represented a trajectory of a point in time, the tangent vector would be a unit vector in the direction of the instantaneous velocity **u̇** of the point.

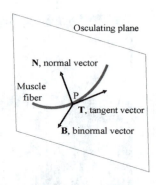

Figure 1.11 Definition of a Frenet frame. The frame is defined at point P.

$$T = \frac{\dot{u}}{\|\dot{u}\|}$$

[1.3]

where $\|\dot{u}\|$ is the magnitude of vector **u̇**. For the muscle fiber, the tangent vector defines the fiber direction at a given point.

2. The normal vector **N** is in the direction of the normal acceleration

$$N = \frac{\ddot{u}}{\|\ddot{u}\|}$$

[1.4]

For muscle fibers, the normal vector is perpendicular to the fiber direction. Because it points toward the center of curvature, the normal vector is also called the *curvature vector*. The normal vector indicates the deviance of the curve from a straight line. Recall that curvature is the change in direction per unit of arc (see *Kinematics of Human Motion*, p. 227). Arc is any part of the curve; it can be represented as a product *arc = radius (r) × angle* (in radians). The curvature is then an inverse of the radius, *curvature = 1/r*.

The plane spanned by the tangent and normal vectors is known as the *osculating plane*. The single-plane hypothesis of muscle architecture suggests that each muscle fiber lies in only one osculating plane. To picture this hypothesis, suppose a part of a curved muscle fiber is in the X-Y plane of a certain system of coordinates. Then, none of the fiber parts lies in the X-Z or Y-Z.

3. The *binormal vector* **B** is perpendicular to the tangent and normal vectors. It is found as the cross-product of **T** and **N**.

$$B = T \times N$$

[1.5]

Frenet frames are also called TNB frames. The main limitation of Frenet frames is that they cannot be computed for straight (noncurved) lines. The reader is invited to explain why this happens. A hint: Recall the definition of the vector cross-product.

1.3 FIBER ARCHITECTURE IN THE FASCICLES

Representative publications: Trotter et al. 1995; Huijing 1999; Fukunaga et al. 1997b; Hijikata and Ishikawa 1997

In contrast to sarcomeres, whose dimensions and shape are almost identical in all skeletal muscles, muscle fibers occur in a variety of forms. There are two main types of muscle fibers: (a) those that span along the entire fascicle, from the tendon of origin to that of insertion, and (b) those that end somewhere in the middle of the fascicle. They are characteristically named *spanning* and *nonspanning fibers*, respectively.

Spanning fibers connect to tendons or aponeuroses at the ends of the fascicle. The relatively short muscles with a small distance between the tendons, such as intrinsic hand muscles or some pennate muscles (e.g., the medial and lateral gastrocnemii and soleus), consist almost entirely of spanning fibers. The trapezius and pectoralis minor also belong to this group. In these muscles, the fibers run along the whole fascicle, and the muscle fiber length is similar to the fascicle length. Of course, in pennate muscles the fascicle length and the muscle length are different.

The sites where muscle fibers connect with collagen fibers of the tendon are called *myotendinous junctions*. The junctions are arranged in series with fibers and serve as sites of force transmission. The force transmission along the fiber axis to the tendon at the myotendinous junctions is called the *myotendinous force transmission, serial force transmission,* or *longitudinal force transmission*. A comment on terms: longitudinal force transmission requires a serial connection of muscle fibers and the corresponding tendon; muscle fibers in fascicles are themselves arranged in parallel.

Muscles with nonspanning fibers are *series fibered*. The series-fibered muscles should be distinguished from the muscles with tendinous intersections (TI), such as rectus abdominis and semitendinosus. In the muscles with TIs, fibers are linked end to end through the TIs, which serve as places of force transmission. Hence, the fibers go either from a tendon to a TI or from a TI to a TI. In the series-fibered muscles, the fibers are too short to reach from the tendon of origin to the tendon of insertion; they terminate somewhere in the fascicle and are serially linked. This arrangement is typical for the long muscles such as the sartorius, gracilis, and brachioradialis. For instance, the sartorius at a length of 50 to 60 cm has the longest span among human muscles, but the majority of its fibers are only 5 to 18 cm long. When a fiber is much shorter than the fascicle, at least one end of the fiber is not connected to the myotendinous junction. The fiber force is transmitted to the tendon through other mechanical connections. The force exerted by a nonspanning fiber can be transmitted to other fibers and only then to the tendon. It is also possible that the forces exerted by nonspanning fibers are transmitted first in the lateral direction to the surrounding connective tissue (endomysium) via shear force. This type of force transmission is called *lateral force transmission* or, because muscle fascia is involved in the process, *myofascial force transmission*.

Nonspanning fibers usually taper near the free end that terminates within the fascicle; fiber cross-sectional area decreases at the end (figure 1.12). The ends are very narrow: the angle between the longitudinal fiber axis and the fiber surface in the vicinity of the fiber free end can be 1°. The area of such fiber tips is very small. If a fiber force were transmitted longitudinally through such a sharp fiber end, it would cause membrane damage because of the enormous pressure in the contact area. For this reason, at the free ends of nonspanning fibers, serial force transmission is not feasible. Nevertheless, tapering is advantageous for force transmission (explained in section **3.1.5**).

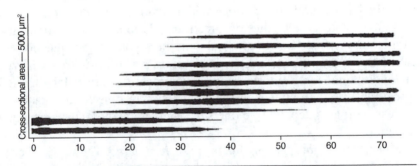

Figure 1.12 Fibers within a motor unit from one fascicle. All fibers taper at one or both ends. Cat tibialis anterior muscle.

From M. Ounjian, R.R. Roy, E. Eldred, et al., 1991, "Physiological and developmental implications of motor unit anatomy," *Journal of Neurobiology* 22(5): 547-559. © 1991 John Wiley & Sons. Reprinted by permission of John Wiley & Sons, Inc.

■ ■ ■ *FROM THE LITERATURE* ■ ■ ■

Tension Along the Muscle Is Transmitted Using Both In-Series and In-Parallel Arrangements

Source: Paul, A., J. Rodda, M. Duxson, and P. Sheard. 2000. Examination of intrafascicular muscle fiber terminations: Implications for tension delivery in series-fibered muscles. *Journal of Morphology* 245(2):130-145

Mammalian skeletal muscles with long fascicle lengths are predominantly composed of short muscle fibers that terminate midbelly with no direct connection to the muscle origin or insertion. The manner in which these short fibers terminate and transmit tension through the muscle to their tendons is poorly understood. The authors made an extensive morphological study of a series-fibered muscle of a guinea pig. Intrafascicularly terminating fibers end about equally often in either a long progressive taper or a series of small or larger blunt steps. However, when analyzed at higher resolution, the apparently smooth progressive tapers appear also to be predominantly composed of a series of fine stepped terminations. Tension from intrafascicularly terminating fibers is likely to be passed along the muscle to the tendon using both in-series and in-parallel arrangements.

It is not presently known what portion of the fiber's force is transmitted to the tendon and bone through the myotendinous path.

● ● ● *ANATOMY AND PHYSIOLOGY REFRESHER* ● ● ●

Motor Units

A *motor unit* (MU) is a collection of muscle fibers innervated by the same *motoneuron*, a neural cell in the ventral part of the spinal cord or in the brain (in the case of facial muscles). Motoneuron activity excites all fibers in the MU.

(continued)

Anatomy and Physiology Refresher *(continued)*

Motor units differ in their properties. Slow, or slow-twitch, MUs are specialized for prolonged use at relatively slow velocities. They consist of (a) small, low-threshold motoneurons with low discharge frequencies; (b) axons with relatively low conduction velocities; and (c) muscle fibers adapted to lengthy aerobic activities. Fast, or fast-twitch, MUs are specialized for brief periods of activity characterized by large power outputs, high velocities, and high rates of force development. They consist of (a) large, high-threshold motoneurons with high discharge frequencies; (b) axons with high conduction velocities; and (c) muscle fibers adapted to explosive or anaerobic activities. A more detailed classification includes three types of muscle fibers: Type I, slow; Type IIA, fast but fatigue resistant; and Type IIX, fast with low resistance to fatigue. All human muscles contain all three types of MUs in various proportions. The maximal shortening velocity of fast muscle fibers is almost 4 times greater than the maximal velocity of slow muscle fibers. The difference in force producing capacity among the MUs can be 100-fold.

To control muscle force, the central nervous system activates the MUs according to the size of their motoneurons, from small motoneurons at low forces to large motoneurons at high forces (the *size principle*). Because small motor neurons innervate slow-twitch fibers whereas large motor neurons innervate fast-twitch fibers, the size principle implies that at low forces only slow muscle fibers are active. Force gradation within one MU is accomplished through firing rate changes (*rate coding*).

The fibers from one motor unit may be distributed in a fascicle quite differently. The distribution of muscle fibers affects the capacity of motor units to generate force. The total force of a group of fibers is essentially the sum of forces produced by the fibers arranged in parallel to one another, not in series. Therefore, if motor units in the sartorius (nonspanning fibers) and the trapezius (spanning fibers) have the same number of fibers, it can be expected that a single motor unit in the sartorius would produce a smaller force than a unit with the same number of fibers in the trapezius.

Two extreme fiber arrangements are shown in figure 1.13. Two motor units, A and B, are shown. In the top panel, all fibers from one unit are connected through myotendinous junctions to the tendon at one side, either at the origin or at the insertion. In the bottom panel, the fibers of the same motor unit are equipped with myotendinous junctions at both the origin and insertion. In the first case, each motor unit has the largest number of fibers at one level; see the top panel of figure 1.13. Such an arrangement allows each MU to generate a large force. The force, however, should be transmitted from the free end of the fiber to the tendon. It can be done, for instance, via the second motor unit. This requires simultaneous coordinated activation of both motor units, A and B, and myofascial force transmission. Otherwise, the force generated by one MU will not be transmitted to the second tendon. In the arrangement shown in the bottom panel, the fibers are distributed along the long axis of the fascicle. The force exerted by each MU is smaller than in the first case—because of the smaller cross-sectional area—but the forces are transmitted to both tendons via the myofascial and myotendinous pathways. In fact, a mixture of various fiber arrangements is found in different muscles.

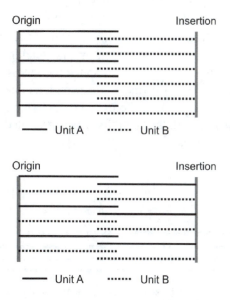

Figure 1.13 Two extreme arrangements of the muscle fibers of a motor unit. Fibers from two motor units, A and B, are shown. An arrangement shown at the top allows each motor unit to generate larger force, but the transmission of the force both to the origin and to the insertion is not possible without activating other motor units or using myofascial pathways. In the arrangement at the bottom panel, each motor unit exerts a smaller force transmitted to both tendons through the myotendinous pathways.

Muscles with spanning fibers are innervated through one motor endplate located near the middle of the muscle fibers. Muscles with nonspanning fibers are innervated through two or more motor endplates distributed along the muscle length. Muscles of newborns have several motor endplates that become more widely spaced with growth.

1.4 MUSCLE AS A FIBER-REINFORCED COMPOSITE

Representative publication: Huijing 1999

Muscles possess mechanical properties of composite materials that are common in engineering. This is especially true when muscles are considered together with adjoining connective tissues. The composite materials (*composites* for short) are made from two or more materials. They display the best characteristics of each of the components. The components can be put together in different ways. In the so-called *fiber-reinforced composites*, the fibers made of a material with high tensile strength are embedded in another material (called *matrix*), which glues the fibers together and transfers internal loads (stresses). The matrix is more extensible than the fibers. Fiber-reinforced composites are used in aircrafts, racing cars, and advanced sports equipment like racing bicycles and some tennis rackets. The most and familiar natural composite material is wood; its fibers are made of cellulose. Fiber-reinforced composite materials may have either continuous or discontinuous fibers. In the latter, force is transmitted longitudinally via the matrix (figure 1.14).

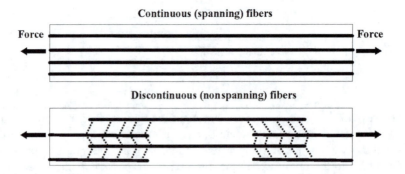

Figure 1.14 Force transmission in the composites with continuous (spanning) and discontinuous (nonspanning) fibers. Oblique dotted lines represent the matrix connections.

In muscles, intramuscular connective tissues (i.e., the endomysium and perimysium) form an amalgamated flexible frame, or lattice, the *muscle matrix*. The endomysium that surrounds each muscle fiber is a part of this matrix that holds the muscle structure together. The muscle can be considered a composite consisting of an extracellular matrix enforced by (a) an active component (i.e., fibers that generate force and bear loads) and two passive components—(b) the basal lamina and (c) the muscle fascias. The myofascial pathways of force transmission exist at three levels of muscular organization: within muscle fibers, within a muscle, and within a muscle compartment (figure 1.15).

Not only a single muscle but also a group of muscles can be considered a fiber-enforced composite. Because neighboring muscles in a compartment are interconnected via connective tissue, they may interact mechanically with each other. This fact inspires questions regarding the limitations of the anatomical nomenclature used for muscles. The traditional nomenclature is based on the epimysial system. In anatomy classes, muscle anatomy is studied by dissecting one muscle from another using the epimysium as a criterion for separation. With such an approach, the muscle is considered a morphological unit. It is not evident, however, whether the muscle is only a morphological unit or also a functional unit used by the central nervous system to control movements. This intriguing issue is discussed later, but we have to admit right away that we cannot answer this question.

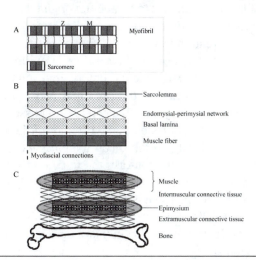

Figure 1.15 Myofascial pathways of force transmission. *(a)* Within muscle fibers. *(b)* Within a muscle. *(c)* Within a fascial compartment.

Reprinted, by permission, from H. Maas, 2003, "Myofascial force transmission" (PhD dissertation, Vrije Universiteit, Amsterdam).

After Tendon Transfer, Rectus Femoris Becomes an Antagonist to Itself

Source: Riewald, S.A., and S.L. Delp. 1997. The action of the rectus femoris muscle following distal tendon transfer: Does it generate knee flexion moment? *Dev Med Child Neurol* 39(2):99-105

Rectus femoris transfer surgery involves detaching the rectus femoris from the patella and reattaching it posterior to the knee. This procedure is thought to convert the rectus femoris from a knee extensor to a knee flexor. However, prior to this study, moments generated by this muscle after transfer had not been measured. The authors stimulated the rectus femoris via intramuscular electrodes in four subjects, two after transfer to the semitendinosus and two after transfer to the iliotibial band, while measuring the resultant knee moment. Electromyographic activity was monitored in the quadriceps, hamstrings, and gastrocnemius muscles to verify that the rectus femoris was the only muscle activated by the stimulus. It was found that the rectus femoris generated a knee extension moment in all of the subjects tested. This finding suggests that transfer surgery does not convert the rectus femoris to a knee flexor and that a mechanism exists that transmits rectus femoris forces anterior to the knee joint center after distal tendon transfer.

The following explanation has been offered (Huijing 1999). After surgery, the distal part of the rectus femoris transmits force via the longitudinal pathway, whereas the proximal part—which was not separated during the surgery from the neighboring tissues—transmits force via the lateral transmission path. The part of the muscle force transmitted longitudinally to the distal tendon generates a flexion moment at the knee joint. In contrast, the force transmitted laterally is transmitted through the connective tissue to the knee extensors. Hence, part of the muscle force is used for knee flexion and the other for knee extension. The force recorded externally is the resultant of these two forces. It can be said that in this situation, the muscle works as a self-antagonist.

1.5 FIBER, FASCICLE, AND MUSCLE LENGTH: LENGTH–LENGTH RATIOS

Muscle excursion is proportional to the excursion of its fibers. Hence, fiber length affects the feasible muscle excursion.

1.5.1 Fiber and Fascicle Length

Representative publications: Friden et al. 2004; Fukunaga et al. 1997a; Herbert et al. 2002

Muscle fibers are long and thin. Their *aspect ratio* (length/diameter) is very large, commonly more than 1,000. For instance, for a fiber of 20 cm length (even longer fibers have been found in human sartorius) and 50 μm diameter, the aspect ratio is huge: 4,000. In such long fibers there should be about 100,000 serially connected sarcomeres (assuming a sarcomere length of 2μm). In some muscles (e.g., tibialis anterior) the fibers are the same length throughout, whereas in other muscles (e.g., sartorius, see table 1.1) the length of individual fibers can be sharply different. Muscles with fibers of similar length are at times called *homogeneous muscles* or *muscles with uniform architecture.*

■ ■ ■ *FROM THE LITERATURE* ■ ■ ■

Contradictory Data on the Length of Muscle Fibers

Sources: (1) Maganaris, C.N., V. Baltzopoulos, and D. Tsaopoulos. 2006. Muscle fibre length-to-moment arm ratios in the human lower limb determined in vivo. *J Biomech* 39(9):1663-1668. (2) Agur, A.M., V. Ng-Thow-Hing, K.A. Ball, E. Fiume, and N. McKee,. 2003. Documentation and three-dimensional modelling of human soleus muscle architecture. *Clin Anat* 16(4):285-293. (3) Huijing, P.A. 1985. Architecture of the human gastrocnemius muscle and some functional consequences. *Acta Anat (Basel)* 123(2):101-107

The first study reports that fibers obtained from measurements in various regions of such muscles as the gastrocnemius medialis, gastrocnemius

(continued)

From the Literature   *(continued)*

lateralis, vastus lateralis, vastus intermedius, and soleus have similar length within the muscle. These muscles are claimed to be homogeneous. In contrast, the second study reports that the soleus is not a homogeneous muscle. The length of muscle fibers in different regions of the soleus ranges from 16 to 45 mm. Both studies cited literature data that supported their views. Large variations in fiber length within the human gastrocnemius were also found in the third study. Unfortunately, readers are left to guess which conclusion is correct.

Separating individual fibers from a muscle tissue is technically difficult and requires special skills. Therefore, fascicle (bundle) length rather than the fiber length is typically measured. However, in muscles with nonspanning fibers, the bundle length and the fiber length can be quite different. When interpreting the literature, the reader should pay attention to what was actually measured in a given study: fiber length or fascicle length. The ubiquity of nonspanning fibers has been recognized only recently; in the past researchers did not appreciate the difference between fascicle and fiber lengths.

Fiber length depends on muscle length, fiber pennation, and the spanning pattern of the fibers in the fascicles, whether the fibers are spanning or not. Fascicle length depends only on muscle length and fiber pennation.

1.5.2 Length–Length Ratios

Representative publications: Woittiez et al. 1985; Zajac 1989

The following ratios are used to characterize muscle architecture.

• The fascicle length can be much smaller than the entire length of the fleshy part of the muscle–tendon complex. The ratio of mean fascicle length to the muscle belly length at optimal muscle length is known as the *index of architecture*. The index of architecture is an indirect measure of the fiber pennation. The *optimal muscle length* is the length at which maximal force can be produced. For muscles with spanning fibers, the fiber and fascicle length are the same and the index of architecture can be defined as the ratio: mean fiber length–muscle belly length. Because maximal muscle excursions are proportional to fiber length, muscles with high fiber length–muscle length ratio (e.g., sartorius, semitendinosus, gracilis muscles) are well suited for large excursions and hence a high velocity of shortening.

■ ■ ■ *FROM THE LITERATURE* ■ ■ ■

Muscle Fiber Versus Muscle Length

Source: Friedrich, J., and R.A. Brand. 1990. Muscle fiber architecture in the human lower limb. *J Biomech* 23(1):91-95

The authors studied the architecture of 47 muscles and muscle parts in two cadavers. They compared their data with the data of Wickiewicz et al. (1983; 3 cadavers) and Weber and Weber (1846). Within a given muscle, fiber lengths were remarkably similar. Only a few muscles had standard deviations greater than 10%. For a given muscle, the ratios of the fiber length to the muscle length (i.e., the index of architecture) were consistent among specimens and studies. Among the muscles, the ratios varied substantially (table 1.3).

Table 1.3   Fiber Length to Muscle Length Ratios for Some Leg Muscles

Muscle	Mean	Standard deviation
Rectus femoris	0.20	0.02
Vastus lateralis	0.23	0.04
Vastus medialis	0.23	0.03
Gastrocnemius, medialis	0.16	0.02
Gastrocnemius, lateralis	0.25	0.05
Sartorius	0.88	0.09

Combined data from Friedrich and Brand (1990) and Wickiewicz et al. (1983).

• The ratio of tendon slack length to optimal fiber length is often computed (*normalized tendon length*, $L_{T,Slack} / L_o$). The *tendon slack length* is the threshold length at which a stretched tendon begins to develop force. The *optimal fiber length* is the length at which the active fiber generates maximal force. For muscles with zero pennation and spanning fibers, the fiber length equals the muscle belly length. Hence, for these muscles the ratio represents slack tendon length normalized by muscle belly length. Tendons are elastic: when extended, they store the potential energy of deformation that can later be used for movement. For a given muscle–tendon unit length, the longer the tendon, the larger the amount of the stored potential energy.

Table 1.4 The Normalized Tendon Length for Human Muscles (after Zajac 1989)

Muscle	Normalized tendon length
Plantar flexors	
Soleus	11
Gastrocnemius	9
Others	7
Dorsiflexors	3
Quadriceps	
Vasti	3
Rectus femoris	5
Hamstring	
Semitendinosus	2
Semimembranosus	7
Hip one-joint muscles	0.2
Biceps and triceps brachii	1-2
Wrist and extrinsic finger flexors	3-8

Hence, normalized tendon length is an indirect measure of the ability of the muscle–tendon complex to store and recoil the potential energy of deformation. The biomechanical aspects of such energy transformations are considered in chapters 4 and 7. The normalized tendon length of some human muscles is presented in table 1.4.

• The larger the normalized tendon length, the longer the tendon and the shorter the muscle fibers. Such architecture is typical for the proximally located muscles that serve distal joints. The proximal muscle location allows for the distal parts to be slim. It reduces the mass and moments of inertia of the distal body segments (e.g., the fingers and feet). It also allows for large elastic potential energy accumulation in the tendons. Animals that are good runners, such as horses, have bulky muscles near the torso and slender legs with very long elastic tendons. This arrangement reduces inertial resistance and thus helps horses to run faster.

• The ratio of muscle belly length to muscle–tendon unit length is used to compute the speed of muscle shortening or lengthening in human movements. In force–velocity equations (discussed in chapter 3), muscle velocity is understood to be the rate of change of the muscle belly length, not the muscle–tendon unit length. Therefore, the muscle–tendon unit length is commonly multiplied by this ratio.

■ ■ ■ *FROM THE LITERATURE* ■ ■ ■

Regular Wearing of High Heels Changes the Tendon-to-Fascicle Length Ratio

Source: Csapo, R., C.N. Maganaris, O.R. Seynnes, and M.V. Narici. 2010. On muscle, tendon and high heels. *J Exp Biol* 213(15):2582-2588

Wearing high heels places the calf muscle–tendon unit in a shortened position. Because muscles and tendons are highly malleable tissues, chronic use of high heels might induce structural and functional changes in the calf muscle–tendon unit. To test this hypothesis, the authors recruited 11 women who regularly wore high heels and a control group of 9 women. Gastrocnemius medialis fascicle length, the Achilles tendon length, and some other variables were assessed in both groups. Shorter gastrocnemius medialis fascicle lengths were observed in the group who wore high heels (49.6 ± 5.7 mm vs. 56.0 ± 7.7 mm), resulting in greater tendon–fascicle length ratios.

1.6 MUSCLE PATH: MUSCLE CENTROIDS

Representative publications: Jensen and Davy 1975; Jensen and Metcalf 1975; Tsuang et al. 1993; Gao et al. 2002

Forces exerted by the muscle are transmitted from one attachment site to another along a certain line of action, usually referred to as a *muscle path.* The knowledge of muscle path allows us to determine muscle length, the speed of muscle shortening or lengthening, the force that the muscle exerts in the lateral direction on adjacent structures, and muscle moment arm. These variables are all important in muscle mechanics.

Based on the path, muscles can be categorized as (a) muscles with a straight line path or (b) curved muscles. Muscles with a curved path wrap around bones or other tissues. Straight muscles exert forces along their lines of action on the bones of origin and insertion only. Curved muscles also exert forces toward the center of curvature. For instance, when the upper arm is abducted below 90°, the deltoid exerts an inward force on the humerus. Such transverse forces may either stabilize or destabilize the joint.

Determining muscle path is challenging even for a single-joint configuration: muscle geometry changes when muscles contract, muscles interact with neighboring structures (bones, other muscles that can change shape when active), and lateral forces and friction can affect the muscle path. These difficulties are amplified when we attempt to determine muscle path for many joint configurations. Several approaches are used either separately or in combination to determine muscle paths, including the (a) straight-line model, (b) centroid model, (c) via-point model, and (d) Cosserat model.

1.6.1 Straight-Line Representation of Muscle Path

The simplest way to represent the muscle path is to model it as a straight line joining the geometric centers of the muscle attachment areas, the origin and insertion. Such a straight-line approximation, however, is not accurate for many muscles. Muscles have complex shapes that change during contraction. On their way from the origin to insertion, the muscles may wrap over bones, pass through a tendon sheath, and change their direction. Only some muscles can be accurately modeled as straight lines from the origin to insertion.

▪ ▪ ▪ *FROM THE LITERATURE* ▪ ▪ ▪

Straight-Line Representation of Hip Musculature

Source: Dostal, W.F., and Andrews, J.G. 1981. A three-dimensional biomechanical model of hip musculature. *J Biomech* 14(11):803-812

The authors assumed that the force generated by a muscle lies along a straight line connecting the approximate center of the origin on the pelvis and the approximate center of the attachment on the femur. The straight-line model was selected because it is conceptually simple and easy to use. (The authors also claimed that the model is "reasonably accurate.") The centers of muscle attachments were marked on the bony pelvis of a male dry bone specimen. Two-joint muscles crossing both the hip and knee joints were given fictitious distal attachment points on the femur. Two muscles, the obturator internus and iliopsoas, that wrapped around the underlying bony pelvis prior to crossing the hip were also assigned fictitious pelvis attachments at the points of muscle contact with the bony pelvis.

To test model validity, the investigators fixed elastic strings at the muscle attachment points and moved the femur manually through a full range of hip joint angles. A string was considered to validly represent the muscle path if no significant interaction or contact was observed between the string and the underlying bone or other strings. The most significant interaction occurred for the string representing the gluteus maximus when the hip was flexed more than 5° from the zero joint configuration. At these angles, the string was wrapped around the underlying bony pelvis, the femur, and the external rotator. The model was used to determine the moment arms of the hip muscles in three dimensions.

Straight-line representations of muscle paths allow for uncomplicated determination of muscle moment arms with the classic methods of vector analysis (the details are provided in chapter 5).

1.6.2 Centroid Model of Muscle Path

Representative publications: Jensen and Davy 1975; Tsuang et al. 1993

In this method, the *muscle centroid*—the locus of the geometric centers of the muscle transverse cross-sections—is determined. The method involves a three-stage procedure: (a) cross-sectional images (slices) of the muscle are obtained at small intervals along the muscle; (b) for each thin slice, the location of the geometric centers of cross-sections is determined; and (c) the locus of geometric centers along the muscle, the muscle centroid, is viewed as the muscle path. The centroid represents the muscle line of action provided that the force per unit of cross-sectional area is the same for all regions of a muscle slice. The centroids are occasionally modeled as frictionless elastic bands. Examples of the muscle centroids are presented in figure 1.16.

The centroid method allows for a more realistic representation of a muscle's line of action than does the straight-line model (figure 1.17). Unfortunately, for some muscles the method is difficult to use in practice: the location of the geometric centers of the muscles cross-sections is not easy to obtain. If the cross-sections are determined on a cadaver specimen, the measurements correspond to only one muscle length. Hence, the variation of the muscle centroid with the joint position remains unknown.

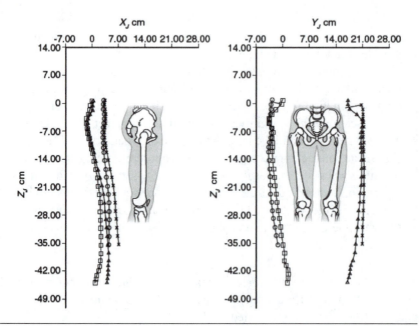

Figure 1.16 Rectus femoris centroids in two projections. The rectus femoris inserts into the quadriceps tendon in the distal two-thirds of the thigh, and at this point the tendon outline was difficult to define. Therefore, the centroid in the graph terminates a short distance above the knee. As compared with the curved centroid, the moment arms of the muscle forces for the straight line from the origin to insertion are less by 1% to 12%.

Reprinted from *Journal of Biomechanics* 8(2), R.H. Jensen and D.T. Davy, "An investigation of muscle lines of action about the hip: A centroid line approach vs. straight line approach," 103-110, copyright 1975, with permission from Elsevier.

■ ■ ■ *FROM THE LITERATURE* ■ ■ ■

Centroids of Trunk Muscles Displace During Twisting

Source: Tsuang, Y.H., G.J. Novak, O.D. Schipplein, A. Hafezi, J.H. Trafimow, and G.B. Andersson. 1993. Trunk muscle geometry and centroid location when twisting. *J Biomech* 26(4-5):537-546

The trunk muscles of the lumbar region were studied using magnetic resonance imaging. The cross-sectional areas of the muscles were measured from the transverse scans at 1 cm intervals from L2 to S1 and the muscle centroids determined. All muscle centroids were displaced when twisting. Thus, the locations and orientations of all muscle lines of action changed. The changes in the muscle moment arms were sometimes as great as twofold.

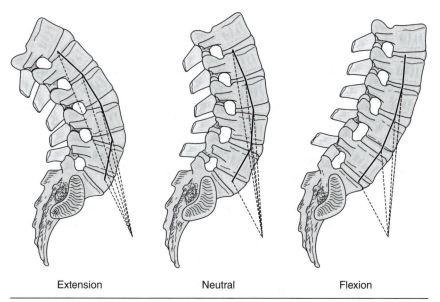

Extension	Neutral	Flexion

Figure 1.17 Line of actions of the psoas major muscle, sagittal view. The centroid line of action (the thick line) is compared with straight lines of action (the thin lines) in three lumbar postures. The authors concluded that the mechanics of psoas cannot be adequately represented with a series of straight-line vectors from vertebral origins to insertion. When compared with the straight-line approach, the centroid path yields larger compression forces and increasing shear forces from upper to lower lumbar levels.

Reprinted from *Journal of Biomechanics* 28(3), P.L. Santaguida and S.M. McGill, "The psoas muscle: A three-dimensional geometric study," 339-345, copyright 1995, with permission from Elsevier.

A serious limitation of centroid models is that they neglect muscle pennation. In pennated muscles the muscle fiber force acts at a certain angle to the cross-sectional slices. These oblique forces produce not only a pulling effect along the centroid but also shear forces and, depending on aponeurosis attachment location, moments of force. Regrettably, the existing centroid models ignore these effects.

1.6.3 Curved and Wrapping Muscles

Representative publications: Charlton and Johnson 2001; Carman and Milburn 2005

Most muscles are curved to some extent. Some of them, such as the gluteus maximus, psoas major, extensor pollicis longus, extensor digitorum communis, sartorius, obturator internus, serratus anterior, and oblique and transverse abdominal muscles, are curved substantially. Usually, curved muscles wrap around passive structures, but they can also wrap around other muscles.

When a curved muscle develops tension, the muscle tends to (a) bring the origin and insertion closer together and (b) straighten. Curved muscles exert forces not only along the muscle path but also laterally, toward the inside. Such compression forces act orthogonally to the muscle path. The mechanism is similar to what is happening when a bent rope is pulled at both ends: The rope tends to straighten. If external objects on the concave side of the rope resist the straightening, the rope exerts force on them. In the same way, curved muscles exert lateral forces acting inward. Curved muscles coiled about a central rod induce torsion on it. In particular, muscles that encircle bones when at rest (e.g., supinator and pronator teres, the long head of biceps brachii) induce untwisting torsional effects when active (the so-called *law of detorsion*).

▪ ▪ ▪ *FROM THE LITERATURE* ▪ ▪ ▪

Difference in Force Production Between Linear and Curved Muscle Models Can Be as Large as 37%

Source: Gatton, M., M. Pearcy, and G. Pettet. 2001. Modelling the line of action for the oblique abdominal muscles using an elliptical torso model. *J Biomech* 34:1203-1207

The torso was represented as an elliptical cylinder, and the internal and external oblique muscles were assumed to wrap around the torso. To achieve anatomic similarity, the major and minor axes of the cylinder varied linearly with height from the base of the torso (figure 1.18). Compared with a model of linear muscle path, the torso model resulted in a 15% increase in the axial twist moment and 5% and 2% decreases in the lateral bend and extension moments, respectively. When the torso is flexed, extended, or twisted, the forces predicted by the linear and curved muscle models differed as much as 37%.

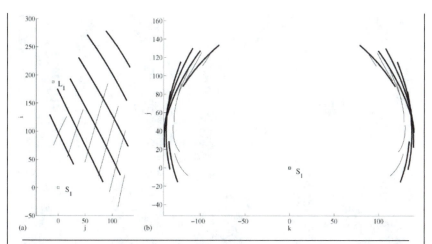

Figure 1.18   Modeling the path of the internal (light line) and external (dark line) oblique abdominal muscles. *(a)* The sagittal plane and *(b)* the transverse plane. L_1 and S_1 represent the centers of the corresponding vertebral bodies.

Reprinted from *Journal of Biomechanics* 34(9), M. Gatton, M. Pearcy, and G. Pettet, "Modelling the line of action for the oblique abdominal muscles using an elliptical torso model," 1203-1207, copyright 2001, with permission from Elsevier.

The paths of curved muscles cannot be described accurately by the coordinates of their origin and insertion alone. Additional information is necessary. Curved muscles are typically represented either as arches or as sequences of straight-line or curved segments connected at via points (figure 1.19). When muscle length changes, as occurs during concentric contraction, muscle segment length also changes. The coordinates of some via points with respect to the skeleton can also change. Such points are called *dynamic points* to distinguish them from fixed points, which are constant with respect to their bony attachment: for instance, the origin and insertion. The plane that contains all fixed and dynamic points is called the *muscle–tendon plane* or *dominant muscle plane* (when Frenet frames are used, this plane is the osculating plane). During movement, the orientation of the muscle–tendon plane either remains constant—as in hinge joints with 1 degree of freedom—or changes. Muscle movements within a muscle–tendon plane with constant orientation can be studied with the methods of planar kinematics;

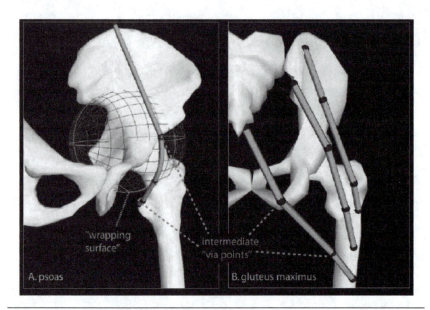

Figure 1.19 Modeling of the psoas and gluteus maximus as sequences of straight-line or curved segments connected at via points. The muscles wrap around the pelvic edge, hip joint capsule, and femoral neck. The via points are shown in black, and the wrapping surface that is used for the modeling of the psoas is shown as a wire frame. For the gluteus maximus, only the paths of some representative muscle fascicles are shown. Note that the muscles wrap around more than one surface. This makes modeling the muscle path more difficult.

With kind permission from Springer Science+Business Media: *Annals of Biomedical Engineering,* "Three-dimensional representation of complex muscle architectures and geometries," 33(5), 2005, 661-673, S.S. Blemker and S.L. Delp, fig. 1.

movements involving change of orientation of the muscle–tendon plane require three-dimensional analysis.

Finding the path of a wrapping muscle is challenging, requiring the use of sophisticated mathematical methods and imaging techniques. One of the guiding ideas in this endeavor is that muscle paths follow a geodesic. In its original sense, a *geodesic* is the shortest route between two points on the earth's surface (finding such routes is important for airline companies striving to reduce costs). The concept was generalized in mathematics and in the present context can be understood as the shortest muscle path length that is achievable without penetrating the neighboring objects. If a muscle follows a geodesic, the muscle length is minimal.

■ ■ ■ *FROM THE LITERATURE* ■ ■ ■

Modeling Musculotendinous Paths

Source: Marsden, S.P., and D.C. Swalles. 2008. A novel approach to the prediction of musculotendon paths. *Proc Inst Mech Eng H* 222(H1):51-61

The majority of experts accept the premise that muscle wrapping minimizes muscle length. This presumption is realized in several biomechanics-specific software packages, such as SIMM (Software for Interactive Musculoskeletal Modeling, MusculoGraphics, Inc., Santa Rosa, CA), its free version OpenSim 1.1, and AnyBody (AnyBody Technology A/S, Aalborg, Denmark). Muscles and muscle fascicles are modeled as strings, and the anatomical structures around which the strings are wrapped are modeled as simple geometric shapes, such as cylinders, spheres, and ellipsoids. The present study suggests a method of calculating muscle path as a smooth line of minimal length that does not penetrate the wrapping object. The parts of the string that are not in contact with the wrapping surface will be straight lines, whereas the part of the string lying on the surface must follow a geodesic path that connects smoothly with the straight line segments.

If the muscle path is established, kinematic analysis of via point models is straightforward. Dynamic analysis (i.e., force computation) presents difficulties.

With *kinematic analysis,* when a centroid is represented as a series of n line segments, the muscle length L is computed as the sum of the point-to-point distances

$$L = \sum_{i=0}^{n} \left\| \mathbf{P}_{i+1} - \mathbf{P}_i \right\|, \qquad [1.6]$$

where $\mathbf{P}_i$ and $\mathbf{P}_{i+1}$ are the 3-D positional vectors of the via points i and $i+1$, respectively; and the doubled vertical lines signify the vector magnitude (its norm). The rate of the change of the muscle length is determined as

$$\dot{L} = \sum_{i=0}^{n} \left[(\dot{\mathbf{P}}_{i+1} - \dot{\mathbf{P}}_i) \cdot \frac{(\mathbf{P}_{i+1} - \mathbf{P}_i)}{\left\| \mathbf{P}_{i+1} - \mathbf{P}_i \right\|} \right], \qquad [1.7]$$

where $(\dot{\mathbf{P}}_{i+1} - \dot{\mathbf{P}}_i)$ is the relative velocity of point $\mathbf{P}_{i+1}$ with respect to point $\mathbf{P}_i$, $(\mathbf{P}_{i+1} - \mathbf{P}_i)/\|\mathbf{P}_{i+1} - \mathbf{P}_i\|$ is the unit vector in the direction from $\mathbf{P}_i$ to $\mathbf{P}_{i+1}$, and the entire expression in the square brackets—which is a dot product of the above two vectors—represents the component of the relative velocity along the $\mathbf{P}_i$-to-$\mathbf{P}_{i+1}$ line segment.

Dynamic analysis depends on mechanical interaction of the muscle with the adjacent objects (bones, other muscles, connective tissues). Two extreme cases are (a) the muscle contacts without friction—the muscle can glide over the contact area without any resistance, and (b) the muscle is fixed to the contacting object. The intermediate variants are innumerable (different friction, various elastic resistance, and others). Without knowing the interaction mechanics, we cannot conduct accurate dynamic analysis.

■ ■ ■ *FROM THE LITERATURE* ■ ■ ■

Linear Versus Nonlinear (Centroidal) Models of the Shoulder Muscles

Sources: Wood, J.E., S.G. Meek, and S.C. Jacobsen. 1989. Quantitation of human shoulder anatomy for prosthetic arm control: I. Surface modeling. *J Biomech* 22(3):273-292; II. Anatomy matrices. *J Biomech* 22(3):309-325

Using a specially designed measuring device, the authors recorded the three-dimensional coordinates of the 30 shoulder muscles of one cadaver specimen. The muscle paths were determined in two ways, as straight lines from the origin to insertion and by the centroid method. The muscle moment arms in 3-D were then computed for both models. The muscles were then classified as (1) geometrically linear (those well approximated as a taut string); (2) functionally linear (geometrically nonlinear but functionally can be modeled as linear; moment arm differences between nonlinear and linear models were small and could be neglected); and (3) curvilinear. Only four muscles were considered geometrically linear: brachialis, lower head (anatomy texts usually do not distinguish two parts of the brachialis, but in the studied specimen two distinct parts of the muscle were seen); pectoralis minor; subclavius; and trapezius, lower part. These muscles are also functionally linear. Thirteen muscles were classified as functionally linear: biceps, long head; biceps, short head; brachialis, upper head; coracobrachialis; deltoid, acromial part; infraspinatus; supraspinatus; subscapularis; teres major; triceps, lateral head; triceps, medial head; triceps, long head; trapezius, middle part. The other 13 muscles were classified as curvilinear: brachioradialis; deltoid, clavicular part; deltoid, scapular part; latissimus dorsi; pectoralis major, abdominal part; pectoralis major, clavicular part; pectoralis major, sternocostal part; rhomboid major; rhomboid minor; serratus anterior, lower part; serratus anterior, upper part; teres minor; and trapezius, upper part.

1.6.4 Twisted Muscles

Representative publication: Pai 2002

Curved muscle centroids can be nontwisting or twisting. In the majority of muscles, the centroids are not twisted. To analyze the path of the centroids of these muscles, the Frenet frames (described in section **1.2**) can be used. However, in some muscles the bundles are not only bent but also twisted about each other. The pectoralis major serves as an example. In the anatomical position, the parts of the pectoralis major that arise from the clavicle and those from the lower part of sternum are twisted and crisscrossed; the clavicular fibers that have the highest origin have the lowest insertion (figure 1.20*a*). In this posture, the clavicular head is a prime mover for flexion, whereas the sternocostal portion assists in arm extension. Both parts work as agonists in horizontal flexion, especially when the arm is elevated and the muscle fibers are untwisted (figure 1.20*b*). If the two parts of the pectoralis major are represented by separate centroids, in anatomical position the centroids would be twisted.

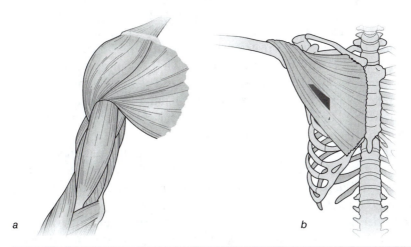

a   *b*

Figure 1.20   Anterior view of the right pectoralis major *(a)* in anatomical position and *(b)* when the arm is elevated. In anatomical position, the two parts of the pectoralis major are twisted with respect to each other.

For twisting centroid descriptions, the so-called *Cosserat models* (also called *Cosserat rods*) are occasionally used. They are named after French scientists and brothers Eugène (1866-1931) and François Cosserat (1852-1914). The Cosserat rods are used to describe fiberlike deformable strands that not only bend and change their length but also twist. The examples are surgical sutures, DNA strands, and some muscle centroids. Cosserat rods are not described further in this book. An example of a Cosserat rode is presented in figure 1.21.

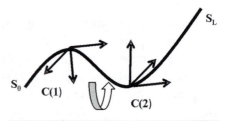

Figure 1.21 A Cosserat rod. A thin strand is bent and twisted. The Cosserat representation is similar to the Frenet representation (described in section **1.2**), but in addition to rod bending the rod twisting is also considered. The Cosserat rod in space is analogous to motion of a rigid frame C if coordinate S is interpreted as time. Hence the "time" changes from S_0 to S_L. On the way from C(1) to C(2), the strand twisted 180°. Note the different axes orientations at C(1) and C(2).

1.6.5 Muscles Attaching to More Than Two Bones

Representative publication: Hogfors et al. 1987

Some muscles, or more precisely, the structures described in anatomy textbooks as a single muscle, have more than one origin or insertion. For instance, the deltoid has its origins at the clavicle and scapula. The pectoralis major arises at the clavicle, the sternum, and the cartilages of all the true ribs. The trapezius arises at no less than three locations—the occipital bone, the spinous process of the seventh cervical, and several dorsal vertebrae. Its superior fibers proceed downward whereas the inferior fibers proceed upward; the middle fibers are directed horizontally. The extrinsic muscles of the hand (e.g., flexor digitorum profundus) fan out into four tendons that attach to the fingers, from the second to fourth.

Muscle parts attached to different bones produce different mechanical effects. For instance, the action of the upper part of the trapezius includes pulling on the humeral head, whereas the inferior fibers that do not attach to the head do not generate this action. A different effect is produced by muscle fibers going in different directions. The upper trapezius elevates the scapula and clavicle whereas the lower trapezius depresses them. Evidently such muscles cannot be represented by a single muscle path. Biomechanically, they are modeled as two or more muscles. For instance, the trapezius is usually modeled as either three or even four different muscles.

1.7 CROSS-SECTIONAL AREA, PHYSIOLOGICAL AND ANATOMICAL

Representative publications: Fukunaga et al. 1992b; Morse et al. 2005

Because many muscles have irregular shape, determining a muscle cross-section is not always a straightforward operation. It is simple for muscles with parallel fibers. For these muscles, the circumference and cross-section can be determined at any distance between muscle origin and insertion, but the same cannot be done with convergent muscles. How would you determine the cross-section of the infraspinatus (shown in figure 1.9), pectoralis major (figure 1.20), or trapezius (figure 1.26)? It seems that for such muscles the only valid cross-section is the one through all muscle fibers. Such a cross-section—orthogonal to the long direction of all the fibers in the muscle—is called the *physiological cross-section* and the corresponding area is known as the *physiological cross-sectional area* (PCSA). If all the fibers in a muscle were arranged in parallel, the PCSA would equal their total cross-sectional area. Because muscle forces arise from the summation of forces exerted in parallel by myofibrils in muscle fibers and muscle fibers in fascicles, it is commonly expected that the maximal force that all the muscle fibers together can exert scales with the PCSA.

In addition to the PCSA, the area of the muscle slices cut perpendicularly to the longitudinal muscle direction or to the muscle centroid can be determined (figure 1.22). This area is called the *anatomical cross-sectional area*

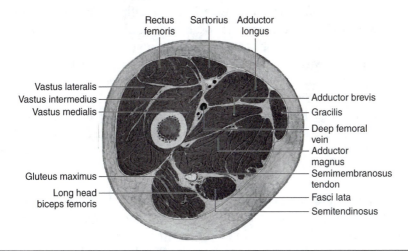

Figure 1.22 Cross-section of the thigh. Anatomical cross-sections of the muscles are seen. The image is obtained from a cadaver specimen dissected perpendicularly to the long axis of the body at the regular distances of 1 cm.

Adapted from A.C. Eycleshymer and D.M. Schoemaker, 1911, *A cross-section anatomy* (New York: D. Appleton and Company).

(ACSA) or *simply cross-sectional area* (CSA). The ACSA usually varies from slice to slice and is only valid for one slice. The ACSA estimated from limb circumference measurements is not very reliable because of different contributions of subcutaneous fat and bone across subjects. The shortest distance between the aponeuroses (internal tendons) of origin and insertion measured in the plane of CSA is called the *muscle thickness* (see figure 1.6*a*).

■ ■ ■ *FROM THE LITERATURE* ■ ■ ■

Anatomical and Physiological Cross-Sectional Area of Elbow Flexors and Extensors

Source: Kawakami, Y., K. Nakazawa, T. Fujimoto, D. Nozaki, M. Miyashita, and T. Fukunaga. 1994. Specific tension of elbow flexor and extensor muscles based on magnetic resonance imaging. *Eur J Appl Physiol Occup Physiol* 68(2):139-147

Series cross-section images of the upper extremity were obtained for four men by MRI and ACSA of elbow flexor muscles (biceps brachii, brachialis, brachioradialis) and extensor muscles (triceps brachii) were measured. PCSA was calculated from the muscle volume and muscle fiber length, the volume from the series of the ACSA, and the fiber length from the muscle length multiplied by previously reported fiber–muscle length ratios. Elbow flexion–extension torque was measured using an isokinetic dynamometer, and the force at the tendons was calculated from the torque and moment arms of muscles measured by MRI. Maximal ACSA of triceps brachii was comparable to that of total flexors, whereas PCSA of triceps brachii was greater by 1.9 times. Within flexors, brachialis had the greatest contribution to torque (47%), followed by biceps brachii (34%) and brachioradialis (19%).

For parallel-fibered muscles, the PCSA and ACSA are similar (figure 1.22). For pennate muscles, the PCSA is larger than the ACSA. At a given muscle volume, fiber pennation increases the PCSA. For some pennate muscles, such as shown in figure 1.23, the PCSA does not correspond to a single muscle cross-section in any physical plane. It is rather a set of several slices.

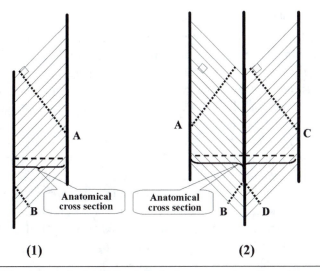

(1)   **(2)**

Figure 1.23 Anatomical and physiological cross-sections for unipennate (1) and bipennate (2) muscles. Schematics, a planar case. Physiological cross-sections are represented by dotted lines. They are perpendicular to the fiber directions. The letters A, B, C, and D designate the segments of the physiological cross-sections. The physiological cross-section for the unipennate muscle equals A + B; for the bipennate muscles it is equal to the sum A + B + C + D. The anatomical cross-sections are shown as the dashed lines. They are perpendicular to the tendons. The shortest distance between the tendons of origin and insertion measured in the plane of anatomical cross-section is muscle thickness.

▪ ▪ ▪ *FROM THE LITERATURE* ▪ ▪ ▪

A Muscle PCSA Can Be Eightfold Larger Than Its ACSA

Source: Fukunaga, T., R.R. Roy, F.G. Shellock, J.A. Hodgson, M.K. Day, P.L. Lee, H. Kwong-Fu, and V.R. Edgerton. 1992b. Physiological cross-sectional area of human leg muscles based on magnetic resonance imaging. *J Orthopaed Res* 10(6):928-934

Magnetic resonance imaging techniques were used to determine the PCSAs of the major muscles or muscle groups of the lower leg. The boundaries of each muscle or muscle group were digitized from images

(continued)

From the Literature   *(continued)*

taken at 1 cm intervals along the length of the leg. Muscle volumes were calculated from the summation of each anatomical CSA and the distance between each slice (i.e., 1 cm). Muscle length was determined as the distance between the most proximal and distal images in which the muscle was visible. The PCSA of each muscle was calculated as muscle volume times the cosine of the angle of fiber pennation divided by fiber length (i.e., the projected PCSA was determined, explained subsequently in the text). To estimate fiber lengths, the published fiber length–muscle length ratios were used. The mean PCSAs of the soleus, medial gastrocnemius, and tibialis posterior were 230, 68, and 36.8 cm², respectively. These PCSA values were almost 8 (soleus), 4 (medial gastrocnemius), and 7 (tibialis posterior) times larger than their respective maximum ACSA (table 1.5).

Table 1.5   Anatomical and Physiological Cross-Sectional Areas of Some Human Ankle Plantar Flexors

Muscle	ACSA, cm²	PCSA, cm²	PCSA/ACSA
Medial gastrocnemius	16.49	68.34	4.14
Lateral gastrocnemius	11.24	27.78	2.47
Soleus	29.97	230.02	7.68
Flexor hallucis longus	4.85	19.32	3.98
Tibialis posterior	5.40	36.83	6.82
Flexor digitorum longus	1.59	9.12	5.73

Group averages, based on the study by Fukunaga et al. (1992b). ACSA, anatomical cross-sectional area; PCSA, physiological cross-sectional area.

Determining the PCSA is not simple. When such techniques as sonography and MRI were not available, PCSA was determined on cadavers using one of the following three techniques:

1. Muscle circumference was measured and the PCSA was computed assuming a cylindrical and parallel-fibered muscle.
2. Muscle cuts were made perpendicular to the fibers, and their area was measured.
3. Muscle volume and fiber length were measured, and PCSA was computed from these measurements.

With modern imaging techniques, the third method has become the most popular. None of these above methods, however, is very accurate.

The PCSA is often estimated from the following equation:

$$PCSA = \text{Muscle Volume / Fiber Length} \qquad [1.8a]$$

or

$$PCSA \ [mm^2] = \frac{M \ [g]}{\rho \ [g/mm^3] \cdot l_f \ [mm]} \qquad [1.8b]$$

where M is muscle mass, l_f is the fiber length, and ρ is muscle density (1.056 g/cm³ in the fresh state, see table 1.1). For muscles with spanning straight fibers of equal length, the derivation of equation 1.8 is straightforward (figure 1.24).

Problems with equation 1.8 arise when the muscle fibers end in the middle of the muscle (see figures 1.12, 1.13, and 1.14), when they are curved, or when the PCSA of a convergent (triangular) muscle, such as is shown in figure 1.9, should be determined. For such muscles, for example, the triangular muscles, the average PCSA can be determined, but the physiological meaning of "average PCSA" is not clear: In particular it is not evident whether the forces exerted at the muscle origin and insertion are proportional to the average PCSA.

Some authors prefer using another definition of the PCSA. They determine a projection of the PCSA (determined as described previously) on the muscle's line

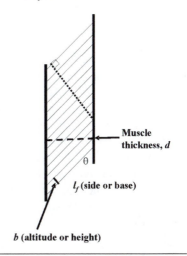

of action. To achieve this, the PCSA is multiplied by the cosine of the pennation angle (equation 1.9). This measure is called *projected PCSA* or *functional PCSA*.

Figure 1.24 Derivation of equation 1.8a for a unipennate muscle with equal fiber length. The parallel fibers form the sides of parallelograms. For a parallelogram with side l_f and the altitude (height) to this side b, area = $l_f \times b$. Hence, $b = $ area $/ l_f$. The sum Σb equals the cross-section of all the fibers. It can be estimated by dividing the total muscle area by l_f. Assuming that the third muscle dimension (not shown in the figure) is constant, the physiological cross-sectional area of the muscle is estimated from the ratio of muscle volume to fiber length.

$$\text{Functional PCSA} \ [mm^2] = \frac{M \ [g]\cos\theta}{\rho \ [g/mm^3] \cdot l_f \ [mm]}, \qquad [1.9]$$

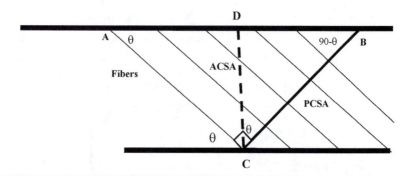

Figure 1.25 Computing a projected PCSA. Consider a triangle with vertices A, B, and C, where side AC represents a muscle fiber and side BC represents the PCSA. Note that the triangle ABC is a right triangle and the angle at C is a right angle. If the pennation angle equals θ, then the other two angles in triangle ABC are θ and 90 − θ, as shown in the figure. From the vertex C, draw the altitude to point D and consider the right triangle CDB. In this triangle the angle at vertex C is θ. Hence, the projected PCSA equals the product PCSA × cos θ. In the presented example, the projected PCSA equals ACSA; however, they can differ from each other. The reader is challenged to explain the reason behind this discrepancy (hint: peruse figure 1.23).

where θ is the pennation angle. Figure 1.25 explains the reason behind this computation.

When interpreting the literature, the reader is advised to pay attention to how the PCSA was determined: (a) as an area of the cross-section of muscle fibers or (b) as the projection of the above area (what is described in [a]) on the plane of the anatomical cross-section (functional PCSA).

When only the muscle volume, V, muscle thickness, d, and angle of pennation θ are known, the PCSA is determined in the following way. The tendon area to which the muscle fibers attach equals $T = V/d$ (muscle volume divided by thickness, see figure 1.23). Each muscle fiber of cross-sectional area a occupies $t = a / \sin \theta$ of this area. Hence, the cross-sectional area of the individual fiber is $a = t \sin \theta$, and the total physiological cross-sectional area of all the fibers is

$$\text{PCSA} = T \sin \theta = V \sin \theta / d. \qquad [1.10]$$

In the ideal case shown in figure 1.23 (two parallel tendons, muscle fibers of equal length), equation 1.10 is equivalent to equation 1.8a. The projected PCSA in this case equals

$$\text{Projected PCSA} = \frac{2V \sin\theta \cos\theta}{d} = \frac{2V}{d} \sin 2\theta. \qquad [1.11]$$

In addition to the CSA and PCSA, measures that are expressed in units of area, mm^2 or cm^2, *normalized muscle area* (NMA) is also used. To compute NMA, the area of the muscle is divided by the total area of all the muscles in the studied specimen. The NMA is sometimes called the *tension fraction* (of all the muscle tensions that can be exerted by a given specimen).

■ ■ ■ *FROM THE LITERATURE* ■ ■ ■

Alternative Methods of Estimating the Physiological Cross-Sectional Areas of the Muscles

A. Methods Based on Muscle Geometry

Sources: (1) Fick, R. 1911. *Handuch der Anatomie und Mechanik der Gelenke, Teil 3.* Jena, Germany: Gustav Fischer. (2) Howell, S.M., A.M. Imobersteg, D.H. Seger, and P.J. Marone. 1986. Clarification of the role of the supraspinatus muscle in shoulder function. *J Bone Joint Surg* 68-A:398-404. (3) Wood, J.E, S.G. Meek, and S.C. Jacobsen. 1989. Quantification of human shoulder anatomy for prosthetic arm control: I. Surface modelling. *J Biomech* 22:273-292

To estimate PCSA, some authors have used computational techniques different from those presented in equations 1.8 and 1.10. Instead of fiber length, other measures of length may be used. In particular, the PCSA has been estimated as (1) product of muscle mass and inverse density divided by bundle (fascicle) length, (2) muscle volume divided by muscle length, (3) muscle volume divided by bundle length, and (4) maximum cross-sectional area perpendicular to the centroid. Because muscle fiber length can differ substantially from the muscle length (see table 1.3 as an example), PCSA estimates obtained by these methods can sharply differ from each other.

B. Regression Equations (for Estimating Trunk Muscle Areas)

Sources: (1) McGill, S.M., N. Patt, and R.W. Norman. 1988. Measurement of the trunk musculature of active males using CT scan radiography: Implications for force and moment generating capacity about the L4/L5 joint.

(continued)

From the Literature   *(continued)*

J Biomech. 21(4):329-341. (2) Tracy, M.F., M.J. Gibson, E.P. Szypryt, R. Rutherford, and E.N. Corlett. 1989. The geometry of the muscles of the lumbar spine determined by magnetic resonance imaging. *Spine* 14(2):186-193. (3) Wood, S., D.J. Pearsall, R. Ross, and J.G. Reid. 1996. Trunk muscle parameters determined from MRI for lean to obese males. *Clin Biomech* 11(3):139-144. (4) Chaffin, D.B., M.S. Redfern, M. Erig, and S.A. Goldstein. 1990. Lumbar muscle size and locations from CT scans of 96 women of age 40–63 years. *Clin Biomech* 5(1):9-16

In principle, the cross-sectional areas of the muscles can be estimated from regression equations in which anthropometric measures, such as body height, weight, circumferences, and fat percentage, are used as input variables. Presently, such regression equations use different input variables; muscle area estimates from different studies do not agree well with each other. For instance, Wood et al. (1996) failed to predict the CSA of the psoas muscle from anthropometric measurements (in 26 male subjects), whereas other authors using different methods obtained dissimilar results. The regression equations for the CSA psoas (cm^2) were as follows:

   A. Tracy et al. (1989): = 2.5 + 0.195 (body mass, kg)

   B. McGill et al. (1988): = 1320 − 7.47 (height, cm) − 14.1 mass, kg) + 0.08 [(height)(mass)]

   C. Chaffin et al. (1990): = 5 + 0.007 (trunk area, cm^2)

   Evidently, more work is necessary to obtain more accurate regression equations that will be accepted by the scientific community as a gold standard.

In some muscles, cross-sectional areas depend on body posture. For instance, anatomical cross-sectional areas of the lower lumbar–level trunk muscle decrease with torso flexion.

1.8 MUSCLE ATTACHMENT AREA

Representative publication: Van der Helm and Veenbaas 1991

Muscle fibers generate force along their longitudinal axes. In the parallel-fibered muscles, the forces exerted by the individual fibers are parallel. For these muscles, the Varignon theorem is valid (see section **1.1.6** in *Kinetics of Human Motion*) and the fiber force action can be replaced by a single resultant force (you may also say that the muscle produces a zero-pitch wrench, i.e., a pure force without moment). For such a muscle, the moment of force about the point of application of the resultant is zero (it is not, however, zero about other centers). If all the muscle fibers produce equal force, the point of application of the resultant is at the geometric center of the attachment area.

Some muscles attach to bones over a small area. For such muscles, the attachment can be modeled as a point at which the tendon resultant force is applied. These muscles exert a force on the bone: They are pure *force actuators*. Other muscles have a broad area, or a line, of contact. If force is exerted at only one part of the contact area, such a muscle can exert a moment of force with respect to the attachment center. Hence, muscles with broad attachment area can act not only as force actuators but also as *moment actuators*. Muscles with the pointlike contact are mainly located at the extremities whereas the muscles of the trunk, for instance, the shoulder muscles, commonly have large attachment areas at the origin.

It may happen that a muscle is too wide to fit within the available body space although the space is sufficiently large to accommodate the muscle's volume. To fit onto the available space, the muscle should become more compact. The presumption is that during evolution, the muscles adapted in size, shape, and bone attachment and consequently in function. One consequence of this theory is that the distinctions between muscle types— fusiform, unipennate, and bipennate—are not very strict. The muscle types are not separate entities but rather the fragments of a morphological continuum (figure 1.26).

Muscle attachments can be represented as

- a point, when the attachment area is small;
- a line, when an aspect ratio of the attachment (i.e., the ratio of length to width) is large; or
- a surface.

When an attachment is modeled as a line, two variants—straight line and curved line approximations—are distinguished. In the first case, all the points of contact are on a straight line and hence muscle fibers cannot exert a moment of force about this line, for example, about the muscle aponeurosis. A moment, however, can be exerted about any axis perpendicular to the attachment line (figure 1.27). A curved line approximation is required when at least three points in the attachment area are not collinear. A curved line approximation yields the same mechanical effects as a surface approximation; for example, moments of force can be exerted about any axis lying in

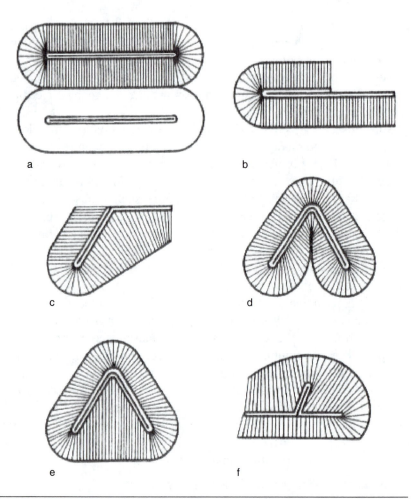

Figure 1.26 Cross-sections of different muscle types. *(a)* Bipennate muscles stacked on each other. The end tendons are double-folded unipennate tendons. The aponeuroses of origin have grown together in bipennate intermuscular septa. *(b)* Partly folded unipennate model. *(c)* Model *(b)* with end tendon folded at 120°. The end tendon is partly unipennate and partly bipennate. *(d)* Bipennate model *(a)* folded at an angle of 60°. *(e)* Model with slack removed from the outer aponeurosis. *(f)* Cross-section of a very compact muscle with the end tendon with an extra fold.

From J.N.A.L. Leijnse, 1997, "A generic morphological model of a muscle group: Application to the muscles of the forearm," *Acta Anatomica* 160: 100-111. Reprinted with permission of S. Karger AG, Basel.

the plane tangent to the contact (provided that the individual fibers produce force independently from each other).

The number of actions of a muscle equals the number of independent motions that can be impressed by this muscle on an unconstrained bone. The concept of the number of muscle actions is closely related to the concept of degrees of freedom (see section **2.2.1** in *Kinematics of Human Motion*). A free solid body in space has 6 degrees of freedom (DoFs); it can move in three directions and rotate about three mutually orthogonal axes. When a muscle acts on an unconstrained bone, it could induce bone motion along 1 or several DoFs. For instance, a muscle with two single-point attachments can exert force in only one direction, along the line between the origin and exertion.

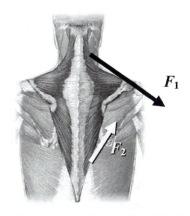

Figure 1.27 The muscle fibers of the trapezius are in different directions. Forces F_1 and F_2 exert not only a resultant force but also a moment of force on the spine. The moment is about an axis perpendicular to the attachment line.

Its number of actions equals one. Depending on the combination of the attachment sites, muscles can have a variety of number of actions. In total, nine attachment–attachment combinations are possible (table 1.6). The maximal number of DoFs that a muscle can affect is 6.

In the present context, bones are assumed to be constrained only by muscles that are directly connected to them. Joint constraints have not yet been considered. To visualize this concept, imagine that one of the bones is

Table 1.6   Number of Actions of Muscles With Different Attachments at the Origin and Insertion

Attachments at the insertion	Attachment at the origin		
	Point	Straight line	Surface
Point	1	2	3
Straight line	2	4	5
Surface	3	5	6

The idea for this table comes from Van der Helm and Veenbaas (1991).

perturbed by an external force or moment. The question is whether a given muscle, if properly activated, can resist the perturbation. A muscle with a point–point attachment can resist the perturbation in only one direction; it cannot, for instance, resist perturbations in the direction perpendicular to its line of action.

In research, muscles with broad attachment areas, as well as muscles that connect to more than two bones (discussed in **1.6.6**), are often modeled by two or several muscle parts (bundles) with point-to-point attachments (see figure 1.18, right panel, for an example). Such muscle parts, "muscles within muscles," are pure force generators. They are conveniently called *actons* (a term introduced by Polish researchers Morecki, Ekiel, and Fidelus [1971]). Muscles with point-to-point attachments are also considered actons.

There is a certain analogy between muscle fibers and ideal engineering prismatic joints. Prismatic joints are served by linear actuators that exert force along an axis. Muscle attachments in this case are modeled as spherical joints that can rotate about two or three axes (in the latter case, fiber twisting is allowed). In the model, linear actuators can exert a pulling force whereas revolute joints are passive. With such a model, muscle–bone systems are analogous to parallel robots consisting of rigid bodies connected by prismatic joints (a so-called *Stewart platform*). This approach allows one to analyze muscle architecture with the methods developed in robotics.

Consider as an example two bones, the occipital bone of the skull and scapula, connected by an acton consisting of superior fibers of the trapezius. We are interested in the number of DoFs and the constraints in such a system. To compute the number of DoFs, the Gruebler formula can be used (described in section **2.2.1** of *Kinematics of Human Motion*). According to the formula, the DoFs are computed as the difference between the total number of DoFs of the links (for two solid bodies, DoFs = 12) and the number of constraints imposed by the joints (do not forget that in the present context the muscle fibers are modeled as prismatic joints).

$$\text{DoF} = 6N - \sum_{i=1}^{k}(6 - f_i) = 6(N - k) + \sum_{i=1}^{k} f_i, \qquad [1.12]$$

where N is the number of links (=2), k is the number of joints, and f_i is the number of DoFs in the ith joint. For the prismatic joints, $f_i = 1$. The formula is only valid when the constraints imposed by the joints are independent. In the muscle, the "joints" (i.e., individual fibers modeled as prismatic joints) are kinematically constrained by bone positions. For instance, in planar line-to-line contacts, the length of any two fibers, such as the length of two reference fibers shown as the dotted lines in figure 1.28, determines the length of all other fibers. For a surface attachment, the number of reference fibers equals three. With these conventions, the number of DoFs of the two-bone–one-muscle system can be computed with the Gruebler formula where the muscle actions play the role of constraints. The results will be the same as in table 1.6.

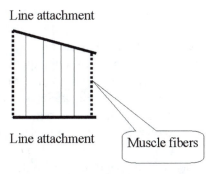

Figure 1.28 Constraints imposed on the muscle fibers (prismatic joints) with planar line-to-line attachments. See explanation in the text.

▪ ▪ ▪ *From the Literature* ▪ ▪ ▪

Dividing Muscles in Their Functional Parts (Actons)

Source: Van der Helm, F.C., and R. Veenbaas. 1991. Modelling the mechanical effect of muscles with large attachment sites: application to the shoulder mechanism. *J Biomech* 24(12):1151-1163

In the literature, the muscles with large attachment sites are often replaced by sets of muscle bundles comprising the muscle (actons). This division is commonly made without a thorough mechanical analysis, based only on personal opinion. A method is suggested for determining the necessary number of muscle lines of action experimentally.

In a cadaver experiment, the complete attachment sites and bundle distribution of 16 muscles of the shoulder mechanism were recorded. For each muscle, the resulting force and moment vector for a large number (up to 200) and a reduced number of muscle lines of action were computed (maximal 6). Reduced representations with minimal deviation from the complete representation were selected and recommended for future use. The number of muscle lines of action in the reduced representation depended on the shape of the attachment site and muscle architecture.

1.9 SUMMARY

Muscle architecture is the internal structure of the muscles. Muscles exist in variety of forms. Although all the muscles are three-dimensional, given technical complexities they are often studied in two dimensions.

The organization of fascicles relative to the axis of force generation is used to classify muscles as parallel-fibered, fusiform, pennate, convergent, and circular. In fusiform muscles, muscle fascicles are arranged along the line of muscle force action. In pennate (or pinnate) muscles, fascicles are attached to the tendon at an angle. In such muscles, the directions of fiber shortening and tendon movement are different. Depending on the number of fiber directions, pennate muscles are classified as unipennate, bipennate, and multipennate. The pennation angle is determined as the angle between the direction of muscle fibers and either (a) the line of muscle force action (external portion of the tendon) or (b) the aponeurosis (internal tendon). The first method is preferred. The angle increases with muscle hypertrophy and growth. The angle decreases when muscle size decreases, for example, with advanced age or disuse atrophy. Convergent muscles are broad at the origin and converge at one attachment point. Individual fibers of a convergent muscle pull in dissimilar directions, and hence their action at the joint can be different.

Many muscles are curved and have complex three-dimensional geometry. For their architectural analysis, a muscle fiber (or fascicle, muscle line of action, or ligament) is represented as a curved line in space whose geometry can be described with Frenet frames. These methods have the advantage that line curvature is computed simply, as the inverse of the radius. If a force is transmitted along the band, this method allows one to compute the lateral forces acting normally from the band on neighboring tissues (provided that the force acting along the band is known).

There are two main types of muscle fibers: (a) those that span along the entire fascicle, from the aponeurosis of origin to that of insertion (spanning fibers); and (b) those that end somewhere in the middle of the fascicle (nonspanning fibers). Nonspanning fibers usually taper near the end that terminates within the fascicle. The forces exerted by nonspanning fibers are transmitted first to the connective tissue matrix and then to the tendon. This type of force transmission is called lateral force transmission or myofascial force transmission.

Muscles possess mechanical properties of fiber-reinforced composite materials that are made from two or more materials and are commonly used in engineering. The intramuscular connective tissue forms the muscle matrix. The muscle is a composite consisting of an extracellular matrix enforced by myofibers that generate force and bear loads.

Muscle excursion is proportional to the excursion of its fibers. Hence, fiber length affects the feasible muscle excursion. Fiber length depends on (a) muscle length, (b) fiber pennation, and (c) the spanning pattern of the fibers in the fascicles, whether the fibers are spanning or not. The fascicle length depends only on (a) and (b). The following length–length ratios are used to characterize muscle architecture. (A) The ratio of mean fiber length to muscle belly length at a muscle's optimal length. This ratio is known as the index of architecture. (B) The ratio of tendon slack length to optimal fiber length. The tendon slack length is the threshold length at which a stretched tendon begins to develop force. This ratio represents the slack tendon length normalized by muscle belly length (normalized tendon length). (C) The ratio of muscle belly length to the muscle–tendon unit length. This ratio is used to compute the speed of muscle shortening or lengthening in human movements.

Forces exerted by the muscle are transmitted from one attachment site to another along a certain line of action, usually referred to as a muscle path. Muscles can be categorized as (a) muscles with a straight-line path, (b) curved muscles, and (c) wrapping muscles. Several approaches are used either separately or in combination to determine muscle paths. They are (a) straight-line model, (b) centroid model, (c) via-point model, (d) obstacle-set model, and (e) Cosserat model. The muscle centroid is the locus of the geometric centers of the muscle transverse cross-sections. The centroids of the curved muscles can be twisting or nontwisting. Twisting centroid descriptions can use Cosserat models, not described in detail in this book. Some muscles have more than one origin or insertion. They are typically modeled as two or more muscles.

The physiological cross-section is orthogonal to the muscle fibers direction. The corresponding area is known as the physiological cross-sectional area (PCSA). Because the muscle force arises from the summation of the forces exerted in parallel by myofibrils in muscle fibers and muscle fibers in fascicles, it is commonly expected that the maximal force that a muscle can exert scales with PCSA. In addition to the PCSA, the area of muscle slices cut perpendicularly to the longitudinal muscle direction or to the muscle centroid can be determined. It is called the anatomical cross-sectional area (ACSA). The ACSA usually varies from slice to slice and it is only valid for one slice. The PCSA and the ACSA are similar for parallel-fibered muscles and different for pennate muscles. The projected PCSA is a projection of the PCSA onto the plane of the ACSA. As a rule, the projected PCSA and ACSA are not equal.

Depending on their attachment area and shape, muscle attachments are modeled as points, lines, or surfaces. Muscles with broad attachment areas can act not only as force actuators but also as moment actuators. The number of actions of a muscle equals the number of independent spatial motions that can be impressed to an unconstrained bone. A muscle's number of actions depends on its particular combination of attachment sites. Actons are muscles or muscle parts (muscles within muscles) that work as single force actuators.

1.10 QUESTIONS FOR REVIEW

1. Explain the terms *in vivo*, *in situ*, and *in vitro*.

2. Define the terms *muscle architecture* and *muscle morphometry*.

3. How does one make the gastrocnemius muscle contract eccentrically? How about the soleus muscle? Describe a movement for which the gastrocnemius contracts eccentrically while the soleus contracts concentrically.

4. What muscles are called fusiform? How do they differ from parallel-fibered muscles? Give examples of fusiform muscles.

5. Discuss pennate muscles. Why are they called pennate? How are pennation angles determined? How can pennated muscles be further classified? What is a pennation plane? Give examples of pennated muscles.

6. What is an aponeurosis?

7. Discuss how curved muscle fibers are analyzed. What is fiber curvature? Explain Frenet frames. What is a binormal vector?

8. What are spanning and nonspanning fibers? Explain the term *myofascial force transmission*.

9. Define and compare myotendinous force transmission and myofascial force transmission.

10. Muscles are sometimes considered composite materials. Explain. Provide an example of a nonorganic composite material comparable to muscle.

11. Describe length ratios used to characterize muscle architecture. What is the index of architecture (muscle architecture)? What is normalized tendon length?

12. Discuss the concept of a muscle path.

13. Explain how muscle centroids are determined. Discuss the advantages and limitations of this method.

14. Define and compare physiological cross-sectional area (PCSA) and anatomical cross-sectional area of a muscle. How is the PCSA of a muscle determined?

15. What is projected PCSA or functional PCSA?

16. Some muscles act as pure force generators whereas other muscles exert not only a force but also a moment on the bone to which they attach. Explain what causes this difference.

17. What is the number of actions of a muscle?

18. Explain the term *acton* (or *muscle acton*).

1.11 LITERATURE LIST

Agur, A.M., V. Ng-Thow-Hing, K.A. Ball, E. Fiume, and N.H. McKee. 2003. Documentation and three-dimensional modelling of human soleus muscle architecture. *Clin Anat* 16(4):285-293.

Benard, M.R., J.G. Becher, J. Harlaar, P.A. Huijing, and R.T. Jaspers. 2009. Anatomical information is needed in ultrasound imaging of muscle to avoid potentially substantial errors in measurement of muscle geometry. *Muscle Nerve* 39(5):652-665.

Binzoni, T., S. Bianchi, S. Hanquinet, A. Kaelin, Y. Sayegh, M. Dumont, and S. Jequier. 2001. Human gastrocnemius medialis pennation angle as a function of age: From newborn to the elderly. *J Physiol Anthropol Appl Human Sci* 20(5):293-298.

Blemker, S.S., D.S. Asakawa, G.E. Gold, and S.L. Delp. 2007. Image-based musculoskeletal modeling: Applications, advances, and future opportunities. *J Magn Reson Imaging* 25(2):441-451.

Blemker, S.S., and S.L. Delp. 2005. Three-dimensional representation of complex muscle architectures and geometries. *Ann Biomed Eng* 33(5):661-673.

Blemker, S.S., P.M. Pinsky, and S.L. Delp. 2005. A 3D model of muscle reveals the causes of nonuniform strains in the biceps brachii. *J Biomech* 38(4):657-665.

Carman, A.B., and P.D. Milburn. 2005. Dynamic coordinate data for describing muscle-tendon paths: A mathematical approach. *J Biomech* 38(4):943-951.

Chaffin, D.B., M.S. Redfern, M. Erig, and S.A. Goldstein. 1990. Lumbar muscle size and locations from CT scans of 96 women of age 40–63 years. *Clin Biomech* 5(1):9-16.

Charlton, I.W., and G.R. Johnson. 2001. Application of spherical and cylindrical wrapping algorithms in a musculoskeletal model of the upper limb. *J Biomech* 34(9): 1209-1216.

Csapo, R., C.N. Maganaris, O.R. Seynnes, and M.V. Narici. 2010. On muscle, tendon and high heels. *J Exp Biol* 213(Pt 15):2582-2588.

Dostal, W.F., and J.G. Andrews. 1981. A three-dimensional biomechanical model of hip musculature. *J Biomech* 14(11):803-812.

Eycleshymer, A., and D. Schoemaker, eds. 1911. *A Cross-Section Anatomy.* New York: Appleton. (http://web.mac.com/rlivingston/Eycleshymer/Welcome.html).

Fick, R. 1911. *Handuch der Anatomie und Mechanik der Gelenke, Teil 3.* Jena, Germany: Gustav Fischer.

Friden, J., R.M. Lovering, and R.L. Lieber. 2004. Fiber length variability within the flexor carpi ulnaris and flexor carpi radialis muscles: Implications for surgical tendon transfer. *J Hand Surg Am* 29(5):909-914.

Friedrich, J.A., and R.A. Brand. 1990. Muscle fibre architecture in the human lower limb. *J Biomech* 23:91-95.

Fukunaga, T., K. Funato, and S. Ikegawa. 1992a. The effects of resistance training on muscle area and strength in prepubescent age. *Ann Physiol Anthropol* 11(3):357-364.

Fukunaga, T., R.R. Roy, F.G. Shellock, J.A. Hodgson, M.K. Day, P.L. Lee, H. Kwong-Fu, and V.R. Edgerton. 1992b. Physiological cross-sectional area of human leg muscles based on magnetic resonance imaging. *J Orthop Res* 10(6):928-934.

Fukunaga, T., Y. Ichinose, M. Ito, Y. Kawakami, and S. Fukashiro. 1997a. Determination of fascicle length and pennation in a contracting human muscle in vivo. *J Appl Physiol* 82(1):354-358.

Fukunaga, T., Y. Kawakami, S. Kuno, K. Funato, and S. Fukashiro. 1997b. Muscle architecture and function in humans. *J Biomech* 30(5):457-463.

Fung, L., B. Wong, K. Ravichandiran, A. Agur, T. Rindlisbacher, and A. Elmaraghy. 2009. Three-dimensional study of pectoralis major muscle and tendon architecture. *Clin Anat* 22(4):500-508.

Gans, C. 1982. Fibre architecture and muscle function. *Exerc Sport Sci Rev* 10:160-207.

Gao, F., M. Damsgaard, J. Rasmussen, and S.T. Christensen. 2002. Computational method for muscle-path representation in musculoskeletal models. *Biol Cybern* 87(3):199-210.

Gatton, M., M. Pearcy, and G. Pettet. 2001. Modelling the line of action for the oblique abdominal muscles using an elliptical torso model. *J Biomech* 34(9):1203-1207.

Herbert, R.D., A.M. Moseley, J.E. Butler, and S.C. Gandevia. 2002. Change in length of relaxed muscle fascicles and tendons with knee and ankle movement in humans. *J Physiol* 539(Pt 2):637-645.

Hijikata, T., and H. Ishikawa. 1997. Functional morphology of serially linked skeletal muscle fibers. *Acta Anat (Basel)* 159(2-3):99-107.

Hogfors, C., G. Sigholm, and P. Herberts. 1987. Biomechanical model of the human shoulder: I. Elements. *J Biomech* 20(2):157-166.

Howell, S.M., A.M. Imobersteg, D.H. Seger, and P.J. Marone. 1986. Clarification of the role of the supraspinatus muscle in shoulder function. *J Bone Joint Surg Am* 68(3):398-404.

Huijing, P.A. 1985. Architecture of the human gastrocnemius muscle and some functional consequences. *Acta Anat (Basel)* 123:101-107.

Huijing, P.A. 1999. Muscle as a collagen fiber reinforced composite: A review of force transmission in muscle and whole limb. *J Biomech* 32(4):329-345.

Huxley, H.E. 1966. The fine structure of striated muscle and its functional significance. *Harvey Lect* 60:85-118.

Jensen, R.H., and D.T. Davy. 1975. An investigation of muscle lines of action about the hip: A centroid line approach vs the straight line approach. *J Biomech* 8(2):103-110.

Jensen, R.H., and W.K. Metcalf. 1975. A systematic approach to the quantitative description of musculo-skeletal geometry. *J Anat* 119(Pt 2):209-221.

Kardell, T., and P. Maquet, eds. 1994. *Steno on Muscles: Introduction, Texts and Translation (Transactions of the American Philosophical Society)*. Philadelphia: American Philosophical Society.

Kawakami, Y., T. Abe, and T. Fukunaga. 1993. Muscle-fiber pennation angles are greater in hypertrophied than in normal muscles. *J Appl Physiol* 74(6):2740-2744.

Kawakami, Y., K. Nakazawa, T. Fujimoto, D. Nozaki, M. Miyashita, and T. Fukunaga. 1994. Specific tension of elbow flexor and extensor muscles based on magnetic resonance imaging. *Eur J Appl Physiol Occup Physiol* 68(2):139-147.

Klimstra, M., J. Dowling, J.L. Durkin, and M. MacDonald. 2007. The effect of ultrasound probe orientation on muscle architecture measurement. *J Electromyogr Kinesiol* 17(4):504-514.

Leijnse, J.N. 1997. A generic morphological model of a muscle group: Application to the muscles of the forearm. *Acta Anat (Basel)* 160(2):100-111.

Lieber, R.L. 2009. *Skeletal Muscle Structure, Function and Plasticity: The Physiological Basis of Rehabilitation*. 3rd ed. Philadelphia: Lippincott, Williams & Wilkins.

Maas, H. 2006. Myofascial force transmission. PhD dissertation, Vrije Universiteit Amsterdam, The Netherlands.

Maganaris, C.N., V. Baltzopoulos, and D. Tsaopoulos. 2006. Muscle fibre length-to-moment arm ratios in the human lower limb determined in vivo. *J Biomech* 39(9):1663-1668.

Marsden, S.P., and D.C. Swailes. 2008. A novel approach to the prediction of musculotendon paths. *Proc Inst Mech Eng H* 222(1):51-61.

Matthews, G.G. 2002. *Cellular Physiology of Nerve and Muscle*. 4th ed. New York: Wiley.

McGill, S.M., N. Patt, and R.W. Norman. 1988. Measurement of the trunk musculature of active males using CT scan radiography: Implications for force and moment generating capacity about the L4/L5 joint. *J Biomech* 21(4):329-341.

Morecki, A., J. Ekiel, and K. Fidelus. 1971. *Bionika Ruchu [Bionics of Movement]* (in Polish). Warszawa, Poland: Panstwowe Wydawnictwo Naukowe.

Morse, C.I., J.M. Thom, N.D. Reeves, K.M. Birch, and M.V. Narici. 2005. In vivo physiological cross sectional area and specific force are reduced in the gastrocnemius of elderly males. *J Appl Physiol* 99(3):1050-1055.

Muramatsu, T., T. Muraoka, Y. Kawakami, A. Shibayama, and T. Fukunaga. 2002. In vivo determination of fascicle curvature in contracting human skeletal muscles. *J Appl Physiol* 92(1):129-134.

Otten, E. 1988. Concepts and models of functional architecture in skeletal muscle. *Exerc Sport Sci Rev* 16:89-137.

Ounjian, M., R.R. Roy, E. Eldred, A. Garfinkel, J.R. Payne, A. Armstrong, A.W. Toga, and V.R. Edgerton. 1991. Physiological and developmental implications of motor unit anatomy. *J Neurobiol* 22(5):547-559.

Pai, D.K. 2002. STRANDS: Interactive Simulation of Thin Solids using Cosserat Models. Paper read at EUROGRAPHICS, Saarbrücken, Germany.

Pappas, G.P., D.S. Asakawa, S.L. Delp, F.E. Zajac, and J.E. Drace. 2002. Nonuniform shortening in the biceps brachii during elbow flexion. *J Appl Physiol* 92(6):2381-2389.

Riewald, S.A., and S.L. Delp. 1997. The action of the rectus femoris muscle following distal tendon transfer: Does it generate knee flexion moment? *Dev Med Child Neurol* 39(2):99-105.

Santaguida, P.L., and S.M. McGill. 1995. The psoas major muscle: A three-dimensional geometric study. *J Biomech* 28(3):339-345.

Scott, S.H., C.M. Engstrom, and G.E. Loeb. 1993. Morphometry of human thigh muscles: Determination of fascicle architecture by magnetic resonance imaging. *J Anat* 182(Pt 2):249-257.

Tracy, M.F., M.J. Gibson, E.P. Szypryt, A. Rutherford, and E.N. Corlett. 1989. The geometry of the muscles of the lumbar spine determined by magnetic resonance imaging. *Spine* 14(2):186-193.

Trotter, J.A., F.J.R. Richmond, and P.P. Purslow. 1995. Functional morphology and motor control of series-fibered muscles. *Exerc Sport Sci Rev* 23:167-214.

Tsuang, Y.H., G.J. Novak, O.D. Schipplein, A. Hafezi, J.H. Trafimow, and G.B. Andersson. 1993. Trunk muscle geometry and centroid location when twisting. *J Biomech* 26(4-5):537-546.

Van der Helm, F.C., and R. Veenbaas. 1991. Modelling the mechanical effect of muscles with large attachment sites: Application to the shoulder mechanism. *J Biomech* 24(12):1151-1163.

Ward, S.R., C.M. Eng, L.H. Smallwood, and R.L. Lieber. 2009. Are current measurements of lower extremity muscle architecture accurate? *Clin Orthop Relat Res* 467(4):1074-1082.

Weber, W., and E. Weber. 1846. *Mechanics of the Human Walking Apparatus.* Translated by P. Maquet and R. Furlong. Berlin: Springer.

Wickiewicz, T.L., R.R. Roy, P.L. Powell, and V.R. Edgerton. 1983. Muscle architecture of the human lower limb. *Clin Orthop Relat Res* (179):275-283.

Woittiez, R.D., R.H. Rozendal, and P.A. Huijing. 1985. The functional significance of architecture of the human triceps surae muscle. In *Biomechanics IX-A,* edited by D. Winter, R.W. Norman, R.P. Wells, K.C. Hayes, and A.E. Patla. Champaign, IL: Human Kinetics, pp. 21-26.

Wood, J.E., S.G. Meek, and S.C. Jacobsen. 1989. Quantitation of human shoulder anatomy for prosthetic arm control: I. Surface modelling. *J Biomech* 22(3):273-292.

Wood, J.E., S.G. Meek, and S.C. Jacobsen. 1989. Quantitation of human shoulder anatomy for prosthetic arm control: II. Anatomy matrices. *J Biomech* 22(3):309-325.

Wood, S., D.J. Pearsall, R. Ross, and J.G. Reid. 1996. Trunk muscle parameters determined from MRI for lean to obese males. *Clin Biomech (Bristol, Avon)* 11(3):139-144.

Zajac, F.E. 1989. Muscle and tendon: Properties, models, scaling, and application to biomechanics and motor control. *Crit Rev Biomed Eng* 17(4):359-411.

2

PROPERTIES OF TENDONS AND PASSIVE MUSCLES

The chapter deals with the mechanical properties of tendons and passive muscles. Tendons are biological organs of complex structure whose main function is to transmit muscle forces to bones. In doing this, the tendons accumulate and release potential energy of elastic deformation. Tendons also assist in energy transfer between body segments. To transmit forces between muscles and bones, tendons must be strong yet compliant to prevent damage to the muscle tissues. The strength and elastic properties of animal tendons have been known since ancient times, when animal tendons were used as bowstrings, and the Bible describes that Samson was fastened with tendon cords.

Properties of passive muscle are a benchmark for comparison with the properties of active muscles. During movements, not all muscles and muscle fibers are active. When the muscles are relaxed, their bellies are moved by other muscles, both joint antagonists and joint agonists. For instance, when the elbow joint flexes, the joint extensors are stretched. If the force exerted during elbow flexion is small, not all fibers of elbow flexors are active; the passive fibers are deformed by adjacent active muscle fibers.

The chapter covers the following topics. Section **2.1** is devoted to mechanical properties of tendons and aponeuroses, and section **2.2** addresses the mechanical properties of passive muscles. In the both cases the elastic properties are described first and the viscoelastic properties second. Section **2.1.1** deals with elastic behavior of tendons and covers the following issues: **2.1.1.1**—stress-strain relations, **2.1.1.2**—tendon forces, **2.1.1.3**—tension and elongation in tendons and aponeuroses, and **2.1.1.4**—constitutive equations for tendons and ligaments. Viscoelastic behavior of tendons is described in section **2.1.2**. The basic concepts of viscoelasticity are explained in subsection **2.1.2.1**, and viscoelastic properties of tendons are discussed in subsection **2.1.2.2**. Section **2.1.3** addresses interaction of tendons with surrounding tissues. It covers the following issues: **2.1.3.1**—intertendinous shear force and lateral force transfer, **2.1.3.2**—interfinger connection matrices, **2.1.3.3**—gliding resistance between the tendons and surrounding tissues, **2.1.3.4**—tendon wrapping, **2.1.3.5**—bowstringing, **2.1.3.6**—interaction of tendon properties and muscle function, and **2.1.3.7**—musculotendinous architectural indices. Section **2.2** addresses mechanical properties of passive muscles and muscle–tendon complexes, highlighting the concept of muscle tone in **2.2.1**, mechanical properties of relaxed muscles and muscle fibers in **2.2.2**, elastic properties in **2.2.2.1**, and viscoelastic properties in **2.2.2.2**. A brief discussion on joint flexibility is provided in section **2.3**. Subsections **2.1.3.2** (interfinger connection matrices), **2.1.1.4** (constitutive equations for tendons and ligaments), and **2.1.2.2.1** (computational models of the viscoelastic properties of tendons) are considered advanced material. These sections are marked with 《 》.

••• ANATOMY REFRESHER •••

Tendons

Tendons are made up of longitudinally arranged collagen fibers oriented mainly along the tendon length. Some of the fibers are oriented at an angle to the main fiber direction and may cross each other. Relaxed fibers follow a sinusoidal wave pattern known as *crimp*. The fibers lie within the tendon matrix of watery gel. Tendons have a hierarchical structure that resembles the structure of parallel-fibered muscles (figure 2.1 and table 2.1).

The tendons and their collagen fibers are surrounded by three gel-like matrices of connective tissues. *Paratenon*, which is the loose connective tissue between a tendon and its sheath (the descriptive modifier *para-* means "by the side of"), allows a tendon to move freely. *Epitenon* (*epi-* means "outside") covers a tendon just beneath the paratenon. Each fiber bundle is surrounded by the *endotenon* (*endo-* means "within").

Muscle forces are transmitted to the tendons via the *myotendinous junctions*. The junctions are usually the weakest points of the

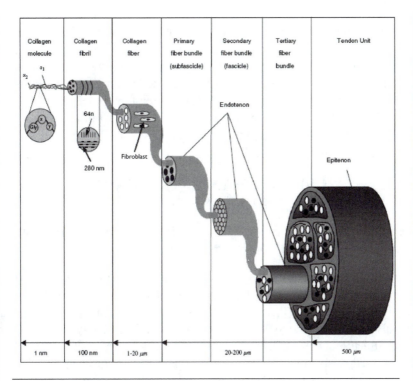

Figure 2.1 Hierarchical structure of a tendon.

Reprinted from *Journal of Biomechanics* 39(9), J.H.-C. Wang, "Mechanobiology of tendon," 1563-1582, 2006, with permission from Elsevier. Previously adapted from *Journal of Biomechanics* 36(10), F.H. Silver, J.W. Freeman, and G.P. Seehra, "Collagen self-assembly and the development of tendon mechanical properties," 1529-1553, copyright 2003, with permission from Elsevier.

Table 2.1 Muscle and Tendon Structure

Muscles	Tendons	
	Structure	Typical size (diameter)
Muscle fascicle	Fascicle	20-200 nm
Muscle fiber	Fibril (crimped)	50-500 nm
Myofibril	Subfibril	10-20 nm
Myofilament (filament)	Microfibril (filaments)	~3.5 nm
Sarcomere (functional unit)	Collagen molecule (functional unit)	

Adapted from M.J. Alter, 2004, *Science of flexibility,* 3rd ed. (Champaign, IL: Human Kinetics), 37.

(continued)

Anatomy Refresher *(continued)*

muscle–tendon units and thus a common site of injury. The tendon–bone junction is known as the *enthesis*. Besides containing collagen fibers, tendons include elastic fibers that are made of elastin. Elastin makes up about 2% of the dry weight of the tendon and contributes to fiber crimp pattern recovery following tendon stretch.

As was already mentioned in chapter 1, in pennate muscles tendons consist of an external part and an internal part called the *aponeurosis*.

2.1 BIOMECHANICS OF TENDONS AND APONEUROSES

Representative publications: Woo et al. 1997; Butler et al. 1978; Wang 2006; Magnusson et al. 2008; Benjamin et al. 2008

Tendon biomechanics research has three main subfields: (a) *mechanobehavior*—mechanical aspects of tendon function, (b) *mechanotransduction*—mechanisms governing the conversion of mechanical stimuli into biochemical signals, and (c) *mechanobiology*—the interdisciplinary study of changes in tissue structure and function in response to systematic mechanical loading (adaptation and deadaptation). We discuss here only the purely mechanical behavior of tendons, using both elastic and viscoelastic (time-dependent) models.

2.1.1 Elastic Behavior

Representative publications: Zajac 1989; Woo et al. 1997

In everyday life, tendons are exposed to large tensile forces. When a tendon is subjected to an external stretching force, the tendon length increases. When the force is removed, the tendon returns to its original length. If the force exceeds a certain critical magnitude, the tendon will rupture.

••• MECHANICS REFRESHER •••

Deformable Bodies in Extension, Elasticity

In mechanics, material bodies are classified as either rigid or deformable. In a *rigid body*, the distance between any two points within that body is always the same. Such bodies are mathematically ideal abstractions that do not actually exist, but if deformation is very small relative to the object size, the body can be considered rigid for computational purposes. In a *deformable body*, in contrast, distances between points within the body can change. When an external force is applied to a deformable body, the body changes shape. Deformable bodies can be classified further as either *plastic* or *elastic*. When external force is removed, elastic bodies return to their original size and shape whereas plastic materials (e.g., clay) retain a deformed shape. *Elasticity*, or *resilience*, is the capacity of objects or materials to resist external forces and resume their normal shape (e.g., tendon length, after force removal).

We consider here only extension forces, similar to those that act on tendon ends along the tendon. Many deformable bodies behave linearly when subjected to small forces. That is, elongation (Δl) is linearly proportional to the force $\Delta l = cF$, and the proportionality coefficient c is called *compliance*. Compliance is measured in m/N. Often it is more convenient to use the inverse of compliance, *stiffness*. Stiffness has dimensionality, newtons per meter (N/m). The expression $F = -k\,\Delta l$, where F is the force resisting the deformation and k is stiffness, or spring constant, is known as *Hooke's law* after Robert Hooke (1635-1703).

Extension forces in nature do not act on single points but instead are distributed over an area; force magnitude divided by the area on which it acts is called *stress*. When a stress plane is perpendicular to the force direction, the stress is called *normal stress* (tensile or compressive stress); stress in planes parallel to the force direction is termed *shear stress*. Stress, similar to pressure, is measured in newtons per square meter (N/m^2), pascals, or derivative units. One pascal (Pa), named after French scientist Blaise Pascal (1623-1662), represents very small biomechanical loading, so megaspascals (MPa, 1 million pascals) or gigapascals (GPa, 1 billion pascals) are more commonly used. The elongation per unit object length is called *strain* and is dimensionless, usually expressed as a percentage (i.e., meters per meter). Strain can be computed as the ratio of the change in length of a line segment to either (a) its initial length (*Lagrangian strain*, or simply strain) or (b) its final length (*Eulerian strain*).

In research, forces are usually normalized with respect to the cross-sectional area of the object, normal stress, and elongation is expressed per unit of the object length (l) (i.e., as strain). The stress–strain relation is $\sigma = E\Delta l\,/\,l = E\varepsilon$, where σ is stress (N/m^2), ε is dimensionless strain, and E is a proportionality constant, *Young's modulus* (N/m^2), named after Thomas Young (1773-1829). Young's modulus has the same dimensions as stress; it is the stress required to elongate the body to double its original length. Usually maximal elongation is much smaller than that: real objects would break well before doubling their lengths.

(continued)

Mechanics Refresher *(continued)*

Note that elastic modulus is not the same as stiffness. Stiffness is a property of a deformable body, whereas elastic modulus is a property of the constituent material.

The mathematical relations between stresses and strains are called *constitutive equations*. Hooke's law, for example, is a linear constitutive relation that is valid for linear materials.

When a tendon is forcibly stretched, its cross-sectional area decreases. The stress measured relative to the instantaneous area is called the *Cauchy stress* (after French mathematician A.L. Cauchy, 1789-1857) or *true stress*. However, for simplicity in small deformations, the stress is estimated using the original cross-sectional area of the tendon. This stress measure is sometimes called the *Lagrangian stress* (after J-L. Lagrange, 1736-1813, an Italian mathematician who lived the most of his life in France). For large deformations, changes in cross-sectional area cannot be neglected, so instantaneous area should be used. In large deformations, therefore, stress increases are caused by two interacting factors: (a) force increases and (b) cross-sectional area decreases. Consequently, analysis of large deformations requires complex mathematics and is beyond the scope of this book. The ratio of transverse strain (normal to the applied load) divided by the relative extension strain (in the direction of the applied load) is known as *Poisson's ratio*.

2.1.1.1 Stress–Strain Relations

Representative publications: Frisen et al. 1969; Diamant et al. 1972

The elastic behavior of tendons and other objects is represented as either a force–deformation or a stress–strain relation. In the first case, the main variable of interest is tendon stiffness (N/m) or compliance (m/N); in the second case, elastic behavior is characterized by Young's modulus. The value of Young's modulus, also called the *tangent modulus of elasticity* or *elastic modulus*, equals the slope of the stress–strain curve, $d\sigma_T / d\varepsilon_T$, where σ_T is the tendon stress (N/m²), that is, the ratio of tendon force to tendon cross-sectional area, and ε_T is the strain, the ratio of the tendon elongation to the tendon length at rest (dimensionless). Young's modulus for tendons can be regarded as the stress required making $\Delta l_T / l_T = 1$, where l_T is the length of the tendon prior to the deformation. It can also be represented by the product of tendon stiffness and the ratio of the original length to the cross-sectional area.

In some joint configurations, the tendon may be slack: it does not resist force when elongated. For such cases, elongation Δl_T is defined relative to the slack length, $\Delta l_T = l_{Te} - l_{Ts}$, where l_{Te} is the length of the elongated (deformed) tendon and l_{TS} is its *slack length*, the length at which the tendon begins to resist external force. The slope of the stress–strain curve, in general, depends on tendon elongation. The curve has three regions (figure 2.2): (a) an initial toe region where the slope increases, (b) a linear region where the slope is constant, and (c) a failure region.

When in vivo testing is performed on humans, loads are below the injury threshold and thus the failure region is not reached (figure 2.3).

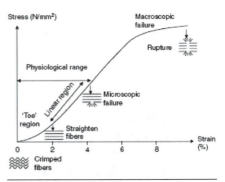

Figure 2.2 Stress–strain curve of a tendon, schematics.

Reprinted from *Journal of Biomechanics* 39(9), J.H.-C. Wang, "Mechanobiology of tendon," 1563-1582, 2006, with permission from Elsevier.

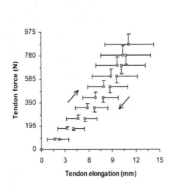

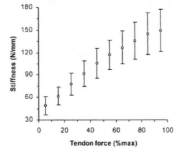

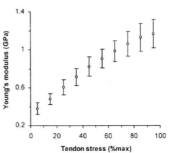

Figure 2.3 Mechanical properties of the gastrocnemius tendon. Measurements were taken in vivo in six men. Tendon forces were calculated from the moment of force generated during isometric plantar flexion contraction. Left panel: The gastrocnemius tendon force–elongation relation. Arrows indicate loading and unloading directions. Values are means ± standard deviations. The elongation of the tendon increased in a curvilinear manner with the force acting upon it, from 1.7 ± 1 mm (0.8% ± 0.3% strain) at 87.5 ± 8.5 N to 11.1 ± 3.1 mm (4.9% ± 1% strain) at 875 ± 85 N. Top right panel: Tendon stiffness as a function of tendon load. Bottom right panel: Young's modulus at different loads. The tendon Young's modulus and hysteresis were 1.16 ± 0.15 GPa and 18% ± 3%, respectively (hysteresis is explained in section **2.1.2.1**).

Reprinted from *Journal of Biomechanics* 35(12), C.N. Maganaris and J.P. Paul, "Tensile properties of the in vivo human gastrocnemius tendon," 1639-1646, copyright 2002, with permission from Elsevier.

2.1.1.1.1 Stress–Strain Relations in the Toe Region

Representative publications: Frisen et al. 1969; Diamant et al. 1972

In the toe region, the tendon is strained to approximately 2% (range 1.5%-4%). The only known exception is the human psoas tendon, where this region extends to 10% strain. Under these low strain conditions, collagen fibers straighten and lose their crimp pattern but the fiber bundles (figure 2.2) themselves are not stretched. The modulus of elasticity increases with strain until it reaches a constant value at the start of the linear region. The average slope of the stress–deformation curve in the toe region is at times called the *toe-region modulus*.

Nonlinear elasticity in the toe region is explained by two mechanisms. The first model assumes consecutive recruitment of individual collagen fibrils at varying degrees of crimp. With increasing tensile deformation, new load bearing fibrils jump into action and resist deformation. A nonlinearity arises because of the increasing number of fibrils contributing to the tendon resistance (figure 2.4, *a* and *b*). This model explains how a large number of linear elements with different resting lengths can produce a globally nonlinear force–deformation curve (figure 2.4*c*).

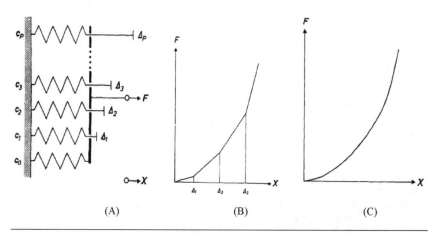

(A)   (B)   (C)

Figure 2.4 Model of the nonlinear elasticity of the tendon in the toe region. *(a)* Linear elements with spring constants $\Delta_0, \Delta_1, \Delta_2 \ldots \Delta_p$ come into action at certain stages of deformation ($\Delta_1, \Delta_2, \Delta_3 \ldots \Delta_p$). *(b)* Load-deformation relation of the model. *(c)* Continuous representation of the model.

Reprinted from *Journal of Biomechanics* 2(1), M. Frisén, M. Mögi, I. Sonnerup, and A. Viidik, "Rheological analysis of soft collagenous tissue: Part I. Theoretical considerations," 13-20, copyright 1969, with permission from Elsevier.

Another model for toe-region nonlinear elasticity is the flexible hinge model of tendon crimp (figure 2.5). This model assumes that crimped collagen fibers are lengthened until they become straight, similar to stretching a helical spring until it becomes a straight wire. Stiffness increases from a low value for the spring to a much larger value for the straight wire.

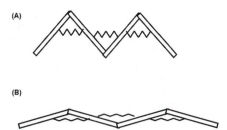

Figure 2.5 The flexible hinge model of the tendon crimp. In the model, the rigid collagen fibers are connected by flexible hinges. The unbending occurs by stretching the flexible hinges.

Reprinted, by permission, from J. Diamant, A. Keller, E. Baer, M. Litt, and R.G.C. Arridge, 1972, "Collagen; Ultrastructure and its relation to mechanical properties as a function of ageing," *Proceedings of the Royal Society of London, Series B: Biological Sciences* 180(1060), p. 307, fig. 12.

Following this model, two kinds of axial strain for tendon are distinguished: (a) *apparent strain* (macroscopic strain), which is defined in the usual way, as the amount of elongation per unit of object length, $\varepsilon = \Delta l / l_0$, where Δl is the elongation and l_0 is the initial tendon length, and (b) a *material strain* (microscopic strain), which is the relative fiber length change (tendon fibers considered elastic material), $\varepsilon^f = \Delta l^f / l_0^f$, where superscript f refers to the tendon fiber and subscript zero represents the length of the uncrimped (straight) tendon fiber at minimal load (figure 2.6).

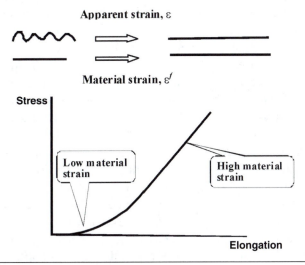

Figure 2.6 Apparent and material strain of a tendon, schematic.

2.1.1.1.2 Stress–Strain Relations in the Linear Region

Representative publication: Maganaris and Paul 1999

The linear region begins at approximately 2% of the strain for most tendons. In this region, tendon elongation results from stretching prealigned collagen fibers, and hence the stiffness is indicative of the collagen fiber stiffness (material stiffness). The slope of the stress–strain curve in the linear region is usually referred as the Young's modulus of the tendon (*linear-region modulus*). If the tendon is stretched beyond 4% of its original length, microscopic tearing of the tendon fibers can occur (with the exception of the psoas tendon, mentioned previously).

Macroscopic failure occurs when the tendon is forcibly elongated by more than 10% of its original length. To determine tendon elongation, the tendon length that corresponds to zero strain should be established. During in vivo measurements on human subjects, this length is usually not known precisely because at any joint position antagonist muscles can exert nonzero forces that balance each other at the joint but nevertheless cause tendon lengthening. This limitation causes decreased accuracy during noninvasive measurements (e.g., with ultrasound imaging methods) but can be overcome when noninvasive imaging is used in conjunction with more invasive methods, like tendon force transducer implants.

The linear region begins when the tendon is stressed 5 to 30 MPa and ends at the beginning of the failure region, at about 70 to 100 MPa. The Young's modulus in the linear region is approximately 0.6 to 1.7 GPa. Some examples from the literature are presented in table 2.2. In experiments performed on human subjects, the tendon cross-sectional area is not always known and hence only stiffness or compliance values are reported.

The aforementioned mechanical behavior of tendons arises from both their size (cross-sectional area) and the biological properties of the tendon material itself. For example, the Achilles tendon has higher resistance to deformation than other tendons in the body, but this can be attributed to its larger cross-sectional area. In some tendons, material properties vary across tendon portions. For example, the anterior portion of the supraspinatus tendon is mechanically stronger than other portions (has a larger Young's modulus), and it is largely this portion that plays the main functional role of the tendon.

The strain patterns in some tendons are not uniform along their lengths. They may exhibit stress-shielded areas and even areas subjected to compressive loading, especially at the enthesis (tendon–bone junction). Additional work is needed to clarify whether these findings are important for tendon pathology and rehabilitation techniques (see section **5.1.2**).

Table 2.2 Examples of Elastic Properties of Human Tendons Determined From in Vivo Ultrasound Measurements

Tendon	Method, subjects	Stress, MPa (1 MPa = 1 N/mm²)	Strain, %	Stiffness, N/mm	Young modulus, GPa (1 GPa = 1,000 N/mm²)	Author, year
Patellar	Maximal voluntary contraction. Two groups: (a) high range of knee motion, and (b) small range of knee motion.	(a) 22.4 ± 6.5 (b) 34.0 ± 17.6	(a) 6.5 ± 1.6 (b) 7.2 ± 1.9	(a) 3,269 ± 1,591 (b) 3,185 ± 1,457	(a) 0.81 ± 0.35 (b)1.22 ± 0.52	Bojsen-Moller et al. 2007
Patellar[a]	10 s ramp isometric knee extension. Young males.	30.2 ± 2.2	6.9 ± 0.6	4,334 ± 562	1.09 ± 0.12	Hansen et al. 2006
Patellar[a]	In vivo, senior adults, before (a) and after (b) 14 weeks of strength training.	(a) 40 ± 11 (b) 42 ± 11	(a) 9.9 ± 2.2 (b) 5.9 ± 2.4	(a) 2,187 ± 713 (b) 3,609 ± 1220	(a) 1.3 ± 0.3 (b) 2.2 ± 0.8	Reeves et al. 2003
Tibialis anterior tendon[a]	Transcutaneous electrical stimulation from (a) 25 V to (b) 150 V. Male subjects.	(a) 3.4 ± 0.5 (b) 25 ± 2.5	(a) 0.7 ± 0.1 (b) 2.5 ± 0.4	(a) 62 ± 12 (b) 161 ± 26	(a) 0.45 ± 0.06 (b) 1.2 ± 0.15	Maganaris and Paul 1999
Tibialis anterior tendon and aponeurosis	Maximal voluntary contraction, male subjects. (a) Tendon, (b) aponeurosis, (c) aponeurosis distal region, (d) aponeurosis proximal region.		(a) 3.1 ± 0.2 (b) 6.5 ± 0.6 (c) 3.5 ± 0.3 (d) 9.2 ± 1.0			Maganaris and Paul 2000c
Gastrocnemius tendon[a]	Male subjects. Isometric plantar flexion contraction.		(a) 0.8 ± 0.3 at 87.5 ± 8.5 N (b) 4.9 ± 1.0 at 875 ± 5 N		1.16 ± 0.15	Maganaris and Paul 2002
Gastrocnemius tendon and aponeurosis	Maximal plantar flexion efforts.		5.0 at 140 Nm			Arampatzis et al. 2005
Achilles tendon	Maximal voluntary 10 s isometric plantar flexion.	41.6 ± 3.9	4.4-5.6		1.048-1.474	Magnusson et al. 2001
Achilles tendon[b]	One leg hopping.		8.3 (average peak)	145-231 (interquartile range)	0.67-1.07 (interquartile range)	Lichtwark and Wilson 2005
Achilles tendon	Maximal voluntary plantar flexion. (a) Runners, (b) nonrunners	(a) 26.3 ± 5.1 (b) 38.2 ± 9.8	(a) 4.1 ± 0.8 (b) 4.9 ± 0.8		(a) 306 ± 61 (b) 319 ± 42	Rosager et al. 2002

[a]A curvilinear relation between the stress and strain has been reported.

[b]The force–length relation is relatively linear, particularly at high strains.

2.1.1.2 Tendon Forces

Representative publications: Komi 1990; Fleming and Beynnon 2004

Tendon forces can be quite large. Because of the proximity of tendons to joint axes of rotation, tendon forces can be several times larger than the forces exerted on the environment. For instance, forces on the flexor digitorum superficialis tendon during static efforts can be 5 to 6 times larger than the finger contact force; during keyboard tapping maximum tendon forces can be 4 to 7 times larger than fingertip forces; forces on the Achilles tendon can be 10 to 12 times body weight.

In everyday life, the tendons usually operate in the toe region. However, during extreme force exertions, tendon forces can venture into the linear region of the force–deformation curve.

■ ■ ■ *FROM THE LITERATURE* ■ ■ ■

Achilles Tendon Forces During Walking, Running, and Jumping Can Be 12.5 Times Body Weight

Source: Komi, P.V. 1990. Relevance of in vivo force measurements to human biomechanics. *J Biomech* 23(Suppl 1):23-34

The author developed a method to directly record in vivo forces from the human Achilles tendon. A "buckle" force transducer was fixed directly to the Achilles tendon under local anesthesia. (In the best traditions of good science, the author volunteered to be the first subject to undergo the surgical implantation. In later studies, instead of the bulky force transducer, less traumatic optic fiber sensors were used.) Achilles tendon forces during walking, running, and jumping were recorded. In some cases the loading of Achilles tendon reached values as high as 9 kN, corresponding to 12.5 times the body weight or, when expressed per cross-sectional area of the tendon, 1,100 N/cm^2 (11.1 MPa).

Achilles Tendon Forces During Cycling

Source: Gregor, R.J., P.V. Komi, and M. Jarvinen. 1987. Achilles tendon forces during cycling. *Int J Sports Med* 8(Suppl 1):9-14

The forces of the triceps surae muscles were recorded in vivo directly from the Achilles tendon of a healthy male subject (the first author of the paper). The subject pedaled on a standard bicycle ergometer at varying loads (88, 176, and 265 W). Peak Achilles tendon forces increased from 489 N to 661 N. These peak forces, which were recorded at 115° of the pedaling cycle, did not change with increases in the pedaling cadence.

Tendons can sustain and thus transmit very large forces. Their in vitro tensile strength is about 50 to 100 MPa. The values for ultimate tensile strain at failure are around 15% ± 5%. A tendon with a cross-sectional area of 1 cm² is capable of supporting a weight up to 10 kN. For comparison, specific force of a mammalian muscle—the force per unit of physiological area—is on average only about 25 N/cm² (250 KPa, described in section **3.1.6.1**).

▪▪▪ *FROM THE LITERATURE* ▪▪▪

Human Patellar-Tendon Rupture
During Actual Sports Competitions

Source: Zernicke, R.F., J. Garhammer, and F.W. Jobe. 1977. Human patellar-tendon rupture. *J Bone Joint Surg Am* 59(2):179-183

The authors serendipitously filmed a patellar tendon rupture during a national weightlifting championship. The time course and magnitudes of hip, knee, and ankle joint moments of force and of tensile loading of the patellar tendon before and during tendon trauma were determined. At the instant of rupture, the patellar tendon tension was equal to approximately 14.5 kN, or more than 17.5 times body weight.

2.1.1.3 Tension and Elongation in Tendons and Aponeuroses

Representative publications: Proske and Morgan 1987; Magnusson et al. 2008

Aponeuroses thickness gradually increases with proximity to the external tendon. Therefore, the distal parts of the aponeurosis, those that are closer to the external tendon, can transmit larger forces. Experiments on several animals (frogs, cats) have shown that aponeuroses and tendons have similar mechanical properties. Such data do not exist for humans.

In the simplest models of muscle–tendon units (e.g., figure 1.6*a*), tendons and aponeuroses are connected in series. It can be expected that this structural arrangement will yield similar forces in the tendon and aponeurosis (with the possible exception that the parts of the aponeurosis that are located closer to the external tendon transmit larger forces than the parts that are farther from it). However, the force interaction and transmission between the external and internal parts of the tendon for the majority of human muscles remain unknown.

Tendon elongation during natural human movements can be quite large. For instance, during walking the triceps surae elastic structures elongate up to 4%; during running they elongate up to 7% (hence, forces exerted in running are quite large and the elastic structures likely do not operate in the toe region, where the tendon is strained to only 4%, see section **2.1.1.1.1**). The distribution of strain and elongation between the tendon and aponeurosis is different for different muscle groups. In some muscles (e.g., knee extensors on the whole) aponeurosis strain is larger than tendon strain, whereas in other muscles (e.g., gastrocnemius) the reverse is true, and in still others (e.g., vastus lateralis) the strains are similar.

▪ ▪ ▪ *FROM THE LITERATURE* ▪ ▪ ▪

Forces on Tendon and Aponeurosis May Differ From Each Other

Source: Epstein, M., M. Wong, and W. Herzog. 2006. Should tendon and aponeurosis be considered in series? *J Biomech* 39(11):2020-2025

Muscle fibers, aponeuroses, and tendons are often considered mechanically "in series" in skeletal muscles. However, for complex architecture (e.g., figure 1.6, *b* and *c*), this assumption is most likely invalid. In this article the authors theoretically demonstrated, using examples of increasing complexity, that tendons and aponeuroses are not mechanically serial structures. The authors' explanations for this finding include not only tendon and aponeurosis direction differences but other factors such as the direct muscle tissue force transmission and muscle incompressibility (discussed in chapter 3). For these reasons, modeling the force transmission between the aponeurosis and tendons based on free-body diagrams (described in chapter 2 of *Kinetics of Human Motion*) can result in substantial inaccuracies.

■ ■ ■ *FROM THE LITERATURE* ■ ■ ■

Strain of the Tendon and Aponeurosis of Knee Extensors and Ankle Plantar Flexors Is Different

Source: Kubo, K., H. Kanehisa, and T. Fukunaga 2005. Comparison of elasticity of human tendon and aponeurosis in knee extensors and ankle plantar flexors in vivo. *J Appl Biomech* 21(2):129-142

Tendon and aponeurosis elongation during isometric knee extension ($n = 23$) and ankle plantar flexion ($n = 22$), respectively, were determined using a real-time ultrasonic apparatus during ramp isometric contractions up to voluntary maximum. For the knee extensors, the maximal aponeurosis strain (12.1% ± 2.8%) was significantly greater than that of the patellar tendon (8.3% ± 2.4%) ($p < .001$). On the contrary, the maximal strain of the Achilles tendon (5.9% ± 1.4%) was significantly greater than that of the ankle plantar flexor aponeuroses (2.7% ± 1.4%) ($p < .001$).

Strain of the Vastus Lateralis Aponeurosis and Tendon Is Similar

Source: Stafilidis, S., K. Karamanidis, G. Morey-Klapsing, G. Demonte, G.P. Bruggemann, and A. Arampatzis. 2005. Strain and elongation of the vastus lateralis aponeurosis and tendon in vivo during maximal isometric contraction. *Eur J Appl Physiol* 94(3):317-322

The maximal estimated strains of the vastus lateralis tendon, tendon plus aponeurosis, and aponeurosis showed no statistical differences (8% ± 2%, 8% ± 1%, and 7% ± 2%, respectively, $p > .05$). The authors concluded that the strains of the human vastus lateralis tendon, vastus lateralis tendon plus aponeurosis, and vastus lateralis aponeurosis, as estimated in vivo during maximal isometric contractions, do not differ from each other. Note the difference from the previously cited paper.

Even for the same muscle groups, various authors have reported different findings. The triceps surae is of special interest because of its importance in walking and running and its anatomical complexity. This muscle consists of three separate muscle bodies merging into the Achilles tendon. The relative contribution of these muscles to the Achilles tendon force may be different depending on (a) the level of activation, (b) the physiological cross-sectional area (PCSA), and (c) the knee joint angle, which influences the length of the two-joint medial and lateral gastrocnemius heads but not the soleus length, which is a one-joint muscle. Hence, depending on whether the authors studied individual muscle heads or entire muscle, they obtained different results.

■ ■ ■ *FROM THE LITERATURE* ■ ■ ■

Strain and Lengthening of the Achilles Tendon and Aponeurosis

A. There Is No Significant Difference in Strain Between the Achilles Tendon and Aponeurosis for Medial Gastroc-nemius Muscle

Sources: (1) Muramatsu, T., T. Muraoka, D. Takeshita, Y. Kawakami, Y. Hirano, and T. Fukunaga. 2001. Mechanical properties of tendon and aponeurosis of human gastrocnemius muscle in vivo. *J Appl Physiol* 90(5):1671-1678. (2) Arampatzis, A., S. Stafilidis, G. DeMonte, K. Karamanidis, G. Morey-Klapsing, and G.P. Bruggemann. 2005. Strain and elongation of the human gastrocnemius tendon and aponeurosis during maximal plantarflexion effort. *J Biomech* 38(4):833-841

Load-strain characteristics of tendinous tissues (Achilles tendon and aponeurosis) were determined in vivo for human medial gastrocnemius muscle. The maximal strain of the Achilles tendon and aponeurosis was 5.1% ± 1.1% and 5.9% ± 1.6%, respectively. There was no significant difference in strain between the Achilles tendon and aponeurosis. In addition, no significant difference in strain was observed between the proximal and distal regions of the aponeurosis.

B. The Strain of the Achilles Tendon Is Much Larger Than That of the Aponeurosis

Sources: (1) Maganaris, C.N., and J.P. Paul. 2000. *In vivo* human tendinous tissue stretch upon maximum muscle force generation. *J Biomech* 33:1453-1459. (2) Maganaris, C.N., and J.P. Paul. 2000. Load-elongation characteristics of in vivo human tendon and aponeurosis. *J Exp Biol* 203:751-756. (3) Magnusson, S.P., P. Hansen, P. Aagaard, J. Brond, P. Dyhre-Poulsen, J. Bojsen-Moller, and M. Kjaer. 2003. Differential strain patterns of the human gastrocnemius aponeurosis and free tendon, in vivo. *Acta Physiol Scand* 177(2):185-195

In the first study, the strain in the human tibialis anterior muscle–tendon unit during maximal voluntary contraction was measured. The aponeurosis strain was 6.5%, whereas the tendon strain was only 3.2%. In the second study, the muscle contraction was induced by percutaneous tetanic stimulation. The strain distribution was similar to that found in the first study. In the third study, at a common tendon force of 2,641 ± 306 N, the respective deformation and segment length were 5.85 ± 0.85 and 74 ± 0.8 mm for the free tendon and 2.12 ± 0.64 and 145 ± 1.3 mm for the distal aponeurosis. Longitudinal strain was 8.0% ± 1.2% for the tendon and only 1.4% ± 0.4% for the aponeurosis. The authors concluded that the free Achilles tendon demonstrates greater strain compared with the distal (deep) aponeurosis during voluntary isometric contraction.

《 2.1.1.4 Constitutive Equations for Tendons and Ligaments

Representative publications: Cowin and Doty 2007; Weiss and Gardiner 2001; Woo et al. 1993; Woo 1982

When a tendon is forcibly stretched, resistance comes from three main sources: (1) uncrimping of the fibers, (2) the fiber lengthening (deformation), and (3) the resistance provided by the tendon matrix. The first term of the constitutive equation that describes the straightening of the fibers is usually modeled as an exponential function. The second component is a linear function that represents the fiber lengthening past a critical stretch level when the fibers are straightened. The third component represents the matrix deformation with the matrix material modeled as either completely incompressible or as slightly compressible. Including the matrix deformation in the equation requires considering the deformation in 3-D space. However, to simplify the constitutive equations in different models, some consequences of 3-D consideration (e.g., shear forces in the matrix) are sometimes neglected. For the complete analysis of the constitutive equations, six material coefficients are necessary: these can include the rate at which collagen fibers uncrimp, the modulus of the straightened collagen fibers, and the level of the fiber stretch after which the fiber is straight and the stress–strain relation is linear. The reader interested in the tendon constitutive equations is directed to the representative publications mentioned in the beginning of this section. 》

2.1.2 Viscoelastic Behavior of Tendons

Tendons, as well as connective tissue in general, display time-dependent behavior; they are viscoelastic bodies.

2.1.2.1 Basic Concepts of Viscoelasticity

Viscoelasticity is characterized by the four main phenomena: (a) sensitivity to strain rate, (b) creep, (c) stress relaxation, and (d) hysteresis. *Creep* is an increase in length under a constant load; it is a retarded deformation (figure 2.7*a*). *Stress relaxation* is a decrease of stress over time under constant deformation. *Hysteresis*, or *elastic hysteresis*, is the difference between the loading and unloading curves in the stress–strain cycle. Under cycling loading, the strain magnitude depends not only on the present stress but also on its previous levels—on whether the stress was increasing or decreasing. The hysteresis is due to inability of the system to follow identical paths upon application and withdrawal of a force. The hysteresis is manifested as the hysteresis loop (figure 2.7*c*). When a tendon is forcibly stretched, the area beneath the force–deformation curve represents the mechanical work done to extend the tendon (work = $\int F\,dE$, where F is force acting in the direction of tendon extension E). For the same level of tendon extension, the force during loading is larger than the force during unloading. The hysteresis area represents the energy loss in a deformation cycle, mainly due to converting mechanical work into heat. Hysteresis is often quantified as the ratio of the hysteresis loop divided by the area under the loading curve. The ratio represents the percentage of the mechanical energy loss in the loading–unloading cycle.

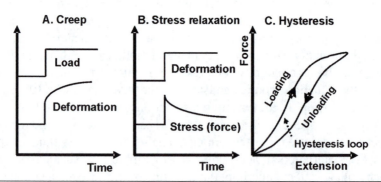

Figure 2.7   Main features of the viscoelasticity. *(a)* Creep. Under constant load, the deformation increases. *(b)* Stress relaxation. Under constant deformation, the stress gradually decreases. *(c)* Hysteresis. The force–deformation curves are different for the loading and unloading periods. The area of the hysteresis loop is representative of the energy loss.

■ ■ ■ *FROM THE LITERATURE* ■ ■ ■

Hysteresis in Intact Human Tendon

Source: Maganaris, C.N., and J.P Paul. 2000. Hysteresis measurements in intact human tendon. *J Biomech* 33(12):1723-1727

Mechanical hysteresis was measured in the human tibialis anterior tendon. With the foot fixed on a dynamometer footplate, the displacement of the tibialis anterior tendon during stimulation and relaxation of the tibialis anterior muscle was recorded by means of ultrasonography in six men. A hysteresis loop was obtained between the load–displacement curves during contraction and relaxation. Measurement of the hysteresis loop area yielded a value of 19%. This value agrees with results from in vitro tensile tests.

Mechanical models of viscoelastic materials commonly include a combination of linear springs and dashpots. The length of a linear spring instantaneously changes in proportion to the load. A dashpot velocity is proportional to the load at any instant (if the velocity is specified, the dashpots provide resistive force in proportion to the loading velocity). Each element is characterized by one parameter, a coefficient of proportionality: (a) for the linear spring, between the load and deformation (stiffness); and (b) for the dashpot, between the load and velocity (damping). Models can include two or more elements, but models with more than three elements are almost never used. Such models are called *models with lumped parameters*. The three prevalent models of viscoelastic behavior are shown in figure 2.8.

The Maxwell model (named after Scottish physicist James Clark Maxwell, 1831-1879) includes a spring and dashpot in series. The force applied to the spring equals the force exerted on the dashpot. The total strain equals the sum of strains of the contributing elements. After a sudden application of a load, the model predicts an immediate deformation of the spring followed by creep of the dashpot. A sudden deformation produces an immediate reaction of the spring, which is followed by stress relaxation. When an object is subjected to repetitive loading and unloading cycles, the model predicts decreasing hysteresis with increasing frequency because the dashpot movement will be reduced.

The Voigt model (after German physicist Woldemar Voigt, 1850-1919) is a combination of a spring and dashpot in parallel. In such a model, a strain applied to the body is the same for the spring and the dashpot. The

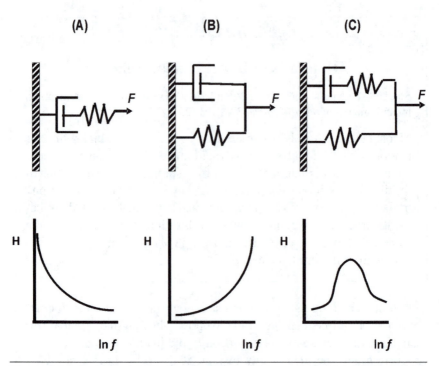

Figure 2.8   Basic models of viscoelasticity. *(a)* Maxwell model, *(b)* Voigt model, and *(c)* Kelvin model (standard linear model). Upper row, the models; bottom row, expected responses of the hysteresis (H) to the increased frequency of the loading and unloading cycles. The frequency *f* is represented in its natural logarithm. The H is defined as the ratio of the area of hysteresis loop divided by the area under the loading curve.

force resisted by the body is the sum of the forces acting on the contributing elements. After a sudden application of a load, the model does not predict immediate spring deformation because the dashpot will not move instantly. The deformation will gradually increase. As with the Maxwell model, a sudden deformation produces an immediate reaction of the spring that is followed by stress relaxation. With increasing frequency of loading and unloading cycles, the hysteresis should increase because the dashpot bears more and more of the load. The reactions of the Maxwell and Voigt models to the increased frequency of oscillation are opposite.

A three-element model (figure 2.8c), commonly called the *standard linear model* or the *Kelvin model* (after an Irish physicist Lord W.T. Kelvin, 1824-1907), includes a dashpot and two springs, one in series with the dashpot and the second in parallel to the dashpot–serial spring assemblage. The model predicts a bell-like dependence of the hysteresis on the logarithm of the frequency. The Hill model of muscle mechanics that is described in chapter 4 is similar to the standard linear model with one significant distinction: instead of including the dashpot, the Hill model includes a contractile element (the element possesses also damping properties).

••• PHYSIOLOGY REFRESHER •••

Viscosity

Viscosity is internal friction in a liquid or a gas. In the International System (SI) of metric units, viscosity is defined as the resistance that a liquid or gaseous system offers to flow when it is subjected to a shear stress (figure 2.9).

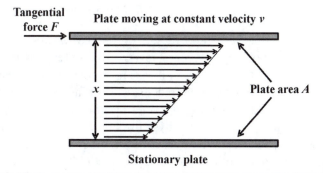

Figure 2.9 Definition of viscosity. Two plates are moving horizontally with respect to each other at velocity *v* at distance *x*. The space between the plates is filled with a viscous liquid. Force *F* is given by equation 2.1.

Viscosity expresses the magnitude of internal friction in a substance, as measured by the force per unit area resisting uniform flow. The governing expression (Newton's equation) is

$$f = \eta A \frac{dv}{dx},\qquad [2.1]$$

where *f* is the force required to maintain a velocity gradient, dv/dx, between planes of area *A*, and η is the viscosity coefficient. The SI metric unit for viscosity is $(Nm^2) \cdot s = Pa \cdot s$ (Pascal-second). The viscosity unit is the force per unit area required to sustain a unit velocity

(continued)

Physiology Refresher *(continued)*

gradient normal to the flow direction. A high viscosity fluid does not flow as easily as a low-viscosity fluid (e.g., honey does not move as easily as water).

The viscosity of some biological fluids, such as blood or synovial fluid, provides important functionality. In healthy people the viscosity of synovial fluid is non-Newtonian; that is, viscosity decreases when the joint angular velocity increases. There is a historical tendency in the biomechanics literature to use the term *viscosity* with a broader meaning, as the resistance of any source proportional to velocity. Such a use of this term should be discouraged. The coefficient in the equation $f = -kv$ should be called the *damping coefficient*.

2.1.2.2 Viscoelastic Properties of Tendons

Representative publications: Fung 1967; Cowin and Doty 2007

Tendons possess viscoelastic properties. However, the hysteresis dependence on the loading–unloading frequency is very small. Tendons (as well as other soft biological tissues) usually have a near-constant hysteresis over a large range of loading–unloading frequencies. For instance, when tendons are subjected to loading–unloading cycles at different rates, a 1,000-fold increase of frequency changes the hysteresis by no more than a factor of 2. The linear viscoelastic models described in the previous section do not explain this fact. Hence, other, more complex models have been suggested.

《 2.1.2.2.1 Computational Models of the Tendons

Representative publications: Fung 1967, 1993; Woo et al. 1993; Atkinson et al. 1997

Detailed description of the contemporary computational models of the tendons is beyond the scope of this book. The most broadly used is the quasi-linear viscoelasticity (QLV) model, suggested by Fung (1967).

••• CALCULUS REFRESHER •••

Convolution and Convolution Integral

Convolution is a mathematical operation on two functions producing a third function. Convolution is similar to cross-correlation: it is the overlap of the two functions as one function is passed (shifted) over the second. Suppose that the system is linear and time-invariant; response of the system to a unit impulse is $h(t)$. If the input signal is $x(t)$, then output signal $y(t)$ is described by the convolution integral where τ is a so-called dummy variable.

$$y(t) = h(t) \otimes x(t) = \int_{-\infty}^{\infty} h(\tau)x(t-\tau)d\tau \qquad [2.2]$$

Initially, both functions are expressed in terms of the dummy variable τ. Then, one of the functions is transposed, $\chi(\tau) \rightarrow \chi(-\tau)$, and the time offset t is added. The function $\chi(t - \tau)$ slides along the axis from $-\infty$ to $+\infty$. Wherever the two functions intersect, the integral of their product is computed. To compute the output $y(t)$ at a specified t, first the integrand $h(\tau)\chi(t - \tau)$ is computed as a function of τ. Then integration with respect to τ is performed, resulting in $y(t)$.

The QLV theory is based on the idea that the stress–strain relation can be separated into two parts: an elastic part and a history-dependent part. The elastic part characterizes the elasticity of the material, that is, its strain–strength relation. The elastic response of a material is the stress in response to an instantaneous step input of strain **E**. If a tendon is suddenly strained to a certain magnitude **E** and maintained constant at this length, the stress will be a function of **E** as well as time. Because of relaxation, stress is usually a monotonically decreasing function of time. The history of stress changes is called the *reduced relaxation function*. The QLV model assumes that the so-called *superposition principle* applies and the stress at time t is the sum of the contributions of the past changes that follow the same reduced relaxation function. The stress resulting from arbitrary stretch history at any time equals a convolution integral of the elastic stress response and the reduced relaxation function. 》

2.1.2.2.2 Factors Affecting Mechanical Properties of the Tendons

Representative publications: Archambault et al. 1995; Magnusson et al. 2008; Kubo et al. 2002, 2003a, 2003b, 2007; Wang 2006

The metabolic activity in human tendon is high, which allows tendons to adapt to changing demands. The tendons adapt to applied forces and motion. The mechanical properties of intact tendons depend on their length (which does not change much in adults), cross-sectional areas, and mechanical properties of the tendon tissue. Consider experimental data on the main factors affecting tendon properties.

• *Gender.* In women, stiffness and Young's modulus are lower whereas the maximum Lagrangian strain ($100 \times L$ / Initial Tendon Length) is greater than in men.

• *Aging.* Aging alters tendon mechanical properties; stiffness and Young's modulus are lower in older adults than in young adults. Maximal strain (L / Initial Tendon Length) and stiffness of the tendon–aponeurosis structures decrease significantly with aging. In contrast, hysteresis increases significantly. Exercise in old age can partly reverse this process.

• *Disuse.* Tendon stiffness decreases with disuse. Complete unloading due to bed rest or spinal cord injury reduces tendon stiffness and modulus, however, only chronic unloading due to spinal cord injury seems to cause tendon atrophy.

• *Exercise and training.* The effects of training on the mechanical properties of tendons are exercise specific. Static stretching systematically applied for 10 min significantly decreases stiffness and hysteresis. Low-load resistance training (squats using body weight, for 6 months) produces no significant change in stiffness and hysteresis but significantly increases the maximal elongation of tendon–aponeurosis structures from 23.3 ± 2.1 mm to 24.8 ± 2.2 mm (Kubo et al. 2003b). Resistance training—plantar flexion at 70% of 1-repetition maximum with 10 repetitions per set (5 sets per day)—increases the stiffness of tendon structures as well as muscle strength and size (Kubo et al. 2002). Increase of the tendon stiffness is usually accompanied by an increase in the rate of force development.

2.1.3 Tendon Interaction With Surrounding Tissues

This section addresses the mechanical interaction of tendons with other tendons, synovial sheaths, laterally located bones, and muscles.

2.1.3.1 Intertendinous Shear Force and Lateral Force Transfer

Representative publications: Bojsen-Moller et al. 2004; Leijnse et al. 1993; Schieber and Santello 2004

When several muscles connect to a common tendon, differential muscle displacements may occur. For instance, during plantar flexion the medial gastrocnemius and soleus aponeuroses deform to different extent; as a result the layers of connective tissue lying in parallel to the acting forces slide with respect to each other (figure 2.10). Such a displacement should be accompanied by shear forces between the adjacent aponeurosis. However, in majority of tendons force transmission occurs mainly along the individual fascicles that act as independent structures, and lateral force transmission between adjacent collagen fascicles is small.

One specific case of intertendinous transfer occurs in the hand, where tendons from extrinsic muscles that serve different fingers usually con-

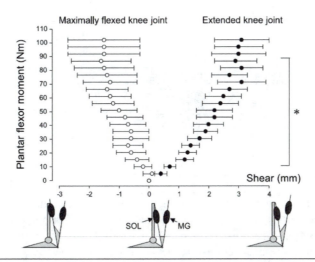

Figure 2.10 Relative displacement of the medial gastrocnemius and soleus aponeuroses proximal to the junction of the Achilles tendon at the different knee joint angles and different voluntary contractions of the plantar flexor muscles. *Statistically significant differences ($p < .05$) in the interaponeuroses shear displacements between two knee positions at plantar flexion moments of 10 to 87 Nm. Moments greater than 87 Nm were achieved by only 4 subjects of 8.

Reprinted, by permission, from J. Bojsen-Møller, P. Hansen, P. Aagaard, et al., 2004, "Differential displacement of the human soleus and medial gastrocnemius aponeuroses during isometric plantar flexor contractions in vivo," *Journal of Applied Physiology* 97(5): 1908-1914.

nect to each other. Tendon interconnections are one of the causes of finger *enslaving*: when one finger exerts force or moves other fingers also move or exert force. You can perform a simple demonstration: place your hand with the palm up and wiggle your ring finger with maximal amplitude as fast as possible. You will see that other fingers also move. In addition to tendon interconnection, neural factors contribute to enslaving phenomena. Neural commands originating from cortical cells may diverge to different muscles, and even if neural commands reach a single muscle their effects may be mechanically divergent because single motor units may connect to tendons that serve different fingers (figure 2.11). When such motor units are active, a constant fraction of motor unit force is transmitted to another finger. The force transmitted via the tendon interconnections depends on several factors such as slackness, elasticity, and connection location. These factors, muscle coactivation and tendon interconnections, hinder finger movement independence, which can have adverse effects in some activities, for instance, piano playing. Some cases of surgical removal of tendon interconnections have been reported in the literature, and following surgery hand dexterity improved.

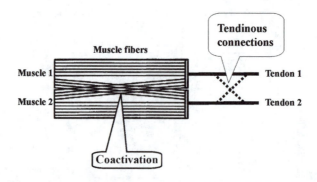

Figure 2.11 A model of two muscles serving different fingers with inter-muscle coactivation and intertendon connections. Coactivation occurs because some fibers from one muscle (or motor unit) attach also to the tendon of another muscle. The connections are between tendons 1 and tendon 2 (dotted lines).

■ ■ ■ *FROM THE LITERATURE* ■ ■ ■

Enslaving in Multifinger Force Production

Source: Zatsiorsky, V. M., Z.M. Li, and M.L. Latash. 2000. Enslaving effects in multi-finger force production. *Exp Brain Res* 131(2):187-195

This study explored the enslaving effects (EEs) in multifinger tasks in which the contributions of the flexor digitorum profundus (FDP), flexor digitorum superficialis (FDS), and intrinsic muscles (INT) were manipulated. A new experimental technique was developed that allows redistribution of the muscle activity between the FDP, FDS, and INT muscles. In the experiment, 10 subjects were instructed to perform maximal voluntary contractions with all possible one-, two-, three-, and four-finger combinations. The point of force application was changed in parallel for the index, middle, ring, and little fingers from the middle of the distal phalanx to the distal interphalangeal joint and then to the proximal interphalangeal joint. It was found that (1) the EEs of similar amplitude were present in various experimental conditions that involved different muscle groups for force production; (2) the EEs were large on average—the slave fingers could produce forces reaching 67.5% of the maximal forces produced by those fingers in a single-finger task; (3) the EEs were larger for neighboring fingers; and (4) the EEs were nonadditive—in most cases, the EEs from two or three fingers were smaller than the EE from at least one finger. EEs among different muscles suggest a widespread neural interaction among the structures controlling flexor muscles in the hand as the main mechanism of finger enslaving.

《 2.1.3.2 Interfinger Connection Matrices

Representative publications: Zatsiorsky et al. 1998; Li et al. 2002; Zatsiorsky and Latash 2008

Finger forces exerted during prehension are constrained by biomechanical and neural finger interdependence. Finger interdependence is manifested as (a) *force deficit* and (b) *finger enslaving*. Force deficit describes the phenomenon whereby peak force generated by a finger during a multifinger maximal voluntary contraction task is smaller than its peak force during a single-finger maximal voluntary contraction task. Deficit increases with the number of explicitly involved ("master") fingers. *Enslaving* describes the phenomenon whereby fingers that are not explicitly required to produce any force (by instruction) are involuntarily activated.

Finger interdependence can be described by *interfinger connection matrices* that relate central commands to individual fingers with actual finger forces via a matrix equation:

$$F = [W]c, \qquad [2.3]$$

where F is a (4×1) vector of the normal finger forces, $[W]$ is a (4×4) interfinger connection matrix whose elements depend on the number of fingers involved in the task, and c is a (4×1) vector of the central (neural) commands. The elements of vector c equal 1.0 if the finger is intended to produce maximal force (maximal voluntary activation) or 0.0 if the finger is not intended to produce force (no voluntary activation).

To compute matrix $[W]$, maximal finger forces exerted in different tasks are recorded: the subjects are instructed to press on force sensors as hard as possible with either one two, three, or four fingers, using all possible finger combinations. Then, the interfinger connection matrices are computed by artificial neural networks. The described approach led to the concept of finger modes, which are arrays of finger forces caused by a single command to one of the fingers. Consider the following matrix that was computed in one of the studies (Zatsiorsky et al. 1998); the matrix is valid for IMRL tasks (I, index finger, M, middle finger, R, ring finger, L, little finger):

$$\mathbf{F} = [\mathbf{W}]\mathbf{c} = \begin{bmatrix} 25.075 & 2.275 & 0.95 & 0.925 \\ 3.55 & 17.35 & 3.4 & 0.875 \\ 2.25 & 5.05 & 13.9 & 2.275 \\ 2.2 & 1.9 & 4.0 & 11.5 \end{bmatrix} \mathbf{c}. \qquad [2.4]$$

When multiplied by commands c_i, the coefficients W represent (a) elements on the main diagonal: the forces exerted by finger i in response to the command sent to this finger (direct, or master, forces); (b) elements in rows: force of finger i due to the commands sent to all the fingers (sum of the direct and enslaved forces); and (c) elements in columns: forces exerted by all four fingers due to a command sent to one of the fingers (a mode).

Force generated by a finger in prehension arises from a command sent to this finger (direct finger force) as well as from the commands sent to other fingers (enslaved force). Direct finger forces can be computed as the product $w_{ii}c_i$ ($i = 1, 2, 3, 4$), where w_{ii} is a diagonal element of the weight matrix and c_i is a command intensity to this finger. To perform these computations, commands c_i to individual fingers should be known. In other words, the command vector **c** should be reconstructed. If matrix [W] is known and actual finger forces (e.g., in a prehension task) are recorded, the vector of neural commands **c** can be reconstructed by inverting equation 2.1:

$$\mathbf{c} = [\mathbf{W}]^{-1}\mathbf{F}. \qquad [2.5]$$

Note that matrix [W] is 4×4 and is always invertible. When the vector **c** is reconstructed, forces generated by individual fingers can be decomposed into components that arise from (a) direct commands to the targeted fingers and (b) enslaving effects, that is, commands sent to other fingers (figure 2.12). »

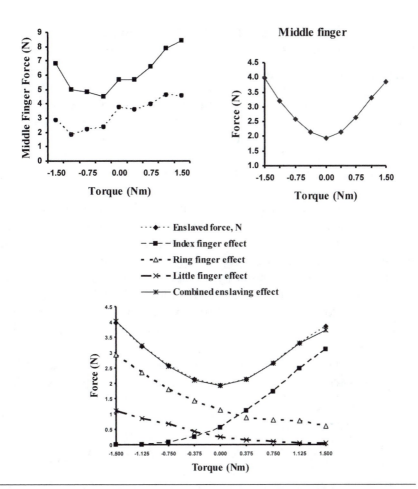

Figure 2.12 Decomposition of the normal forces of the middle finger during holding a 2.0 kg load at different external torques. The data are from a representative subject. Upper left panel: Actual and direct finger forces. The direct forces (dashed line) were computed as the products of the diagonal elements of the interfinger connection matrix times the corresponding finger commands. Upper right panel: Enslaved forces, that is, the difference between the actual and direct forces. Bottom panel: Decomposition of the enslaving effects. Effects of the commands to other fingers on the middle finger force are presented.

Reprinted with kind permission from Springer Science+Business Media: *Biological Cybernetics*, "Force and torque production in static multifinger prehension: Biomechanics and control. II. Control," vol. 87, 2002, pp. 40-49, by V.M. Zatsiorsky, R.W. Gregory, and M.L. Latash, from figs. 5, 6, and 7. © Springer-Verlag 2002.

2.1.3.3 Gliding Resistance Between the Tendons and Surrounding Tissues

Representative publication: An 2007

During joint motion, tendons slide over a certain distance and should overcome gliding resistance. Many tendons are surrounded by a tendon sheath (a tube) that provides nutrition and a low-friction environment to the tendon (figure 2.13). The synovial fluid produced by some cells of the epitenon decreases the friction between the tendon and surrounding tissues. However, some tendons, such as the palmaris longus tendon, are extrasynovial. Intrasynovial tendons provide smaller resistance to movement than extrasynovial. The magnitude of the gliding resistance depends on many factors, such as the proper lubrication, the relative size of the tendon and the surrounding tube or groove, and the level of compression of the tendon by the surrounding tissues (which can increase with some abnormalities). For fingers, the friction coefficient between the flexor tendon and A2 pulley was found to be from 0.020 to 0.063, which is higher than the friction coefficients of diarthrodial joints.

Tendon repair after an injury requires maximal gliding resistance reduction. This is achieved through appropriate suturing, biomaterial selection, and lubrication.

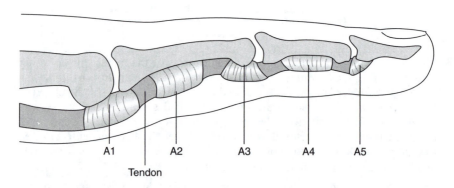

Figure 2.13 The tendon of the flexor digitorum profundus (FDP) and its five pulleys.

■ ■ ■ *FROM THE LITERATURE* ■ ■ ■

Forces During Tendon Gliding Are Not Only Due to Friction

Source: Heers, G., S.W. O'Driscoll, A.M. Halder, C. Zhao, N. Mura, L.J. Berglund, M.E. Zobitz, and K.N. An. 2003. Gliding properties of the long head of the biceps brachii. *J Orthop Res* 21(1):162-166

The gliding resistance of the long head of the biceps brachii tendon during abduction and adduction was measured on nine cadaveric glenohumeral joints. Gliding resistance was calculated as the force difference on the proximal and distal ends of the biceps brachii at five glenohumeral angles (15°, 30°, 45°, 60°, and 75°). The average gliding resistance in abduction at 15°, 30°, 45°, 60°, and 75° for a 4.9 N load was 0.41, 0.40, 0.36, 0.32, and 0.28 N, respectively. At these same angles, but during adduction motion, the force on the proximal tendon end was either identical or less than the distal tendon end, indicating a lack of resistance and even a phenomenon of negative resistance, in which some other force overcame the friction. The difference in gliding resistance between abduction and adduction was significant ($p < .05$). The results indicate that forces opposing biceps tendon gliding are more complicated than simply friction. Tendon deformation inside the bicipital groove produces a direction-dependent effect due to a mechanism of elastic recoil.

2.1.3.4 Tendon Wrapping

Representative publication: Armstrong and Chaffin 1978

Tendons can be classified as direct, those with a straight course between the muscle and bone connections, and wraparound, those that bend around a bone or a pulley (e.g., extrinsic hand muscles). When wraparound (curved) tendons transmit force they also exert a force toward the center of the curvature and tend to straighten. When wraparound tendons slide, they experience larger friction than do direct tendons.

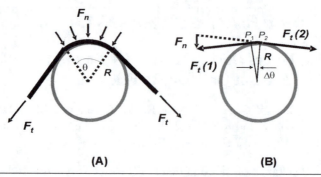

(A)   **(B)**

Figure 2.14 (*a*) A model of a tendon encircling a circular object. *R* is the radius of the cylinder (radius of curvature around supporting object, mm). θ is the angle subtended by the arc of contact (radians). The arc length is $l = R\theta$. F_t is the force along the tendon, and p_n is the normal pressure (N/mm). (*b*) Small part of a tendon in contact with the object. Δθ is the angle subtended at the center by the arc P_1-P_2.

Consider a simple model of a wraparound tendon, a rope encircling a cylindrical object of radius *R* (figure 2.14). The model is analogous to modeling a belt wrapped around a mechanical pulley. We want to determine (a) the normal forces exerted on the pulley per unit of arc length (normal pressure, N/mm) and (b) the friction force.

Normal pressure at rest. Consider a very small part of a tendon wrapped around a cylindrical object (figure 2.14*b*). The tendon is in contact with the object along the arc P_1-P_2 of length *l*. The forces acting at the contact are (a) the forces along the tendon $F_t(1)$ and $F_t(2)$ that are tangent to the supporting surface at points P_1 and P_2, respectively; and (b) the normal pressure p_n. Because $\varDelta\theta$ is small, the normal forces can be considered acting in parallel (approximately) and the normal pressure can be reduced to its resultant F_n.

$$F_n = p_n R\varDelta\theta. \qquad [2.6]$$

Because forces $F_t(1)$, $F_t(2)$, and F_n are in equilibrium, we can form a force triangle with the sides $F_t(1)$, $-F_t(2)$, and F_n, respectively. At equilibrium, $F_t(1) = F_t(2) = F_t$ and the triangle is isosceles. The force triangle and triangle $P_1 O P_2$ are similar and hence the angle at the apex of the triangle equals $\varDelta\theta$. Therefore,

$$\frac{p_n R\varDelta\theta}{2} = F_t \sin\frac{\varDelta\theta}{2} \qquad [2.7]$$

and

$$p_n = \frac{2F_t \sin\dfrac{\varDelta\theta}{2}}{R\varDelta\theta}. \qquad [2.8]$$

At the limit, as $\Delta\theta$ approaches zero, this expression becomes

$$p_n = \frac{F_t}{R},$$ [2.9]

where p_n is the normal force per unit of arc length. Hence, the normal pressure is a function of only the tendon force and the radius of curvature. If F_t is measured in newtons and R in millimeters, p_n is expressed in N/mm.

Friction forces at the wraparound tendons. Consider the case when $F_t(1) \neq F_t(2)$. Let $F_t(1) > F_t(2)$. The difference between the greater force $F_t(1)$ and the smaller force $F_t(2)$ is the friction force. Over a very small arc subtended by the differential angle $d\theta$, the $\Delta F_t \approx dF_t$, where dF_t is the differential of F_t.

$$dF_t = \mu p_n R d\theta,$$ [2.10]

where μ is the coefficient of friction (the coefficient is explained in section **1.2** in the *Kinetics of Human Motion*). Substituting equation 2.9 into equation 2.10 and transposing, we have

$$\frac{dF_t}{F_t} = \mu d\theta.$$ [2.11]

Note that R is not presented in this equation. Integrating equation 2.11 on the interval from $F_t(1)$ to $F_t(2)$ and between the corresponding limits of θ, we obtain

$$\int_{F_t(1)_t}^{F_t(2)} \frac{dF_t}{F_t} = \mu \int_0^\theta d\theta,$$ [2.12]

which yields an equation in natural logarithms

$$\ln F_t(1) - \ln F_t(2) = \mu\theta$$ [2.13a]

or

$$\ln \frac{F_t(1)}{F_t(2)} = \mu\theta.$$ [2.13b]

The corresponding exponential function is

$$\frac{F_t(1)}{F_t(2)} = e^{\mu\theta}.$$ [2.14a]

Equation 2.14 is known as the *Eytelwein formula* (after German scientist Johann Albert Eytelwein, 1764-1848) or *capstan equation*. In biomechanics of human motion equation 2.14a is often written as

$$F_p = F_d e^{\mu\theta},$$ [2.14b]

where F_p and F_d are the tensions at the proximal and distal ends of the tendon, respectively. The force of friction F_{fr} is

$$F_{fr} = F_d(e^{\mu\theta} - 1). \qquad [2.15]$$

The friction coefficient in the preceding equations can be either the static coefficient of friction that corresponds to the point of slipping of the tendon or the kinetic (dynamic) coefficient of friction that is determined when the tendon slides with respect to its sheath (see section **1.2.1** in *Kinetics of Human Motion*).

▪ ▪ ▪ *FROM THE LITERATURE* ▪ ▪ ▪

Friction Between the Flexor Digitorum Profundus Tendon and the A2 Pulley

Source: Uchiyama, S., J.H. Coert, L. Berglund, P.C. Amadio, and K.N. An. 1995. Method for the measurement of friction between tendon and pulley. *J Orthop Res* 13(1):83-89

An experimental setup (figure 2.15) was constructed and tested. The A2 pulley and the piece of the flexor digitorum profundus tendon were dissected from cadavers and fixed with the device as shown. A load $F_1 = 4.9$ N was applied to one end of the tendon while the mechanical actuator pulled the tendon at the other end. The forces at both ends were measured with the force sensors ($F_2 > F_1$).

By varying the angles α and β (see figure 2.15), the five arcs of contact are obtained. With the arc changes from 20° to 60°, the fric-

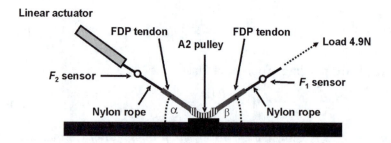

Figure 2.15   Experimental setup, schematic. See explanation in the text.

tion force increased from 0.021 to 0.31 N. Overall, the greater the angle the larger friction force. The friction coefficient was obtained from the equation

$$\mu = \frac{\ln(F_2 / F_1)}{\theta}.$$ [2.16]

The coefficient varied among the studied specimens from 0.022 to 0.063.

Normal forces during tendon gliding. In some movements tendons slide over a curved surface. This happens, for instance, when finger movements are performed with the wrist deviated from the neutral position, flexed or extended. When the wrist is flexed, the tendons of the extrinsic finger flexors are supported by the volar side of the carpal tunnel; when the wrist is extended the tendons are supported by the carpal bones. In both cases normal forces are exerted on supporting structures. The normal pressure is in this case

$$p_n = \frac{F_d e^{\mu\theta}}{R}$$ [2.17]

and the force F_n exerted on the supporting body (e.g., on the flexor retinaculum ligament when the wrist is flexed and the fingers exert force and flex) is

$$F_n = 2F_d e^{\mu\theta} \sin(\theta/2).$$ [2.18]

Equation 2.18 can be deduced from figure 2.14a. We leave this as a student exercise.

2.1.3.5 Bowstringing

Force that arises when a tendon path is not along a straight line is often called the *bowstringing force*. When applied to soft tissues such as finger pulleys or the carpal tunnel, the bowstring force tends to deform and in extreme cases to rupture them.

▪ ▪ ▪ *FROM THE LITERATURE* ▪ ▪ ▪

Physiological Finger Bowstringing in Rock Climbers

Source: Schweizer, A. 2001. Biomechanical properties of the crimp grip position in rock climbers. *J Biomech* 34(2):217-223

(continued)

Rock climbers often use the unique crimp grip position to hold small ledges, whereby the proximal interphalangeal joints are flexed about 90° and the distal interphalangeal joints are hyperextended maximally. In this posture, flexor tendon bowstringing occurs and the tendon exerts high loads on the flexor tendon pulleys. This can cause both acute injuries and overuse syndromes. The bowstringing is well palpated over the course of the tendon sheath as the finger is flexed. The author measured the bowstringing in vivo by (a) the distance from the bone to the flexor tendons and (b) the force that acts perpendicularly to the flexor tendon and causes bowstringing. The largest amount of bowstringing was caused by the flexor digitorum profundus tendon. Loads up to 116 N were measured over the A2 pulley at the fingertip force magnitudes of 30 N (subjects were not able to exert larger forces because of pain) (figure 2.16). Theoretically, at an external resistance of 118 N, the load to the A2 pulley would be almost 400 N, which is very close to the maximum strength of the A2 pulley of 407 N.

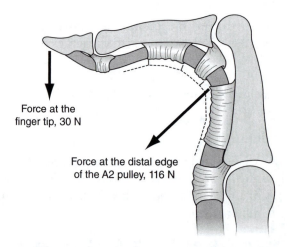

Force at the finger tip, 30 N

Force at the distal edge of the A2 pulley, 116 N

Figure 2.16 The bowstringing force at the A2 pulley is almost four times the force at the fingertip. The force acting normally on the A2 pulley may affect the torque at the metacarpophalangeal joint (this effect is discussed in section **5.3.1.3**).

2.1.3.6 Tendon Properties and Muscle Function

Representative publications: Loren and Lieber 1995; Voigt et al. 1995

Tendon properties are related to muscle function; material properties of tendons are correlated with their physiological functions. Tendon cross-sectional areas correlate well with physiological cross-sectional areas of the associated muscles. Tendons are stiffer in subjects with greater muscle strength, which may reduce the probability of tendon strain injuries. Tendon intrinsic properties accentuate muscle architectural specialization. For instance, among the tendons, strain magnitude in response to a similar stress is positively correlated with the tendon length–fiber length ratio of the muscle–tendon unit, a measure of the compliance of the muscle–tendon unit.

Increased tendon stiffness leads to shorter force transmission durations and higher rates of force development. Decreased hysteresis allows for a smaller energy loss and correspondingly for larger energy storage.

▪ ▪ ▪ *FROM THE LITERATURE* ▪ ▪ ▪

Tendon Biomechanical Properties Correlate With Muscle Specialization

Source: Loren, G.J., and R.L. Lieber. 1995. Tendon biomechanical properties enhance human wrist muscle specialization. *J Biomech* 28(7):791-799

Biomechanical properties of human wrist tendons—extensors carpi radialis brevis, extensor carpi radialis longus, extensor carpi ulnaris, flexor carpi radialis, flexor carpi ulnaris—were measured under loads predicted to be experienced by those tendons under physiological conditions. Loading the corresponding tendons to their maximum tension resulted in significantly different strains among tendons. Strain magnitude was significantly positively correlated with the tendon length–fiber length ratio of the muscle–tendon unit, a measure of the intrinsic compliance of the muscle–tendon unit. These data indicate that tendon intrinsic properties accentuate the muscle architectural specialization already present.

1. Tendon Stiffness Affects Muscle Performance

Source: Bojsen-Moller, J., S.P. Magnusson, L.R. Rasmussen, M. Kjaer, and P. Aagaard. 2005. Muscle performance during maximal isometric and dynamic contractions is influenced by the stiffness of the tendinous structures. *J Appl Physiol* 99(3):986-994

The mechanical properties of the vastus lateralis tendon–aponeurosis complex were assessed by ultrasonography. Maximal isometric knee extensor force and rate of torque development were determined. Dynamic performance was assessed by maximal squat jumps and countermovement jumps on a force plate. From the vertical ground reaction force, maximal jump height, jump power, and force- and velocity-related determinants of jump performance were obtained. Rate of torque development was positively related to the stiffness of the tendinous structures ($r = 0.55$, $p < .05$), indicating that tendon mechanical properties may account for up to 30% of the variance in rate of torque development. A correlation was observed between stiffness and maximal jump height in squat jumps and countermovement jumps ($r = 0.64$, $p < .05$ and $r = 0.55$, $p < .05$). These data indicate that muscle output in high-force isometric and dynamic muscle actions is positively related to the stiffness of the tendinous structures.

2. In Repetitive Jumps, the Tendons Performed 52% to 60% of the Total Work

Source: Voigt, M., F. Bojsen-Moller, E.B. Simonsen, and P. Dyhre-Poulsen. 1995. The influence of tendon Young's modulus, dimensions and instantaneous moment arms on the efficiency of human movement. *J Biomech* 28(3):281-291

The influence of passive tendon work on the gross mechanical efficiency of human whole body movement was examined. The subjects performed repetitive jumps (like skipping) of three different intensities. Metabolic costs and work rates were recorded to obtain mechanical efficiencies. Net joint moments were calculated from film recordings using inverse dynamics. Gross efficiency values of 0.65 to 0.69 and efficiency values of 0.77 to 0.80 at the approximate level of the muscle–tendon complexes were observed. The tendons performed 52% to 60% of the total work. Enhancement of the muscle–tendon efficiency over the maximal theoretical efficiency of the contractile machinery (0.25) could exclusively be explained by the contribution of tendon work.

2.1.3.7 Musculotendinous Architectural Indices

Representative publications: Woittiez et al. 1985; Zajac 1989; Hoy et al. 1990

Several architectural indices, or ratios, characterize tendon performance.

1. *The ratio of resting tendon length to optimal fiber length.* This ratio was discussed in section **1.5.2** and table 1.4. Among the tendons, strain magnitude is positively correlated with the ratio. The ratio is an indirect measure of the ability of the muscle–tendon complex to store and recoil the potential energy of deformation.

2. *The ratio of muscle PCSA to tendon cross-sectional area.* This ratio relates cross-sectional areas of tendons to the muscles that they serve. It is an indirect measure of the stress that can be experienced by the tendon. The higher the ratio, the larger the possible stress.

3. *The ratio of muscle–tendon unit length to muscle moment arm.* The larger the moment arm, the larger the muscle–tendon unit length changes at the same angular displacement at the joint.

■ ■ ■ *FROM THE LITERATURE* ■ ■ ■

A. In Mammals, Tendon Thickness Is Optimized for Maximal Mechanical Energy Storage and Minimal Muscle–Tendon Mass

Source: Ker, R.F., R. McnAlexander, and M.B. Bennett. 1988. Why are mammalian tendons so thick? *J Zool* 216(2):309-324

The maximum stresses to which a wide range of mammalian limb tendons could be subjected in vivo were estimated by considering the relative cross-sectional areas of each tendon and the fibers of its muscle. The majority of the stresses were low. The distribution of

(continued)

From the Literature   *(continued)*

stresses across mammals had a broad peak with maximum frequency at a stress of about 13 MPa, whereas the fracture stress for tendon is about 100 MPa. Thus, the majority of tendons are far thicker than is necessary for rupture safety. Much higher stresses are found among those tendons that act as springs to store energy during locomotion. The authors hypothesized that the thickness of the majority of tendons is co-optimized for maximal mechanical energy storage and minimal musculotendinous mass. A thinner tendon stretches more, so longer muscle fibers would be required to take up this stretch, thereby increasing musculotendinous mass. The predicted maximum stress in tendons of optimal thickness was about 10 MPa, which is quite close to the main peak of the observed stress distribution.

B. Importance of Muscle and Tendon Architecture Indices to a Tendon Transfer

Source: Zajac, F.E. 1992. How musculotendon architecture and joint geometry affect the capacity of muscles to move and exert force on objects: a review with application to arm and forearm tendon transfer design. *J Hand Surg Am* 17(5):799-804

Criteria are given for designing tendon transfer reconstructions to best replicate the biomechanical function of the replaced muscle. To retain the capacities for imparting movement and exerting forces on objects, the donor muscle should have the same moment arm–physiological cross-sectional area product, the same fiber length–moment arm ratio, and the same tendon length–muscle fiber length ratio as the replaced muscle.

2.2 MECHANICAL PROPERTIES OF PASSIVE MUSCLES

The expression "passive muscle" is not precisely defined in the literature; various interpretations of this term exist. Particular care is needed to avoid confusion between different meanings of this expression.

In healthy individuals, muscles are never completely relaxed; even at rest they are partially contracted. During some medical procedures, special drugs are used if necessary to induce complete muscle relaxations. Because in large doses these drugs can force relaxation of the breathing muscles, for safety this procedure is performed only by qualified anesthesiologists under careful control.

> ••• *Pharmacology Refresher* •••
>
> ## Muscle Relaxants
>
> There are two groups of muscle relaxants: peripherally and centrally acting. The peripherally acting relaxants block transmission of neural impulses at the neuromuscular junction. Some tribes in South America used a peripheral muscle relaxant, curare, in hunting. The poison-tipped arrows induced skeletal muscle paralysis. The centrally acting muscle relaxants, commonly called *spasmolytics*, may act at different levels of the central nervous system (e.g., the level of the cortex, brain stem, or spinal cord). They either enhance the level of inhibition or reduce the level of excitation.

2.2.1 Muscle Tone: Equitonometry

Representative publications: Meijer et al. 2001; Wright et al. 2007

At rest, the degree to which muscles can be voluntarily relaxed varies among individuals, a fact well known to physical therapists and masseurs who regularly palpate muscles. Low residual muscle tension—commonly called *muscle tone* or *muscle tonus* (the term was introduced in 1840 by a German physiologist, I. Müller)—always exists in nonparalyzed muscles; complete rest is not physiological.

Unfortunately, there is no commonly accepted method of testing muscle tone or even a commonly accepted meaning of the term. Two main interpretations of the term exist. The first considers muscle tone as residual muscle tension; if this interpretation were generally accepted an ideal method of muscle tone measurement would be measuring the tendon force. The second interpretation of the term is stretch reflex threshold and gain; under this interpretation muscle tone would be assessed by measuring resistance to imposed slow stretch. An additional practical but not very accurate method of diagnosing the state of muscle relaxation is surface electromyographic (EMG) recording; relaxed muscle produce lower or negligibly small electrical activity. However, activity of certain deep motor units may not be detectible by surface EMG.

Muscle tone (tonus) can be explored indirectly by recording the angular positions of multilink kinematic chains under the instruction of complete relaxation and in the absence of the external forces, a method called *equitonometry*. The joint angles in this case represent a balance of tensile forces exerted by the antagonistic muscle groups (e.g., the joint flexors and extensors). For instance, if an arm is placed on a horizontal two-link manipulandum such that the axle of the manipulandum is along the elbow joint axis, in healthy people the forearm and the upper arm always assume the angle 90±2°. At this angle, the elbow flexors and extensors exert joint moments of equal magnitude. In patients with muscle weakness, the forearm displaces in the direction of the healthy muscle group. Relaxed subjects in underwater conditions usually flex their knees and hips to 133° to 134° (figure 2.17). These angular positions represent body postures with equal moments of force exerted by the antagonistic muscles. Patient immobilization forces the equilibrium angle toward the immobilized posture.

Muscle tone is not constant. For instance, it often decreases during sleep, especially during the so-called *REM periods* (rapid eye movement sleep phase, in which dreaming occurs). Muscle resistance to extension increases after fatiguing exercises. For example, in professional baseball pitchers a significant decrease of the passive range of motion (ROM) at the shoulder joint occurs on the first day of spring training.

Some motor disorders result in muscle tone changes, its pathological increase or decrease. *Hypertonia* (*muscle rigidity*) is a condition marked by an abnormal increase in the muscle tone and a reduced ability of a muscle

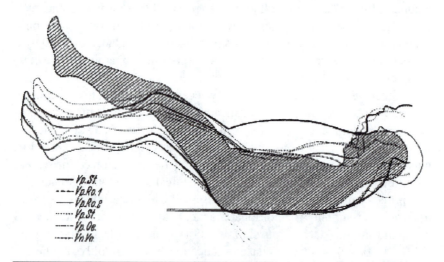

Figure 2.17 Contours of underwater photographs of relaxed subjects. The shaded contour represents a body posture of one subject immediately after submerging into the water.

Reprinted with kind permission from Springer Science+Business Media: *European Journal of Applied Physiology and Occupational Physiology*, "Zur Physiologie des Liegens," vol. 11, 1940, p. 256, B. Lehmann, fig. 2. © 1940, Verlag von Julius Springer.

to stretch. Some clinicians differentiate between *spasticity* (high "tone") and *rigidity*. The former is typical of spinal cord injury, multiple sclerosis, cerebral palsy, and some other illnesses, the latter of Parkinson's disease. Spastic tone strongly depends on velocity of imposed movement; rigidity is not strongly velocity dependent. Muscle hypertonia is believed to be a mainly cortical motor neuron dysfunction (albeit spasticity has also evident spinal mechanisms including impaired presynaptic inhibition). *Muscle hypotonia* is a condition of abnormally low muscle tone. Because hypotonia is commonly diagnosed during infancy, it is also known as "floppy infant syndrome" or "infantile hypotonia."

In the ensuing text, we use the term *passive muscle* with a broad meaning, for both isolated nonstimulated muscles and for muscles in situ at rest, noting that in the latter case muscle responses may depend both on the muscle's mechanical properties and on unspecified tonic activation. People whose muscle responses are being recorded are instructed to "relax" or "not intervene" and to perform muscle stretches very slowly to avoid monosynaptic stretch reflex contribution. Adequate relaxation of the tested muscles is a prerequisite for studying passive muscle properties.

Whole muscles include not only muscle fibers but also connective tissues and blood vessels, which affect muscle mechanical properties. To avoid ancillary effects, research on mechanical properties of isolated muscles is often performed on single muscle fibers or small bundles of fibers.

2.2.2 Mechanical Properties of Relaxed Muscles

2.2.2.1 Elastic Properties

Representative publications: Magnusson 1998; Gajdosik 2001

When a relaxed skeletal muscle is stretched beyond its resting length, it behaves as a deformable body; that is, it deforms and provides *passive resistance* to the stretch. The resistance does not require metabolic energy and, hence, is called passive. The muscle does not obey Hooke's law: the muscle force–deformation curve is not linear. With increased stretching the muscles become stiffer; that is, they demonstrate toe-in mechanical response to lengthening. The behavior of the passive muscle in extension is often compared with the behavior of a knitted stocking: the passive muscle elasticity is mainly attributable to the web of connective tissues within the muscle. During small stretches the web deforms; its threads become progressively taut, and during large stretches the threads themselves may also deform.

The in situ length of animal muscles removed from the body usually is shorter, on average, by about 10%. Because these muscles do not receive any stimulation from the nervous system (no stretch reflexes or tonic activation), this shortening results from the muscle's intrinsic mechanical properties. It indicates that the muscle was stretched while intact in the body. The length of an isolated passive muscle is called its *equilibrium length* (EL). The *rest length* (RL) is the natural muscle length in situ. Muscles with larger relative amount of intramuscular connective tissue are stiffer. They also have a shorter equilibrium length compared with either the RL or the length at which the muscle can generate maximal active tension (figure 2.18). This

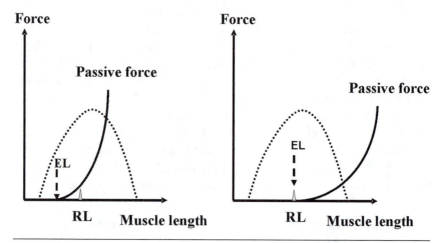

Figure 2.18 Force–length properties of the muscles with different amounts of intramuscular connective tissues (schematics). RL, rest length: the length of the muscle in the body during a natural posture (for humans this would be the anatomical posture—standing upright on a horizontal surface with arms hanging straight down at the sides of the body, head erect). EL, equilibrium length: the length of the isolated muscle without mechanical load. At the EL and below it, the passive force is zero. The dotted curved line is the active force–length relation, which is explained in chapter 3. A solid curved line represents the passive force–deformation relation. Such relations are recorded by fixing one end of the muscle and applying incremental loads to its free end. The load (force) is then plotted versus deformation. Left panel: A muscle with a large amount of connective tissue. Note (a) the passive force–length is shifted to the left (to shorter muscle length), (b) there is a large difference between the EL and RL, and (c) the passive force–length curve exhibits high stiffness. Right panel: A muscle with small amount of intramuscular connective tissue. As compared with the left panel, (a) the passive force–length curve is shifted to the right (to longer muscle length), (b) the EL and RL are closer to each other, and (c) the passive force–length curve is less steep exhibiting smaller stiffness.

is typical for the muscles of lower extremities, where the relative amount of the connective tissues is larger than in the muscles of upper extremities.

When isolated muscle fibers are stretched, resistance to extension is provided by three main structural elements: (1) connective tissues within and around the muscle belly (*parallel elastic components*); (2) stable cross-links between the actin and myosin filaments existing even in passive muscles: the crossbridges resist the stretch a short distance from the resting position before the contacts break and restore at other binding sites; and (3) noncontractile proteins, mainly titin, the giant protein molecules that form elastic connections between the end of the thick filaments and the Z-line. Titin's forces maintain the structural integrity of the sarcomere; they keep thick filaments centered between Z-discs. Mechanical properties of titin are not constant; they can be tuned according to mechanical demands placed on muscles. Actin and myosin filaments slide with respect to each other without visible changes in length.

Sarcomeres that are closer to the fiber ends are less extensible than those at the rest of the fiber. The sarcomere lengths at the middle of the muscle fiber can be increased to approximately 3.65 μm.

At lengths smaller than RL, passive muscles are flaccid. During joint movement the relaxed antagonist muscles do not provide much resistance to movement (e.g., triceps brachii during elbow flexion or biceps brachii during elbow extension). Because the nonstimulated muscles in these conditions exert little tension, they store little energy of elastic deformation. At lengths larger than RL the muscles exhibit resistance to extension (see figure 2.18). In living people, muscle length changes above RL are accompanied by tendon length changes (tendons are in the toe region and provide small resistance to stretching). As a result, changes in muscle–tendon complex length are larger than the change in the muscle itself.

■ ■ ■ *FROM THE LITERATURE* ■ ■ ■

Muscle–Tendon Deformation During Passive Stretch

Source: Herbert, R.D., A.M. Moseley, J.E. Butler, and S.C. Gandevia. 2002. Change in length of relaxed muscle fascicles and tendons with knee and ankle movement in humans. *J Physiol* 539(2):637-645

Ultrasonography was used to measure changes in length of muscle fascicles in relaxed human tibialis anterior and gastrocnemius during passively imposed changes in joint angle. Muscle fascicle length

(continued)

From the Literature *(continued)*

changes were compared with whole muscle–tendon unit length changes calculated from joint angles and anthropometric data. Relaxed muscle fascicles underwent much smaller changes in length than did their muscle–tendon units. On average, muscle fascicles in tibialis anterior account for 55% ±13% (mean ± standard deviation) of the total change in muscle–tendon length. This indicates nearly half of the total muscle–tendon length change resulted from tendon stretch. In the gastrocnemius, which has relatively long tendons, only 27% ± 9% of the total muscle–tendon length change was transmitted to muscle fascicles. The data confirm that length changes seen by muscle fascicles during passive joint movement can be much less than changes in muscle origin–insertion distance. This occurs because tendons undergo significant changes in length, even at very low forces.

Passive resistance in joints results from the interaction of two components, one that depends on joint angle and displacement and one that is independent of these factors (e.g., joint friction). We will call the first component *elastic resistance.* The elastic resistance in the middle range of joint motion is usually small and in the majority of cases can be neglected. The resistance increases exponentially as the joint motion approaches its maximal limit. The elastic resistance is commonly described by the following equation:

$$T = \begin{cases} k(e^{l(\alpha - \alpha_0)} - 1) & \text{if } \alpha \geq \alpha_0 \\ 0 & \text{if } \alpha < \alpha_0 \end{cases}, \qquad [2.19]$$

where T is the passive torque, k and l are experimental constants, α is a joint angle, and α_0 is a joint angle below which the joint structures are slack and do not provide elastic resistance to the movement. Similar equations are used to describe the passive force–muscle length relations in muscles (see figure 2.18), in which case instead of joint angle the muscle length is used and instead of threshold angle EL is used in the equation.

2.2.2.2 Viscoelastic Properties of Passive Muscles: Passive Mechanical Resistance in Joints

Representative publications: Gajdosik 1997, 2001; Magnusson 1998; Esteki and Mansour 1996; Nordez et al. 2008; Proske et al. 1993

Passive muscles and joints possess viscoelastic properties including velocity-dependent resistance, creep, stress relaxation, and hysteresis. Velocity dependence is small. If a passive muscle is forcibly stretched and then allowed to return to its prestretch length, hysteresis is observed. The same happens during rhythmic movements of relaxed joints. The hysteresis area is equivalent to the amount of mechanical energy loss (dissipated to heat). In general, the relation between hysteresis and muscle stretch speed (or rhythmic deformation frequency) can be represented as

$$E = k\omega^n, \qquad\qquad [2.20]$$

where E is the dissipated mechanical energy (the hysteresis area), ω is the oscillation frequency, and k and n are empirical constants. If $n = 1$ the dissipated energy is proportional to velocity, and when $n = 0$ the dissipated energy is velocity independent. In the latter case, the resistance is like friction and is the same at any velocity. In human joints, the exponent n is typically close to zero.

Therefore, in contrast to "classic" viscoelastic bodies, such as rubber, muscle hysteresis is relatively independent of oscillation frequency, and thus the amount of mechanical energy converted to heat is relatively independent of the rate of muscle length change (figure 2.19). This indicates

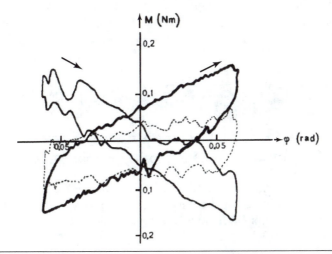

Figure 2.19 Hysteresis loops at various frequencies of oscillation of the relaxed elbow joint. The mechanical resistance (in newton meters) versus the joint angle (radians). The arm was placed horizontally at the mechanical manipulandum and then oscillated in a sinusoidal fashion. The angle between the upper arm and forearm in the midposition was 90°. The frequencies were 2.5 rad/s (thick solid line), 6.0 rad/s (dotted line), and 9.0 rad/s (thin solid line). Note that the hysteresis area is approximately similar at various frequencies; that is, it does not depend much on the speed of muscle length changes.

Reprinted, by permission, from K.L. Boon, A.L. Hof, and W. Wallinga-de Jonge, 1973, The mechanical behavior of the passive arm. In *Biomechanics III: Third International Seminar on Biomechanics, Rome, Italy,* edited by S. Cerquiglini, A. Venerando, and J. Wartenweiler (Basel: S. Karger), 243-248.

that the speed–dependent component does not play a decisive role in the resistance provided by the muscle to its forcible stretching. This behavior is similar to the behavior of other soft biological tissues. The intramuscular mechanisms of the resistance independence on the rate of muscle length change are complex and mainly unknown. If simple mechanical analogies are desired, the resistance more closely resembles dry friction (which at low speeds is independent of speed) than the speed-dependent viscous resistance observed in gels and liquids. Although muscles contain a large amount of water (a viscous liquid), viscous (speed dependent) resistance is apparently quite small.

When a constant load is applied a sufficiently long period of time, the muscle exhibits creep (*lengthening relaxation*). When a constant deformation is applied, the resistance (the tension recorded at the muscle end) declines over time (*stress relaxation*).

Muscles, however, possess thixotropic, or *sticky friction*, properties. *Thixotropy* is a property of a substance to decrease its viscosity when it is shaken or stirred and then solidify when left to stand. Examples of thixotropic materials include bee honey, latex paint, and synovial fluid (found in human joints). In human movements, the muscle and joint thixotropy is manifested mainly as a component of morning stiffness: in the morning

---

■ ■ ■ *FROM THE LITERATURE* ■ ■ ■

In Muscles, Hysteresis Energy Losses Do Not Depend Much on Deformation Rate

Source: Nordez, A., P. Casari, and C. Cornu. 2008. Effects of stretching velocity on passive resistance developed by the knee musculo-articular complex: contributions of frictional and viscoelastic behaviours. *Eur J Appl Physiol* 103(2):243-250

Nine subjects performed passive knee extension–flexion cycles with the hip angle set at 60° on a Biodex dynamometer. A relation between the energy dissipation and the oscillation frequency was determined. A low power value ($n = 0.085$) confirmed that knee muscle–articular complex behavior has little velocity dependence. When stretching angular velocity increased more than 20-fold, increases in passive torque were only 20%.

■ ■ ■ *FROM THE LITERATURE* ■ ■ ■

Muscles Exhibit Thixotropic Properties

Source: Lakie, M., and L.G. Robson. 1988. Thixotropic changes in human muscle stiffness and the effects of fatigue. *Q J Exp Physiol* 73(4):487-500

Stiffness of the relaxed finger musculature after movement followed by various times at rest has been studied. The muscles stiffen considerably as rest duration increases. The stiffening continues at a declining rate for at least 30 min. The increased stiffness observed after rest can quickly be reduced by active or passive movements but not by isometric efforts. These changes characterize a thixotropic system and suggest a long-term molecular rearrangement in relaxed muscle.

hours resistance to movement in some people, especially in seniors and in patients with rheumatoid arthritis, is increased. The resistance decreases after performing several movements at the joint. It increases as a result of muscular fatigue. When only some of the muscle fibers are active, which is typical of movements in which only small muscle tension is generated, the adjacent passive muscle fibers exhibit sticky friction: They resist small changes in length, but after the initial friction is overcome the fibers stretch or shorten smoothly.

2.3 ON JOINT FLEXIBILITY

Representative publications: Alter 2004; Sands et al. 2006; Gajdosik and Bohannon 1987

Joint flexibility is defined as the maximal ROM available at a joint or group of joints. Flexibility is commonly measured in angular units, but in sport and physical education it is oftentimes estimated using linear measures (figure 2.20). Linear measures are less accurate because they are influenced by body proportions. For example, the stand-and-reach test (figure 2.20, right panel) is affected by arm and leg lengths; subjects with longer arms and shorter legs have an advantage in this test.

Maximal ROM achieved during self-movement characterizes active flexibility. *Passive flexibility* is a measure of the ROM achieved due to external forces (e.g., during interaction with a partner or equipment). To

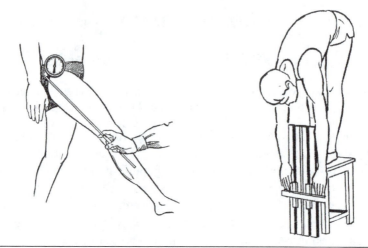

Figure 2.20 Measuring flexibility in angular measures (left figure) and linear measures (right figure). In the left panel, the angle is measured from the vertical.

determine passive flexibility, the direction and point of application of an external force should be known. Passive flexibility is essentially a measure of stretch tolerance, the angular position beyond which further stretching is dangerous or limited by pain.

Active flexibility is determined by the balance between the active forces exerted by the joint agonists and the stretch resistance provided by the joint antagonists. During voluntary active stretching, as a joint approaches its maximal ROM, the joint moment decreases due to both agonist shortening and stretch-induced antagonist resistance. At a certain point the moments of force exerted by the agonists and antagonists become equal and the joint angle does not increase anymore. Hence, active flexibility can be increased (a) by agonist strength training in postures corresponding to their small lengths or (b) by antagonist stretching with the goal of decreasing their resistance to stretching (stiffness). Active flexibility is smaller than passive flexibility (figure 2.21). A large difference between active and passive flexibility is a sign of the agonist weakness. Agonist strength training apparently more effectively reduces this difference than does antagonist stretching.

Muscular resistance to stretching results from (a) neural factors (incomplete muscle relaxation, action of stretch reflexes) and (b) muscle mechanical properties. Both these factors can be affected by training. For instance, if one cannot reach the floor with the fingertips maintaining the leg extended (the stand-and-reach test), the following technique can be used (known in

Figure 2.21 Passive *(a)* and active *(b)* flexibility.

Illustration by Michael Richardson. Reprinted fom M.J. Alter, 2004, *Science of flexibility*, 3rd ed. (Champaign, IL: Human Kinetics), 161, by permission of the author.

yoga as *savasana*): (a) Start bending the trunk with the hands in relaxed fists; (b) when the limit of flexibility is achieved, activate maximally all the muscles for 3 to 4 s while exerting also maximal grasping force and performing forcible exhaling effort with the glottis closed (do not breathe, use the so-called *Valsalva maneuver*); (c) relax while exhaling and immediately continue the trunk bending. After performing this routine several times in a row (usually 3-5 times), the majority of people are able to touch the floor without bending their knees. The described procedure induces muscle relaxation sufficient to temporarily increase flexibility. Because muscle stiffness depends on ambient temperature (at high temperature stiffness decreases), several minutes in a hot sauna may visibly increase flexibility. Systematic stretching training also increases the flexibility.

■ ■ ■ *FROM THE LITERATURE* ■ ■ ■

Mechanical Vibration May Enhance Range of Motion

Source: Sands, W.A., J.R. McNeal, M.H. Stone, E.M. Russell, and M. Jemni. 2006. Flexibility enhancement with vibration: Acute and long-term. *Med Sci Sports Exerc* 38(4):720-725

(continued)

From the Literature *(continued)*

The authors studied whether vibration-aided static stretching could enhance ROM more than static stretching alone in the forward split position. Ten highly trained male volunteer gymnasts were randomly assigned to experimental ($n = 5$) and control ($n = 5$) groups. The test involved a forward split with the rear knee flexed to prevent pelvic misalignment. The height of the anterior iliac spine of the pelvis was measured at the lowest split position. Athletes stretched forward and rearward legs to the point of discomfort for 10 s followed by 5 s of rest and then repeated the procedure 4 times on each leg and each split position (4 min total). The experimental group stretched with a vibration device turned on; the control group stretched with the device turned off. A pretest was followed by an acute phase posttest and then a second posttest measurement was performed after 4 weeks of treatment. The acute phase in the experimental group showed dramatic increases in forward split flexibility for both legs ($p < .05$), whereas the long-term test showed a statistically significant increase in ROM during the right rear leg split only ($p < .05$). The authors concluded that vibration can be a promising means of increasing ROM beyond that obtained with static stretching in highly trained male gymnasts.

2.4 SUMMARY

Tendons are made up of longitudinally arranged collagen fibers oriented mainly along the tendon. Relaxed fibers follow a sinusoidal wave pattern known as crimp. Muscle forces are transmitted to the tendons via myotendinous junctions. The tendon–bone junction is known as an enthesis. Besides containing collagen fibers, tendons include elastic fibers that are made of elastin. In pennate muscles, tendons consist of external and internal (aponeurosis) parts. Tendons are elastic bodies: When a tendon is subjected to an external stretching force the tendon length increases, but when the force is removed the tendon returns to its original length. When forces are not too large, elongation (Δl) is proportional to force. The relation $F = -k\Delta l$, where F is the force resisting the deformation and k is stiffness (N/m), is known as Hooke's law.

A force divided by the area on which it acts is called stress. Stress is measured in newtons per square meter (N/m^2), pascals. The amount of elongation per unit of object length is called strain. Strain is expressed in meters per meter and hence is dimensionless. Strain can be computed as the ratio of the change in length of a line segment to either (a) its initial length (Lagrangian strain, or simply strain) or (b) its final length (Eulerian strain). The stress–strain relation is $\sigma = E\Delta l/l = E\varepsilon$, where σ is stress (N/m^2), ε is dimensionless strain, and E is a proportionality constant, Young's modulus (N/m^2). Young's modulus has the dimension of stress; it is the stress required to elongate the body to double its original length. Usually the maximal elongation is much smaller than that: the object would break long before doubling its length. The mathematical relations between applied stresses or forces and strains or deformations are called constitutive equations.

When the tendon is forcibly stretched, its cross-sectional area decreases. The stress measured relative to the instantaneous area is called the Cauchy stress or "true stress." However, for simplicity in small deformations, the stress is estimated using the original cross-sectional area of the tendon. This stress measure is sometimes called Lagrangian stress. For large deformations, the instantaneous area should be used. In large deformations, stress results from two factors: (a) increasing force and (b) decreasing cross-sectional area. The ratio of the transverse strain (normal to the applied load) divided by the relative extension strain (in the direction of the applied load) is known as Poisson's ratio.

In some joint configurations, the tendon may be slack: it does not resist the force when elongated. For such cases, the elongation is defined relative to the slack length, $\Delta l_\tau = l_{Te} - l_{Ts}$, where l_{Te} is the length of the elongated (deformed) tendon and l_{Ts} is its slack length, the length at which the tendon begins to resist external force.

The tendon stress–strain curve has three regions: (a) the initial toe region where the slope increases, (b) the linear region where the slope is constant, and (c) the failure region. In the toe region the tendon is strained to 2% (range 1.5%-4%). In the toe region collagen fibers are being straightened and lose their crimp pattern but the fibers themselves are not stretched. The slope of modulus increases with strain until it reaches a constant value. In the linear region, tendon elongation results from stretch of the already aligned collagen fibers and hence the stiffness is indicative of collagen fibers stiffness (material stiffness). The slope of the stress–strain curve in the linear region is usually referred as the Young's modulus of the tendon (linear-region modulus). The linear region begins when the tendon is stressed 5 to 30 MPa. The failure occurs at about 70 to 100 MPa. The Young's modulus of tendon is in the range 0.6 to 1.7 GPa. Mechanical properties of different tendons are in part explained by their size (the length and cross-sectional area) and in part by the properties of the tendon material.

Tendons display time- or history-dependent behavior; they are viscoelastic bodies. The viscoelasticity is characterized by four main phenomena: (a) strain-rate sensitivity, (b) creep, (c) stress relaxation, and (d) hysteresis. Creep is an increase in length over time under a constant load. Stress relaxation is a decrease of stress over time under constant deformation. Hysteresis, or elastic hysteresis, is the difference between the loading and unloading curves in the stress–strain cycle. Hysteresis is manifested as the hysteresis loop, which represents energy loss during a deformation cycle, mainly caused by the conversion of mechanical work to heat. Hysteresis is often quantified as the ratio of the hysteresis loop area divided by the area under the loading curve. The ratio represents the percentage of the mechanical energy loss in the loading–unloading cycle. In tendons (and other soft biological tissues), hysteresis is relatively independent of loading–unloading frequency.

Simple mechanical models of viscoelastic materials commonly include a combination of linear springs and dashpots. Such models are called lumped-parameter models. Three prevalent models of the viscoelastic behavior of the tendons are the Maxwell, Voigt, and Kelvin (standard linear) models.

Tendons adapt to applied forces and motion. Tendon properties are related to muscle function; their material properties are correlated with their physiological functions.

Force transmission in tendons occurs mainly along the individual fascicles that act as independent structures, and lateral force transmission between adjacent collagen fascicles is small. However, when several muscles connect to a common tendon, differential displacements (and most probably the lateral force transmission) may occur. It happens, for instance, with the medial gastrocnemius and soleus aponeuroses during the ankle plantar flexion.

One specific case of the lateral force transfer occurs in the hand, where tendons from extrinsic muscles that serve different fingers connect to each other. This tendon interconnection is one of the causes of so-called finger enslaving: when one finger exerts force or moves, other fingers are also involved in the activity.

Tendons can be classified as the direct tendons, those with a straight course between the muscle and bone connections, and wraparound tendons that bend around a bone or a pulley, like seen in extrinsic muscles in the hand. When wraparound (curved) tendons transmit force they also exert a force directed toward the center of the curvature: they tend to straighten. When bent tendons slide they experience larger friction than the direct tendons. Normal force that arises when a tendon path is not along a straight line is often called bowstringing force. When applied to soft tissues such as finger pulleys or the carpal tunnel, the bowstring force tends to deform these structures and in extreme cases can result in tissue rupture.

Several architectural indices, or ratios, characterize the tendon performance.

1. The ratio of resting tendon length to optimal fiber length. This ratio is an indirect measure of the ability of the muscle–tendon complex to store and recoil the potential energy of deformation. Among the tendons, strain magnitude is usually positively correlated with this ratio.

2. The ratio of muscle PCSA to tendon CSA. This ratio relates CSAs of the tendon and the muscle that the tendon serves. It is an indirect measure of the stress that can be experienced by the tendon. The higher the ratio the larger the possible stress.

3. The ratio of muscle–tendon unit length to muscle moment arm. The larger the moment arm, the larger the muscle–tendon unit length changes at the same angular displacement at the joint.

The expression *passive muscle* is not precisely defined in the literature. Particular care is needed to avoid confusion between different meanings of this expression. In healthy people muscles are never completely relaxed; even at rest they are in a partially contracted state, which is usually called muscle tone or muscle tonus. Two main interpretations of the term exist: in the first, muscle tone is the residual muscle tension; in the second, muscle tone is the level (or gain) of the stretch reflexes. Muscle tone (tonus) can be explored indirectly by recording the angular positions of multilink kinematic chains under the instruction of complete relaxation and in the absence of external forces, a method called equitonometry. Some motor disorders result in muscle tone changes, its pathological increase (hypertonia or muscle rigidity) or decrease (muscle hypotonia).

When a relaxed skeletal muscle is stretched beyond its resting length, it behaves as a deformable body; that is, it deforms and provides passive resistance to the stretch. The resistance does not require metabolic energy and hence is called passive. The muscle does not obey Hooke's law: the muscle force–deformation curve is not linear. With increased stretching muscles become stiffer; that is, they demonstrate toe-in mechanical response to lengthening. The behavior of passive muscles in extension is often compared with the behavior of a knitted stocking: passive muscle elasticity is mainly due to the web of connective tissues within the muscle. At small stretches the web deforms and its threads become progressively taut; at large stretches the threads themselves are deformed.

Isolated animal muscles are usually shorter than their intact (in situ) length, on average by about 10%. Because isolated muscles do not receive any neural stimulation, this shortening is representative of a muscle's intrinsic mechanical properties. The length of an isolated passive muscle is called equilibrium length (EL). Rest length (RL) is the natural muscle length in situ. At lengths smaller than RL, passive muscles are flaccid. During joint movements the relaxed antagonist muscles do not provide much resistance. At lengths larger than RL, muscles exhibit resistance to extension. In living people, muscle length changes beyond RL are also accompanied by tendon length changes (the tendons are in the toe region and provide small resistance to stretching). As a result, muscle–tendon complex length changes are larger than the changes in the muscle itself.

When isolated muscle fibers are stretched, their resistance to extension is provided by three main structural elements: (1) connective tissues within and around the muscle belly (parallel elastic components); (2) stable cross-links between the actin and myosin filaments existing even in passive muscles: the cross-bridges resist the stretch a short distance from the resting position before the contacts break and restore at other binding sites; and (3) noncontractile proteins, mainly titin, the giant protein molecules that form an elastic connection between the end of the thick filaments and the Z-line.

Passive muscles, as well as passive joints, possess viscoelastic properties such as velocity-dependent resistance, creep, stress relaxation, and hysteresis. Velocity dependence is very small. Muscles possess thixotropic, or sticky friction, properties. Thixotropy refers to viscosity changes associated with substance disturbance states (e.g., stirred vs. left standing).

Joint flexibility is defined as the maximal range of motion (ROM) available at a joint or group of joints. Maximal ROM achieved by self-efforts characterizes active flexibility. Passive flexibility is a measure of the ROM when subjected to external forces. To determine passive flexibility, the direction and point of application of an applied force should be known. Passive flexibility is essentially a measure of stretch tolerance.

2.5 QUESTIONS FOR REVIEW

1. How are collagen fibers arranged in a resting tendon?

2. Define such terms as *compliance, stiffness, stress, strain, Lagrangian strain, Eulerian strain,* and *Young's modulus.* Name the units used for measuring these variables.

3. Draw a typical stress versus strain plot. Label the toe region, linear region, and plastic region. Describe internal tendon behavior as it

passes through each region. What is the difference between tendon deformation in the toe region and in the linear region?

4. Discuss the functional meaning of the following ratios: (a) resting tendon length to optimal muscle fiber length, (b) muscle PCSA to tendon CSA, and (c) muscle–tendon unit length to muscle moment arm.

5. A cadaver tendon is subjected to a 1,000 N stretching force. The initial cross-sectional area and length are 70 mm² and 50 mm, respectively. The force is then increased by 300 N, resulting in a reduction of cross-sectional area to 50 mm² and increase in length to 55 mm. Calculate the initial and final Cauchy stress in the tendon. Also calculate the final strain.

6. Discuss interfinger connection matrices.

7. Discuss mechanical behavior of wraparound tendons. Explain the Eytelwein formula (equation 2.14).

8. Explain a bowstringing force.

9. Describe the properties of viscoelastic bodies. Explain the meaning of the terms *creep* and *stress relaxation*. Explain how a hysteresis can be recorded. What does a hysteresis loop represent?

10. What is viscosity? What are the SI metric units for viscosity?

11. What are lumped-parameter models? Explain the difference between the Maxwell, Voigt, and Kelvin models.

12. Name a motor disorder that is characterized by muscle hypertonia and one characterized by muscle hypotonia.

13. Explain why during joint movements the relaxed antagonist muscles usually do not provide much resistance to their stretching.

14. What is thixotropy?

15. Define active and passive flexibility.

2.6 LITERATURE LIST

Alter, M. 2004. *Science of Flexibility*. Champaign, IL: Human Kinetics.

An, K.N. 2007. Tendon excursion and gliding: Clinical impacts from humble concepts. *J Biomech* 40(4):713-718.

Arampatzis, A., S. Stafilidis, G. DeMonte, K. Karamanidis, G. Morey-Klapsing, and G.P. Bruggemann. 2005. Strain and elongation of the human gastrocnemius tendon and aponeurosis during maximal plantarflexion effort. *J Biomech* 38(4):833-841.

Archambault, J.M., J.P. Wiley, and R.C. Bray. 1995. Exercise loading of tendons and the development of overuse injuries: A review of current literature. *Sports Med* 20(2):77-89.

Armstrong, T. J., and D.B. Chaffin. 1978. An investigation of the relationship between displacements of the finger and wrist joints and the extrinsic finger flexor tendons. *J Biomech* 11(3):119-128.

Atkinson, T.S., R.C. Haut, and N.J. Altiero. 1997. A poroelastic model that predicts some phenomenological responses of ligaments and tendons. *ASME J Biomech Eng* 119:400-405.

Benjamin, M., E. Kaiser, and S. Milz. 2008. Structure-function relationships in tendons: A review. *J Anat* 212(3):211-228.

Bojsen-Moller, J., K. Brogaard, M.J. Have, H.P. Stryger, M. Kjaer, P. Aagaard, and S.P. Magnusson. 2007. Passive knee joint range of motion is unrelated to the mechanical properties of the patellar tendon. *Scand J Med Sci Sports* 17(4):415-421.

Bojsen-Moller, J., P. Hansen, P. Aagaard, U. Svantesson, M. Kjaer, and S.P. Magnusson. 2004. Differential displacement of the human soleus and medial gastrocnemius aponeuroses during isometric plantar flexor contractions in vivo. *J Appl Physiol* 97(5):1908-1914.

Bojsen-Moller, J., S.P. Magnusson, L.R. Rasmussen, M. Kjaer, and P. Aagaard. 2005. Muscle performance during maximal isometric and dynamic contractions is influenced by the stiffness of the tendinous structures. *J Appl Physiol* 99(3):986-994.

Boon, K.L., A.L. Hof, and W. Wallinga-de Jonge W. 1973. The mechanical behavior of the passive arm. In *Biomechanics III: Third International Seminar on Biomechanics, Rome, Italy,* edited by S. Cerquiglini, A. Venerando, and J. Wartenweiler. Basel, Switzerland: Karger, pp. 243-248.

Butler, D., E. Grood, F. Noyes, and R. Zernicke. 1978. Biomechanics of ligaments and tendons. *Exerc Sport Sci Rev* 6:125-181.

Cowin, S., and S. Doty. 2007. *Tissue Mechanics*. New York: Springer.

Diamant, J., A. Keller, E. Baer, M. Litt, and R.G. Arridge. 1972. Collagen; ultrastructure and its relation to mechanical properties as a function of ageing. *Proc R Soc Edinb Biol* 180(60):293-315.

Epstein, M., M. Wong, and W. Herzog. 2006. Should tendon and aponeurosis be considered in series? *J Biomech* 39(11):2020-2025.

Esteki, A., and J.M. Mansour. 1996. An experimentally based nonlinear viscoelastic model of joint passive moment. *J Biomech* 29(4):443-450.

Fleming, B.C., and B.D. Beynnon. 2004. In vivo measurement of ligament/tendon strains and forces: A review. *Ann Biomed Eng* 32(3):318-328.

Frisen, M., M. Magi, I. Sonnerup, and A. Viidik. 1969. Rheological analysis of soft collagenous tissue: Part I. Theoretical considerations. *J Biomech* 2(1):13-20.

Fung, Y.C. 1967. Elasticity of soft tissues in simple elongation. *Am J Physiol* 213(6):1532-1544.

Fung, Y. 1993. *Biomechanics: Mechanical Properties of Living Tissues.* 2nd ed. New York: Springer-Verlag.

Gajdosic, R. 1997. Influence of age on calf muscle length and passive stiffness variables at different stretch velocities. *Isokinet Exerc Sci* 6:163-174.

Gajdosik, R.L. 2001. Passive extensibility of skeletal muscle: Review of the literature with clinical implications. *Clin Biomech (Bristol, Avon)* 16(2):87-101.

Gajdosik, R.L., and R.W. Bohannon. 1987. Clinical measurement of range of motion: Review of goniometry emphasizing reliability and validity. *Phys Ther* 67(12):1867-1872.

Gregor, R.J., P.V. Komi, and M. Jarvinen. 1987. Achilles tendon forces during cycling. *Int J Sports Med* 8(Suppl 1):9-14.

Hansen, P., J. Bojsen-Moller, P. Aagaard, M. Kjaer, and S.P. Magnusson. 2006. Mechanical properties of the human patellar tendon, in vivo. *Clin Biomech (Bristol, Avon)* 21(1):54-58.

Heers, G., S.W. O'Driscoll, A.M. Halder, C. Zhao, N. Mura, L.J. Berglund, M.E. Zobitz, and K.N. An. 2003. Gliding properties of the long head of the biceps brachii. *J Orthop Res* 21(1):162-166.

Herbert, R.D., A.M. Moseley, J.E. Butler, and S.C. Gandevia. 2002. Change in length of relaxed muscle fascicles and tendons with knee and ankle movement in humans. *J Physiol* 539(2):637-645.

Hoy, M.G., F.E. Zajac, and M.E. Gordon. 1990. A musculoskeletal model of the human lower extremity: The effect of muscle, tendon, and moment arm on the moment-angle relationship of musculotendon actuators at the hip, knee, and ankle. *J Biomech* 23(2):157-69.

Ker, R.F., R. McnAlexander, and M.B. Bennett. 1988. Why are mammalian tendons so thick? *J Zool* 216(2):309-324.

Komi, P.V. 1990. Relevance of in *vivo* force measurements to human biomechanics. *J Biomech* 23(Suppl 1):23-34.

Kubo, K., H. Kanehisa, and T. Fukunaga. 2002. Effects of resistance and stretching training programmes on the viscoelastic properties of human tendon structures in vivo. *J Physiol* 538 (Pt 1):219-226.

Kubo, K., H. Kanehisa, and T. Fukunaga. 2003a. Gender differences in the viscoelastic properties of tendon structures. *Eur J Appl Physiol* 88 (6):520-526.

Kubo, K., H. Kanehisa, and T. Fukunaga. 2005. Comparison of elasticity of human tendon and aponeurosis in knee extensors and ankle plantar flexors in vivo. *J Appl Biomech* 21(2):129-142.

Kubo, K., H. Kanehisa, M. Miyatani, M. Tachi, and T. Fukunaga. 2003b. Effect of low-load resistance training on the tendon properties in middle-aged and elderly women. *Acta Physiol Scand* 178(1):25-32.

Kubo, K., M. Morimoto, T. Komuro, H. Yata, N. Tsunoda, H. Kanehisa, and T. Fukunaga. 2007. Effects of plyometric and weight training on muscle–tendon complex and jump performance. *Med Sci Sports Exerc* 39 (10):1801-1810.

Lakie, M., and L.G. Robson. 1988. Thixotropic changes in human muscle stiffness and the effects of fatigue. *Q J Exp Physiol* 73(4):487-500.

Lehmann, B. 1940. Zur Physiologie des Liegens. *Arbeitsphysiologie* 11:253-261.

Leijinse, J., C. Snijders, J. Landsmeer, J. Bonte, J. van der Meulen, G. Sonneveld, and S. Hovius. 1993. The hand of the musician: The biomechanics of the bidigital finger system with anatomical restrictions. *J Biomech* 26:1169-1179.

Lichtwark, G. A., and A. M. Wilson. 2005. In vivo mechanical properties of the human Achilles tendon during one-legged hopping. *J Exp Biol* 208(Pt 24):4715-4725.

Li, Z.M., V.M. Zatsiorsky, M.L. Latash, and N.K. Bose. 2002. Anatomically and experimentally based neural networks modeling force coordination in static multi-finger tasks. *Neurocomputing* 47:259-275.

Loren, G.J., and R.L. Lieber. 1995. Tendon biomechanical properties enhance human wrist muscle specialization. *J Biomech* 28(7):791-799.

Maganaris, C.N., and J.P. Paul. 1999. In vivo human tendon mechanical properties. *J Physiol* 521(Pt 1):307-313.

Maganaris, C.N., and J.P. Paul. 2000a. Hysteresis measurements in intact human tendon. *J Biomech* 33(12):1723-1727.

Maganaris, C.N., and J.P. Paul. 2000b. Load-elongation characteristics of in vivo human tendon and aponeurosis. *J Exp Biol* 203:751-756.

Maganaris, C.N., and J.P. Paul. 2000c. *In vivo* human tendinous tissue stretch upon maximum muscle force generation. *J Biomech* 33:1453-1459.

Maganaris, C. N., and J. P. Paul. 2002. Tensile properties of the in vivo human gastrocnemius tendon. *J Biomech* 35(12):1639-1646.

Magnusson, S. P. 1998. Passive properties of human skeletal muscle during stretch maneuvers: A review. *Scand J Med Sci Sports* 8(2):65-77.

Magnusson, S. P., P. Aagaard, P. Dyhre-Poulsen, and M. Kjaer. 2001. Load-displacement properties of the human triceps surae aponeurosis in vivo. *J Physiol* 531(Pt 1):277-288.

Magnusson, S.P., P. Hansen, P. Aagaard, J. Brond, P. Dyhre-Poulsen, J. Bojsen-Moller, and M. Kjaer. 2003. Differential strain patterns of the human gastrocnemius aponeurosis and free tendon, in vivo. *Acta Physiol Scand* 177(2):185-195.

Magnusson, S.P., M.V. Narici, C.N. Maganaris, and M. Kjaer. 2008. Human tendon behaviour and adaptation, in vivo. *J Physiol* 586(1):71-81.

Meijer, O.G., Y.M. Kots, and V.R. Edgerton. 2001. Low-dimensional control: Tonus (1963). *Motor Control* 5(1):1-22.

Muramatsu, T., T. Muraoka, D. Takeshita, Y. Kawakami, Y. Hirano, and T. Fukunaga. 2001. Mechanical properties of tendon and aponeurosis of human gastrocnemius muscle in vivo. *J Appl Physiol* 90(5):1671-1678.

Nordez, A., P. Casari, and C. Cornu. 2008. Effects of stretching velocity on passive resistance developed by the knee musculo-articular complex: Contributions of frictional and viscoelastic behaviours. *Eur J Appl Physiol* 103(2):243-250.

Proske, U., and D.L. Morgan. 1987. Tendon stiffness: Methods of measurement and significance for the control of movement: A review. *J Biomech* 20(1):75-82.

Proske, U., D.L. Morgan, and J.E. Gregory. 1993. Thixotropy in skeletal muscle and in muscle spindles: A review. *Prog Neurobiol* 41(6):705-721.

Reeves, N.D., C.N. Maganaris, and M.V. Narici. 2003. Effect of strength training on human patella tendon mechanical properties of older individuals. *J Physiol* 548(Pt 3):971-981.

Rosager, S., P. Aagaard, P. Dyhre-Poulsen, K. Neergaard, M. Kjaer, and S. P. Magnusson. 2002. Load-displacement properties of the human triceps surae aponeurosis and tendon in runners and non-runners. *Scand J Med Sci Sports* 12(2):90-98.

Sands, W.A., J.R. McNeal, M.H. Stone, E.M. Russell, and M. Jemni. 2006. Flexibility enhancement with vibration: Acute and long-term. *Med Sci Sports Exerc* 38(4):720-725.

Schieber, M.H., and M. Santello. 2004. Hand function: Peripheral and central constraints on performance. *J Appl Physiol* 96(6):2293-2300.

Schweizer, A. 2001. Biomechanical properties of the crimp grip position in rock climbers. *J Biomech* 34(2):217-223.

Silver, F.H., J.W. Freeman, and G.P. Seehra. 2003. Collagen self-assembly and the development of tendon mechanical properties. *J Biomech* 36(10):1529-1553.

Stafilidis, S., K. Karamanidis, G. Morey-Klapsing, G. Demonte, G.P. Bruggemann, and A. Arampatzis. 2005. Strain and elongation of the vastus lateralis aponeurosis and tendon in vivo during maximal isometric contraction. *Eur J Appl Physiol* 94(3):317-322.

Uchiyama, S., J.H. Coert, L. Berglund, P.C. Amadio, and K.N. An. 1995. Method for the measurement of friction between tendon and pulley. *J Orthop Res* 13(1):83-89.

Voigt, M., F. Bojsen-Moller, E.B. Simonsen, and P. Dyhre-Poulsen. 1995. The influence of tendon Young's modulus, dimensions and instantaneous moment arms on the efficiency of human movement. *J Biomech* 28(3):281-291.

Wang, J. H. 2006. Mechanobiology of tendon. *J Biomech* 39(9):1563-1582.

Weiss, J.A., and J.C. Gardiner. 2001. Computational modeling of ligament mechanics. *Crit Rev Biomed Eng* 29(3):303-371.

Woittiez, R.D., R.H. Rozendal, and P.A. Huijing. 1985. The functional significance of architecture of the human triceps surae muscle. In *Biomechanics IX-A*, edited by D. Winter, R.W. Norman, R.P. Wells, K.C. Hayes, and A.E. Patla. Champaign, IL: Human Kinetics, pp. 21-26.

Woo, S.L. 1982. Mechanical properties of tendons and ligaments: I. Quasi-static and nonlinear viscoelastic properties. *Biorheology* 19(3):385-396.

Woo, S.L., G.A. Johnson, and B.A. Smith. 1993. Mathematical modeling of ligaments and tendons. *J Biomech Eng* 115(4B):468-473.

Woo, S.L.Y., G.A. Livesay, T.J. Runco, and E.P. Yoing. 1997. Structure and function of tendons and ligaments. In *Basic Orthopaedic Biomechanics,* edited by V.C. Mow and W.C. Hayes. Philadelphia: Lippincott-Raven, pp. 209-251.

Wright, W.G., V.S. Gurfinkel, J. Nutt, F.B. Horak, and P.J. Cordo. 2007. Axial hypertonicity in Parkinson's disease: Direct measurements of trunk and hip torque. *Exp Neurol* 208(1):38-46.

Zajac, F.E. 1989. Muscle and tendon: Properties, models, scaling, and application to biomechanics and motor control. *Crit Rev Biomed Eng* 17(4):359-411.

Zajac, F.E. 1992. How musculotendon architecture and joint geometry affect the capacity of muscles to move and exert force on objects: A review with application to arm and forearm tendon transfer design. *J Hand Surg Am* 17(5):799-804.

Zatsiorsky, V.M., R.W. Gregory, and M.L. Latash. 2002. Force and torque production in static multifinger prehension: Biomechanics and control: II. Control. *Biol Cybern* 87(1):40-49.

Zatsiorsky, V.M., Z.M. Li, and M.L. Latash. 1998. Coordinated force production in multifinger tasks: Finger interaction and neural network modeling. *Biol Cybern* 79(2):139-50.

Zatsiorsky, V.M., Z.M. Li, and M.L. Latash. 2000. Enslaving effects in multi-finger force production. *Exp Brain Res* 131(2):187-195.

Zatsiorsky, V.M., and M.L. Latash. 2008. Multifinger prehension: An overview. *J Mot Behav* 40(5):446-476.

Zernicke, R.F., J. Garhammer, and F.W. Jobe. 1977. Human patellar-tendon rupture. *J Bone Joint Surg Am* 59(2):179-183.

CHAPTER
3

MECHANICS OF ACTIVE MUSCLE

This chapter concentrates on mechanical aspects of muscle contraction occurring at the level above the sarcomere. The "intrasarcomere" mechanisms, such as the actin–myosin interaction, are mentioned only in passing. It is assumed that the readers are familiar with these mechanisms from courses in muscle physiology.

The chapter deals with two main issues: mechanical phenomena occurring in the muscle during muscle contraction (section **3.1**) and the so-called functional relations (section **3.2**). After brief comments on experimental approaches used in studies on muscle mechanics (section **3.1.1**), the chapter starts with a description of the muscle transition from rest to activity (i.e., muscle activation; section **3.1.2**), followed by the transition from activity to rest (i.e., muscle relaxation; section **3.1.3**). Section **3.1.4** describes the fundamental fact of muscle mechanics, the constancy of the muscle volume. Section **3.1.5** addresses force transmission and internal deformations in the active muscles. Section **3.1.6** is devoted to intramuscular stress and pressure, especially intramuscular fluid pressure. The second part of the chapter (section **3.2**) deals with functional relations, force–length relations (section **3.2.1**), and force–velocity relations (section **3.2.2**). Section **3.3** is concerned with the history effects in muscle mechanics.

3.1 MUSCLE FORCE PRODUCTION AND TRANSMISSION

This section addresses the force production and transmission within a muscle. It also describes some intramuscular effects of muscle force changes.

3.1.1 Experimental Methods

The existing knowledge of muscle mechanics was gained from experiments performed on whole muscles, single-fiber preparations, and isolated myofibril preparations. Experiments on whole muscles were conducted in vivo (within a living body), in situ (in the original place but with partial isolation), or in vitro (isolated from a living body). To perform experiments on excised muscles or muscle fibers, the investigator must keep them alive. The muscles (fibers, myofibrils) should be bathed in physiological solution with osmotic pressure and ionic composition similar to those in the tissue fluids in the animal's body. Usually muscles contain plenty of sources of energy, mostly glycogen, that allow them to work for long periods without additional supply of food substances. The crucial issue is adequate oxygen supply. Mammalian muscles have a high metabolic rate. To supply them with oxygen, intact blood circulation has to be preserved. Due to this requirement, experiments on mammalian muscles in vitro are not performed.

In experiments, muscles are typically activated by electric stimuli applied to muscle surface or to the nerve innervating the muscle. If the strength of a single stimulus exceeds a certain threshold, the muscle responds by a brief period of contraction followed by relaxation (*twitch*). If the stimuli are repeated at a sufficiently high frequency, summation occurs and a smooth *tetanus* is observed. Smooth tetanus is characterized by force levels higher than the maximal twitch force. When single fibers (i.e., muscle cells) of mammalian skeletal muscles are stimulated, the fibers follow the all-or-nothing law: the response to any suprathreshold stimulus is maximal and cannot be increased by increasing the strength of the stimulus. In contrast, when the whole muscle is stimulated, the response is graded; with an increasing strength of the stimulus, the muscle force increases because of the increased number of activated fibers. To obtain reproducible results, investigators usually use supramaximal stimuli, which are expected to induce contraction of all the fibers at each presentation.

3.1.2 Transition From Rest to Activity

Representative publications: Hill 1949a, 1949b, 1950; Buller and Lewis 1965

Starting from the arrival of a neural stimulus to a muscle, the muscle needs time to become active, develop force, and start shortening. The period between stimulus arrival and muscle force increase and shortening is known as the *latent period* (figure 3.1).

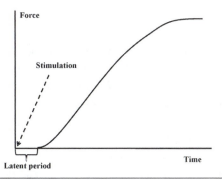

Figure 3.1 Muscle force–time relation, schematic. The time period from the start of stimulation to the start of force development represents the period during which the *active state* of the muscle develops. If judged only by externally manifested force, this is a latent period.

Muscle Activation

Muscle activation involves a complex sequence of events that can be considered at two levels: (1) individual muscle fibers and (2) the entire muscle. Only a brief, simplified overview of these processes is provided here.

At the level of the individual muscle fibers, the process involves the following:

• *Neuromuscular transmission:* the excitation propagates from the motor axon to the muscle fibers. At the neuromuscular junction, the incoming neural impulse opens the calcium channels, which cause local increase of Ca^{2+} ions and release of acetylcholine. The associated events trigger an action potential on the muscle fiber membrane that propagates along its entire length.

• *Excitation–contraction coupling:* excitation propagates from the fiber membranes inside the fibers via the transverse tubular system leading to release of Ca^{2+} ions from the sarcoplasmic reticulum. With an increasing intracellular Ca^{2+} concentration and Ca^{2+} ion–troponin binding along the thin (actin) filaments, the myosin heads establish links (called *crossbridges*) between the myosin and actin filaments. Muscle contraction begins.

(continued)

Physiology Refresher   *(continued)*

At the level of the entire muscle, the process involves the activation (recruitment) of individual motor units and their muscle fibers. The desired level of muscle force is controlled by *recruiting* various numbers and types of motoneurons and associated muscle fibers and by changing the frequency of motoneuronal firing (*rate coding*). A *motor unit* consists of a motoneuron and the muscle fibers it innervates. Motor units possess different properties and are classified as *Type I* (slow, fatigue resistant, their motoneurons are relatively small), *Type IIA* (fast but fatigue resistant), and *Type IIX* (fast with low resistance to fatigue). Fast motor units have motoneurons of relatively large size. During natural contractions, the recruitment follows the *size principle*: with gradually increasing neural drive to a motoneuronal pool, small motoneurons are activated before large ones because of lower activation thresholds. Hence the activation progresses from small to large motoneurons and consequently from the slow to the fast motor units. For a muscle involved in a specific motion, the recruitment order of motor units is relatively fixed.

We first consider the mechanical aspects of muscle activation (excitation–activation coupling) in experimental conditions and then force development in humans. *Excitation* (or *neural excitation*) refers to the motoneuron action potentials or electric stimuli transmitted to the muscle; the term *activation* (or *muscle activation*) designates the set of events linking the excitation with the contractile machinery and resulting in force production. The excitation–contractile machinery linkage is mediated by calcium release (the amount of Ca^{2+} ions released inside the fibers) and its binding to the troponin at the thin (actin) filaments. In vivo muscle excitation is a function of the number of recruited motor units and their firing frequency.

■ ■ ■ *FROM THE LITERATURE* ■ ■ ■

Modeling Neural Control of Gradation of Muscle Force (Neural Excitation)

Source: Tax, A.A., and J.J. Denier van der Gon. 1991. A model for neural control of gradation of muscle force. *Biol Cybern* 65(4):227-234

A model is presented that relates neural control signals linearly to muscle force. The model allows for different relative contributions from the two force-grading mechanisms—the recruitment of motor units and the modulation of their firing frequency. Consider a motoneuron pool that receives input from a nerve bundle. A weighted sum of activities in a nerve bundle I is

$$I = \sum_i u_i e_i , \qquad [3.1]$$

where I is the control signal of the motoneuron pool, e_i is the firing frequency of action potential traveling along each nerve fiber in the bundle, and u_i is the synaptic weight of a nerve fiber i projecting to a motoneuron (in the model, it is assumed that synaptic weights for all motoneurons are equal). When some commonly accepted physiological facts were incorporated into the model (e.g., different recruitment density of the small and large motor units), a linear relation between the control signals I and the muscle force was elucidated. Hence, the control signal I is proportional to the muscle force and can be interpreted as an internal representation of muscle force. The model confirms an intuitive notion that a weighted sum of activities in a nerve bundle can directly represent an externally controlled variable, which in this case is exerted muscle force.

3.1.2.1 Muscle Active State

Representative publications: Gasser and Hill 1924; Hill 1949a, 1949b, 1950; Ritchie and Wilkie 1958; Julian and Moss 1976

During the latent period, the muscle proceeds from rest to the active state. Biochemically, the active state can be characterized by the relative amount of Ca^{2+} ions bound to troponin. The active state develops rapidly, whereas the force at the muscle–tendon junction requires more time to develop.

Mechanically, the active state is manifested via the rapid onset of the ability to resist externally applied stretch. If during a latent period a quick stretch is applied to the muscle, the muscle resists its extension as if it already has developed its maximal force. During the latent period, the muscle can bear large forces without substantial extension (figure 3.2). The increase in muscle force begins later, and if muscle force exceeds weight of the external load, muscle shortening begins. This rapidly achievable ability to resist stretch supports the idea that at the beginning of force increase or muscle shortening, the active state is already fully developed. The rapidly achieved active state is further evidenced by the rapid increase of the temperature in the activated muscle well before the start of shortening. Hence, the chemical processes associated with force

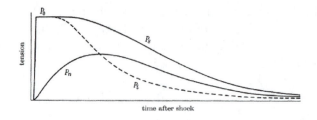

Figure 3.2 Diagram of mechanical changes during a single muscle twitch. P_n, force at the muscle end; P_i, intrinsic force of the contractile component of the muscle (considered to be a mechanical manifestation of the active state); P_s, force in muscle stretched quickly during the latent period; P_0, initial value of P_i and P_s. The quick muscle stretch by a controlled amount lengthens the series elastic component and allows the force at the muscle end to match the force of the contractile component.

Reprinted, by permission, from A.V. Hill, 1949, "The abrupt transition from rest to activity in muscle," *Proceedings of the Royal Society of London, Series B: Biological Sciences* 136(884), p. 410, fig. 7.

and energy production develop well before observable changes in muscle action (force production, shortening).

If during the latent period an experimental device is used to rapidly shorten and hold the muscle at a new length, the muscle develops force more slowly (control–release experiments). This indicates that the internal contractile component of the muscle is able to generate force very fast, whereas the force at the muscle ends is developed with a time delay, necessary to extend the elastic component of the muscle (a more detailed description of the concept of the contractile and elastic muscle components is provided in sections **3.3.2** and **4.4**).

Time course quantifications in the active state are highly dependent on experimental protocol. As a result, in the contemporary literature the active state is frequently treated only qualitatively, without assigning numerical values to its development.

Well-established facts are that during a twitch, the active state reaches maximum and then declines to the resting level. During a fused tetanus, the active state and force are larger than during a twitch contraction because of the summation of responses to individual stimuli. The active state and force are maintained until the end of stimulation or the development of fatigue.

Muscle Force Generation

Muscle force arises from the interaction between the thick (myosin) and thin (actin) filaments in the sarcomeres (see figures 1.2 and 3.7). The interaction consists of cyclic attachments and detachments of the myosin heads (crossbridges) to the active sites on the actin filaments. The interaction results in force generation by the crossbridges and relative sliding of the thick and thin filaments. Throughout relative movement of the actin and myosin filaments with respect to each other (sliding), the lengths of the filaments themselves remain essentially constant (*sliding filament theory*). During active shortening (concentric muscle action), the actin filaments move toward the A-bands whereas during muscle stretch the actin filaments are pulled farther from the arrays of myosin filaments (A-bands), still without length change. Such relative actin–myosin displacements change the extent of overlap of the actin and myosin filaments. If load at the ends of the muscle allows for muscle shortening, the actin filaments will move along the myosin filaments until the sarcomere shortens to the length, called the *sarcomere active slack length*, at which no active force can be generated. The filaments are much stiffer than the crossbridges and are practically nonextensible.

3.1.2.2 Force Development in Humans: Rate of Force Development

Representative publication: Savelberg 2000

When people exert force on the environment under the instruction to exert maximal force as quickly as possible, the magnitude of the attained maximal force (F_m) depends on the resistance, for example, on the load magnitude. The larger the resistance, the larger the maximal force F_{max}. Highest forces are attained against loads with weight exceeding F_{max} in which case the muscle is forced to stretch, so-called *eccentric muscle action* (see chapters 4 and 7). When force is exerted against an unmovable object, for example, in isometric conditions, the force is lower than in eccentric conditions but higher than in concentric conditions. The force cannot be developed instantly. Even for the best athletes it takes usually in excess of 300 to 400 ms to exert maximal force. For isometric force production, the force–time relation is approximately exponential and is oftentimes described by

$$F(t) = F_{max}(1 - e^{-kt}),$$ [3.2]

where $F(t)$ is the force at the instant t, F_{max} is the maximal force achieved in the task, e is the base of the natural logarithm ($e = 2.71828 \ldots$), and k is an experimental parameter. A three-term exponential equation

$$F(t) = A_1(1 - e^{-k_1 t}) + A_2(1 - e^{-k_2 t}) - A_3(1 - e^{-k_3 t})$$ [3.3]

is slightly more accurate than equation 3.2. The first two terms in equation 3.3 describe the time course of the force increase, whereas the third term— $-A_3(1 - e^{-k_3 t})$ —represents the deformation of the subcutaneous soft body tissues in contact with the force measuring device.

When an isometric force development is elicited by electrical stimulation of sufficiently large magnitude and frequency, the time to peak force is shorter and can be around 100 ms (the force magnitude, however, is smaller, which suggests that not all motor units are activated). The difference in the time to peak force between voluntary and electrically elicited contractions is in part explained by the time necessary during voluntary efforts for propagating neural excitation over the motoneuron pools that innervate the involved muscles (recruitment of motor units and increasing their firing rate). The time to peak force is also affected by the time necessary for deformation of the elastic components in the muscle–tendon complexes, in particular the tendons. The force exerted by the tendon on the bone lags behind the force of the contractile components.

One of the factors that seem to affect the rate of force development is muscle composition. Fast, Type II muscle fibers have a shorter twitch rise time than slow, Type I fibers. It is usually expected that people with a large percentage of fast muscle fibers can exert force faster than can people with a low percentage of fast fibers. However, because during maximal force production both fast and slow muscle fibers are activated (recruitment follows the size principle), the force development rate is determined by the slowest muscle fibers (the *weakest-link principle*).

• • • PHYSIOLOGY REFRESHER • • •

Electromechanical Delay

An indirect indicator of force development delay is *electromechanical delay* (EMD), the time period between the first signs of the electrical activity of the muscle (determined by electromyography, or EMG) and the first signs of its mechanical action, such as movement onset or the start of the force increase. Unfortunately, EMD recording methods are not standardized (e.g., what is the threshold EMG value above rest that indicates muscle activity onset?) and depend on the technical details of movement and EMG measurements such as electrode design, placement, and data analysis (e.g., the frequency characteristics of the filters used for processing the movement and the EMG signals). Therefore, the reported EMD values broadly vary among publications, most often quoted between 25 and 100 ms in human movement. Using a sensitive accelerometer can decrease the EMD to about 10 ms.

The frequency of the surface EMG is much larger than the frequency response of the generated muscle force. Therefore, it is said that the muscle works as a low-pass filter of the EMG signals.

In many athletic events, the time available for force development is smaller than the time necessary to produce maximal force. For instance, in well-trained athletes, the time of ground contact during running is around 100 ms, the takeoff time in running long jumps is around 120 ms, and the delivery phase in shot putting (the time of arm extension) is 220 to 270 ms. All these times are shorter than the time necessary to develop maximal voluntary force, $t \geq 300$ to 400 ms. Hence, during rapid movements, athletes have insufficient time to produce maximal force. In activities such as sprinting, this limitation is offset in part by rapid muscle stretch in the support leg during early stance, which (1) stretches the muscle elastic component and thus decreases the EMD and (2) increases the force production in the next push-off phase (the latter mechanism is described in detail in chapters 4 and 7). However, as a rule during fast athletic movements, athletes do not exert the maximal force that they can attain in ideal conditions. (The highest value among maximal force values is termed the *maximum maximorum force*, F_{mm}.) For instance, among the best shot-putters during throws of 20.0 to 21.0 m, the peak force F_m applied to the shot is 500 to 600 N, yet this is only about 50% of the one-arm force that these athletes can generate during bench press of 220 to 240 kg (~1,070-1,176 N per arm). Given these and similar data, investigators have concluded that in rapid athletic events, the rate of force development (RFD) mainly determines the performance results.

The RFD can be computed as the time derivative of the force–time function and is itself a function of time. The average value of the RFD for the initial part of muscular effort is called the *S-gradient* (*S* stands for *start*) and is computed as

$$S\text{-gradient} = F_{0.5m} / T_{0.5m}, \qquad [3.4]$$

where $F_{0.5m}$ is the half of the maximal force and $T_{0.5m}$ the time of attaining $F_{0.5m}$. When measuring S-gradient, investigators ask the subjects to exert the maximal force as fast as possible. In athletes, the force $F_{0.5m}$ and the time $T_{0.5m}$ do not correlate with each other. Hence the ability to exert large forces (muscular strength) and the ability to exert force quickly are independent.

3.1.3 Transition From Activity to Rest: Muscle Relaxation

Representative publications: Hill 1949c; Gillis 1985; Kisiel-Sajewicz et al. 2005

All human movements include phases of muscle activation and *relaxation*, that is, cessation of muscle activity. Relaxation from maximal isometric contraction involves two phases. The first phase—from approximately 100% to 80% of F_{max}—is characterized by slow force decay. The decay is approximately linear so the first phase is usually called the *linear phase*. During this phase the sarcomere lengths do not change. The second phase—from approximately 80% of F_{max} to zero—is exponential, with the fast force decline at the beginning and slow decline at the end. During this phase the force decrease is associated with substantial sarcomere elongation ("give"). The give is not uniform along the muscle and fiber lengths. In experiments performed in situ, it has been demonstrated that the first part of muscle relaxation is governed by reduction of motor unit firing rate, whereas the second part is determined by derecruitment of active motor units.

■ ■ ■ *FROM THE LITERATURE* ■ ■ ■

A Test to Measure the Muscle Relaxation Rate

Source: Lyons, M.F., and A. Aggarwal. 2001. Relaxation rate in the assessment of masseter muscle fatigue. *J Oral Rehabil* 28(2):174-179

It was reported previously that the rate of relaxation from an isometric contraction becomes slower with the onset of local muscle fatigue. The aim of this study was to assess a simple method of measuring relaxation rate during jaw-closing for the purpose of quantifying jaw muscle fatigue. The rates of twitch contraction and relaxation were measured in symptom-free subjects following bilateral direct electrical stimulation of the masseter muscles. The force was recorded via a transducer placed between the anterior teeth. The transducer was held between the teeth with as little force as possible while four single stimuli were delivered at 5 s intervals. The force records of the resulting twitches were averaged, and the half-contraction time, twitch amplitude, and time taken for force to decrease from 75% to 25% of the contraction force (the half-relaxation time) were measured. Half-relaxation time and half-contraction time were independent of twitch amplitude. This method of measuring the relaxation rate of the masseter muscles was found to be practical, and the results were reproducible between sessions.

Muscle relaxation takes longer than activation. In other words, the activation time constant is shorter than the deactivation time constant. A simple demonstration can be used to visualize this claim. Place your right forearm and hand on a table and your left hand on the forearm of the right arm. Tap with your right hand and fingers on the table gradually increasing the frequency of tapping. At the highest frequency you will notice that both wrist flexors and extensors of the right arm are simultaneously active: the muscles have insufficient time to relax.

The rate of muscle relaxation increases if the muscle is allowed to shorten during relaxation, a phenomenon that has important functional implications. This shortening dependence conserves metabolic energy and avoids wasteful simultaneous force production by muscles with opposite mechanical functions (antagonists).

Muscle Shortening Accelerates Relaxation

Source: Barclay, C.J., and G.A. Lichtwark. 2007. The mechanics of mouse skeletal muscle when shortening during relaxation. *J Biomech* 40(14):3121-3129

In locomotion, a considerable fraction of muscle work can be produced by relaxing muscles. Shortening with constant velocity was applied to muscles during relaxation following short tetanic contractions. Curves relating force output to shortening velocity were constructed. Relaxation time was inversely related to shortening velocity during relaxation (figure 3.3).

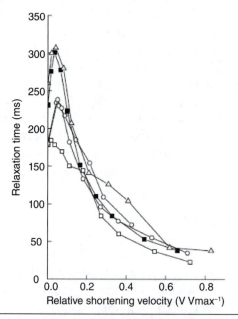

Figure 3.3 The dependence of relaxation time on the speed of muscle shortening. Relaxation time is the time taken for force to decrease from 95% to 5% of the force at the time the last stimulus was delivered. Data are shown for five soleus preparations, each distinguished by symbols joined by lines, for relaxation following a 1 s tetanus. Slow shortening velocities prolonged relaxation time relative to that for isometric contraction (plotted at relative shortening velocity of 0). At velocities greater than $0.1\ V_{max}$, relaxation time decreased as shortening velocity was increased.

3.1.4 Constancy of the Muscle Volume

Representative publications: Baskin and Paolini 1965, 1966, 1967

Active muscles generate force and shorten without substantially changing their volume. The same is true of individual muscle fibers. The recorded changes of muscle volume amount to a small fraction of a percent (between 0.002% and 0.005%). Although these changes may be important for understanding the intrinsic mechanisms of muscle action, they occur on a smaller scale than most issues discussed in this book and, hence, are neglected in the ensuing text.

The constancy of muscle volume during a single contraction was first discovered in 1663 by a Dutch scientist, Jan Swammerdam (1637-1680), who is also credited with being the first to perform experiments on an isolated nerve–muscle preparation of a frog, a classic object of muscle physiology and biomechanics research for more than 300 years (figure 3.4).

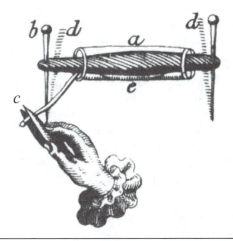

Figure 3.4   Frog nerve–muscle preparation (from J. Swammerdam, *The Book of Nature II*. London, Seyffert, 1758). In Swammerdam's own words, "If we have a mind to observe, very exactly, in what degree the muscle thickens in its contraction, and how far its tendons approach towards each other, we must put the muscle into a glass tube, *a*, and run two fine needles *bb* through its tendons, where they had been before held by the fingers; and then fix the points of those needles, neither too loose nor too firmly, in a piece of cork. If afterwards you irritate, *c*, the nerves, you will see the muscle drawing *dd* the heads of the needles together out of the paces; and that the belly of the muscle itself becomes considerably thicker *e* in the cavity of the glass tube, and stops up the whole tube, after expelling the air. This continues till the contraction ceases, and the needles then move back into their former places."

Adapted from J. Swammerdam, 1758, *The book of nature II* (London: Seyffert).

In one of his experiments Swammerdam placed a nerve–muscle preparation in a glass tube connected to a thin capillary along with a drop of water (figure 3.5). This device is essentially a *plethysmograph* that is used in contemporary science to record volume changes. The nerve was connected to a silver wire that the researcher held in his hand. When the nerve was "irritated" by pulling the wire downward, the muscle contracted. The bubble in the capillary, however, slightly fell rather than moved upward. The fact contradicted an existing theory that the muscle activation is induced by influx of "animal spirits," which were thought to be similar in properties to liquids or air.

The constancy of muscle volume during a muscle contraction is the fundamental fact of muscle mechanics. It greatly affects various mechanical phenomena occurring in the contracting muscle. An example of the effect of constancy of the muscle volume on the force transmission in a unipennate muscle is presented in figure 3.6.

When muscles are activated in situ they exert force on the tendons and extend them. As a consequence, the length of the muscle belly usually changes. In isometric actions of a muscle–tendon complex, the length of the muscle fibers inside the belly initially decreases, while the tendon and aponeurosis elongate, until the muscle develops maximum force and then the fiber length does not change. In concentric action the fibers shorten and in eccentric action they typically elongate. The constancy of the muscle volume requires that the muscle size in the directions orthogonal to the line of muscle force action should change to compensate for the change in muscle belly length. Unfortunately, little is known about the muscle dimension changes in the transverse directions.

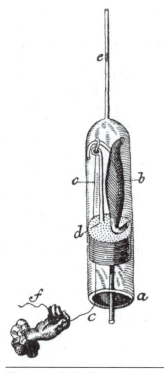

Figure 3.5 Swammerdam's experiment. A frog muscle with a nerve is placed in an air-filled tube sealed at the base. The nerve running through a brass eye is stimulated with a silver wire, by pulling the wire manually. *(a)* The glass tube; *(b)* the muscle; *(c)* a silver wire with a ring on it, through which the nerves passes; *(d)* a brass support with an eye through which the silver eye passes; *(e)* a drop of water in the glass capillary; *(f)* the researcher's hand.

Adapted from J. Swammerdam, 1758, *The book of nature II* (London: Seyffert).

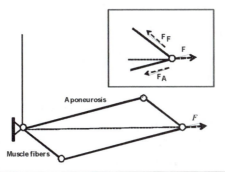

Figure 3.6   A planar model of a unipennate muscle in the pennation plane. The model depicts the muscle as a four-bar linkage with hinge joints. The constancy of the muscle area in the pennation plane is assumed. Without restrictions on the area of the system, under the action of force *F* the system would collapse. However, if the area of the system has to remain constant, the system's behavior would be completely different because of internal resistance provided by the incompressible muscle tissue. The inset shows an incorrect free body diagram with the forces of the muscle fibers *(F_F)* and aponeurosis *(F_A)*. The diagram is incorrect because it neglects the forces necessary to maintain constant area (the constant volume of the muscle). The effects of the constancy of the muscle volume on the force equilibrium and mechanical stability of the muscles were analyzed in several research papers (Otten, 1988; Van Leeuwen and Spoor, 1992, 1993; van der Linden et al. 1998; Epstein and Herzog, 1998, 2003).

Adapted from M. Epstein and W. Herzog, 1998, *Theoretical models of skeletal muscle* (Chichester, UK: Wiley), and M. Epstein and W. Herzog, 2003, "Aspects of skeletal muscle modelling," *Philosophical Transactions of the Royal Society of London: Series B, Biological Sciences* 358(1437):1445-1452.

The rule of constancy of muscle volume during a single muscle contraction does not mean that the muscle volume does not change at all. During and immediately after repeated intensive physical exercise, muscle volume can increase up to 10% over normal conditions due to fluid uptake. Muscle swelling is related to exercise intensity and occurs as a consequence of the accumulation of metabolites and increased osmotic driving force (osmotic pressure). Swelling takes place within seconds, whereas restoration of muscle volume to the rest size is a much longer process. Temporary muscle swelling is used by professional bodybuilders during contests: immediately before posing for judges, contestants perform strength exercises to make their muscles look larger and more impressive. Muscle volume also changes over longer time periods because of strength training or muscle disuse.

3.1.5 Force Transmission and Internal Deformations (Strain)

Representative publications: Huijing 2009; Sandercock and Maas 2009; Yucesoy 2010

The force generated in sarcomeres should be transmitted to the tendons. The transmission and its accompanying events are poorly understood and are the topics of intensive research. As a general principle, the forces are transmitted not only along the longitudinal line of the muscle through the tendons of origin and insertion but also in lateral directions via the adjacent structures (*myofascial force transmission, epimuscular force transmission*). Although the fact of lateral force transmission at all levels of muscle organization is well established, the magnitude of the laterally transmitted force in many cases is unknown. With the exception of the force transmission (a) inside individual muscle fibers and (b) from active to passive muscle fibers inside a muscle, it seems that the lateral force transmission between the muscles in vivo is small and for many practical purposes can be neglected. If this notion receives conclusive experimental confirmation, this would indicate that the muscles are independent force generators; they are united in functional synergistic groups not by lateral mechanical connections but by the motor control processes. If, however, epimuscular force transmission—force transmission via the epimysium, the outer limit of the muscle—is large and cannot be neglected, it would indicate that the central controller cannot control single muscles in isolation. Unfortunately, we do not have sufficient data, especially for human muscles, to answer this question. It follows from several experiments in rats that the epimuscular force transmission increases when the muscles are stretched by external forces. During tendon transfer surgeries in human patients, a tendon is released from its natural position and reattached to another site. This procedure often requires a substantial lengthening of the muscle, which can cause nonintended lateral force transmission.

3.1.5.1 Force Transmission in Muscle Fibers

Representative publications: Monti et al. 1999; Bloch and Gonzales-Serratos 2003

The force generated by the actin–myosin interaction in a sarcomere is transmitted along the myofibril to the neighboring sarcomeres and also laterally to the sarcolemma via transversal riblike structures called *costameres* (in Latin *costa* means "rib"). The costameres overlie the Z-mem-

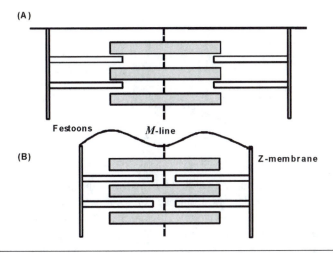

Figure 3.7　Lateral force transmission in the muscle fiber from actin–myosin complexes to the sarcolemma (schematics). *(a)* A sarcomere at rest. *(b)* Sarcomere during contraction. Because of the actin–myosin interaction, the Z-membranes move closer to each other (the length of sarcomere decreases), the force is transmitted laterally to sarcolemma via the costameres (not shown in the figure), and the intercostamer regions of the sarcolemma bulge, forming festoons.

branes as well as M-lines of myofibrils. As a result, the sarcolemma shrivels in sync with the Z- and M-lines of myofibrils, and the intercostameric regions of the sarcolemma slightly bulge, forming so-called *festoons* (figure 3.7).

The force is then transmitted along the muscle fiber to other sarcomeres via both the sarcolemma (approximately 70%-80%) and, longitudinally, the Z-membranes (approximately 20%-30%). The effect of the force transmission via the sarcolemma is a linear relation between the force generated by the muscle fiber and its radius. If the force were transmitted exclusively through the longitudinal pathway, it should be a function of cross-sectional area of the fiber and should be proportional to the radius squared. Instead, because it is also transmitted laterally, it is linearly related to the fiber's perimeter (and hence also to the fiber's radius or diameter; figure 3.8). Despite the linear relations between the fiber force and its perimeter (radius) in the entire muscle, which contains many muscle fibers, the muscle force is still a function of the physiological cross-sectional area. The proof of this statement is left to the readers.

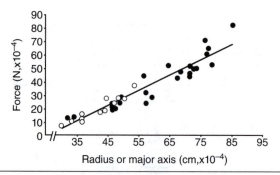

Figure 3.8   Force as a linear function of the radius or the major axis of a muscle-fiber cross-section. The data are from two studies in which the fiber dimensions and force were measured. Closed circles, cylindrical fibers; open circles, elliptical fibers. The data are fitted by a linear regression with $r = 0.95$.

When a muscle fiber develops force in isometric conditions (both ends of the fiber are fixed), the length of the individual sarcomeres can change. Some of the sarcomeres contract while others ("weak sarcomeres") are stretched. The sarcomere strain depends on the position of the sarcomere along the fiber. Sarcomeres near the ends of the fibers may shorten while sarcomeres in the central part of the fibers are stretched. The properties of the sarcomeres also depend on the sarcomere proximity to the motor endplates. Both spatial and temporal distribution of sarcomeres' strain (sarcomere nonuniformity) takes place. The nonuniformity of the sarcomeres in muscle fibers may partly explain the relatively large time for force development in the muscles (see again **3.1.2.2**): when some sarcomeres are shortening other sarcomeres are being forcibly stretched, which slows down the force increase at the muscle ends.

■ ■ ■ *FROM THE LITERATURE* ■ ■ ■

**Spatial and Temporal Heterogeneity
of Muscle Strain During Contraction**

Source: Drost, M.R., M. Maenhout, P.J. Willems, C.W. Oomens, F.P Baaijens, and M.K. Hesselink. 2003. Spatial and temporal heterogeneity of superficial muscle strain during in situ fixed-end contractions. *J Biomech* 36(7):1055-1063

> The spatial and temporal distribution of strain was quantified in fixed-end contracting rat tibialis anterior muscle in situ at optimal muscle length. After termination of stimulation, muscle segments near the motor endplates elongated whereas segments away from the endplates shortened. Muscle plateau shortening strain equaled 4.1%. Maximal plateau shortening of a muscle segment was much larger (9.6%) and occurred distally (at 0.26 of the scaled length of the muscle). During relaxation, distal segments actively shortened at the expense of proximal muscle segments, which elongated. The segments undergoing lengthening were closer to motor endplates than were the segments undergoing shortening. The investigators concluded that local muscle deformation is heterogeneous both temporally and spatially and may be related to proximity to the motor endplates.

When the entire active fiber is forcibly stretched, weak sarcomeres might be pulled beyond actin–myosin overlap and would maintain their length only by passive structural elements (the sarcomere length nonuniformity theory). This theory is discussed in chapter 4.

3.1.5.2 Force Transmission in Muscles: Summation of Muscle Fiber Forces

Representative publication: Patel and Lieber 1997

Muscle fibers and their bundles (fascicles) are arranged in the muscles in the variety of forms briefly described in chapter 1. Muscle architecture has a profound effect on muscle performance.

3.1.5.2.1 Parallel-Fibered and Fusiform Muscles

3.1.5.2.1.1 Nonuniform Shortening of Muscle Fibers

Let us first consider simple models of muscle architecture, two muscle fibers connected either serially or in parallel. In the top panel of figure 3.9, two idealized fibers are connected in series. In static conditions, if the fibers are activated, the force at the end of the series equals the force of a single fiber. If fibers are allowed to shorten, the speed at the free end of the series would be twice the speed of shortening of the individual fiber. In the series-fibered muscles, the serial fibers can be connected in two ways: end to end (serial fibers with serial connections) or side by side (serial fibers with lateral connections). In the fiber series with serial connections, the stress at the connection can be very large and at high forces the fibers can be damaged. To avoid damage, the serially connected fibers often have tapered ends and are connected with each other laterally over large areas, serial fibers with lateral connections (see figure 1.12). In such a fiber arrangement, a local stress at the fiber-to-fiber interface is relatively small. This muscle architecture is typical of the muscles with nonspanning fibers, that is, the muscles whose muscle fibers do not span the entire muscle length and do not pass through from the tendon of origin to the tendon of insertion.

When fibers are connected in parallel, as in the bottom panel of figure 3.9, the force of the fiber set is twice the force of the individual fibers but the speed of shortening is the same as the speed of individual fibers.

In fusiform muscles (e.g., the biceps brachii), the individual fibers are of different length. The fibers located on the periphery of the muscle are longer than the fibers at the muscle center. Hence, when the muscle as a whole shortens, the peripheral fibers shorten more than the central fibers. It is often assumed that the muscle fibers shorten uniformly, that the different muscle parts, such as distal and proximal parts, shorten by the same percentage of their length. This assumption is not valid for all muscles and muscle parts. The muscle is more accurately regarded as a complex

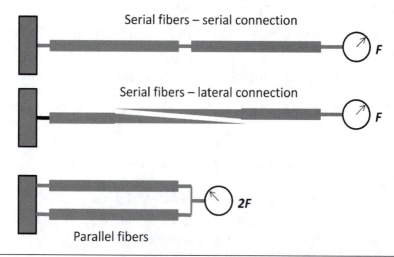

Figure 3.9   Serial and parallel arrangement of two fibers (schematics). Top and middle panels: Two serially arranged fibers with serial connection. Because the cross-sectional area of the first connection is small, the connection can be over-stressed and damaged. The fibers in the middle panel are arranged serially but the force is transmitted from fiber to fiber laterally. Because the area of contact is large, the stress at the interface is small and risk of damage is much lower (compare with figure 1.12). In the two fiber arrangements, the force at the end of the fiber series is the same. Bottom panel: The fibers are arranged in parallel. The total force from two fibers is twice the single-fiber force.

composite of contractile and connective tissue with differently behaving subparts. For instance, the shortening along the anterior boundary of the biceps brachii is uniform whereas the shortening along the centerline of the muscle is not uniform.

■ ■ ■ *FROM THE LITERATURE* ■ ■ ■

Nonuniform Muscle Shortening

Source: Pappas, G.P., D.S. Asakawa, S.L. Delp, F.E. Zajac, and J.E. Drace. 2002. Nonuniform shortening in the biceps brachii during elbow flexion. *J Appl Physiol* 92(6):2381-2389

This study tested the common assumption that skeletal muscle shortens uniformly in the direction of its fascicles during low-load contraction. Cine phase contrast magnetic resonance imaging was used to characterize shortening of the biceps brachii muscle in 12 subjects during repeated elbow flexion against 5% and 15% maximum voluntary contraction (MVC) loads. Mean shortening was relatively constant along the anterior boundary of the muscle and averaged 21% for both loading conditions. In contrast, mean shortening was nonuniform along the centerline of the muscle during active elbow flexion (figure 3.10). Centerline shortening in the distal region of the biceps brachii (7.3% for 5% MVC and 3.7% for 15% MVC) was significantly less ($p < .001$) than shortening in the muscle midportion (26.3% for 5% MVC and 28.2% for 15% MVC). Hence, the shortening was uniform along anterior muscle fascicles and nonuniform along centerline fascicles.

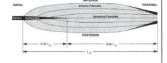

Figure 3.10 Top panel: Schematic line drawing of biceps brachii architecture in sagittal plane. Bottom panel: Centerline and anterior shortening for 15% MVC. Negative length changes indicate muscle shortening with elbow flexion. Vertical error bars indicate ±1 standard deviation.

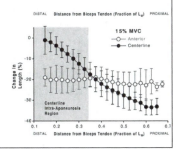

Adapted, by permission, from G.P. Pappas, D.S. Asakawa, S.L. Delp, F.E. Zajac, and J.E. Drace, 2002, "Nonuniform shortening in the biceps brachii during elbow flexion," *Journal of Applied Physiology* 92(6): 2381-2389.

The main sources of nonuniform muscle strain are the differences in fascicle length within the muscle and the curvature of the fascicles.

••• MATHEMATICS AND ENGINEERING REFRESHER •••

Finite Element Method

Finite element method (FEM) was developed in engineering to analyze stress–strain relations in complex mechanical objects, such as a car or an aircraft. In the method, the entire structure is divided into smaller, more manageable (finite) elements. The method uses a system of points, called *nodes*, which make a grid called a *mesh*. The mesh is constructed to represent reaction of a given finite element to loading conditions as manifested in, for instance, the element stiffness. When FEM is used, complex partial differential equations that describe the behavior of a modeled object are often reduced to linear equations that are solved using the standard techniques of matrix algebra. The method requires a large amount of work, especially to generate a realistic mesh correctly representing mechanical properties of the object. The term *finite element modeling* usually refers to the process of meshing: constructing a finite mesh based on a complex geometry, whereas *finite element analysis* refers to the process of solving the equations that govern a mesh behavior.

■ ■ ■ FROM THE LITERATURE ■ ■ ■

Finite Element Modeling of Strain Distribution in a Human Muscle

Source: Blemker, S.S., P.M. Pinsky, and S.L. Delp. 2005. A 3D model of muscle reveals the causes of nonuniform strains in the biceps brachii. *J Biomech* 38(4):657-665

The purpose of this study was to uncover the features of the biceps brachii architecture and material properties that could lead to nonuniform shortening. The authors created a three-dimensional finite element model of the biceps brachii and compared the tissue strains predicted by the model with experimentally measured tissue strains. The finite-element model predicted strains that were within 1 standard deviation of the strains experimentally measured previously with the cine phase contrast magnetic resonance method. Analysis of the model revealed that the variation in fascicle lengths within the muscle and the curvature of the fascicles were the primary factors contributing to nonuniform strains.

3.1.5.2.1.2 Nonlinear Summation of Fiber Forces

Representative publications: Clamann and Shelhorn 1988; Sandercock 2005

When two motor units are experimentally activated, the total force from the units can be smaller or larger than the sum of the forces from these motor units when they are activated separately (*nonlinear summation of motor unit forces*). The difference can be as large as 20%. When the force from two units is smaller than the sum of the individual unit forces the summation is called *subadditive*, and when the force is larger the summation is called *superadditive*. Similar effects, but of smaller relative magnitude, are recorded when individual parts of a whole muscle, rather than single motor units, are activated.

Explanations of this nonlinearity are mainly speculative; that is, they are not based on direct evidence but are rather deduced from general considerations. Explanations for subadditivity include serial arrangement of some muscle fibers, the stretching of common elastic elements, lateral force transmission to neighbor fibers, and change in the angle of pennation (newly recruited motor units may contribute less force because the angle of pennation increases during contraction, discussed later in the chapter). Explanations for superadditivity include thixotropic behavior of passive muscle fibers and increased stiffness of the aponeuroses and tendons with the force increase (motor units recruited in an active muscle exert force on an extended, and hence stiffer, tendon and may experience a steeper force increase and shorter relaxation time).

3.1.5.2.2 Pennate Muscles

We discuss first the force transmission from the fibers to the tendon and then speed transmission.

3.1.5.2.2.1 Force Transmission

Representative publications: Benninghoff and Rollhauser 1952; Otten 1988

In a parallel-fiber muscle, the fascicle force transmits to the tendon without changes (see, however, section **3.1.5.2.1.2** on nonlinear summation of fiber forces). In a pennate muscle, only the fiber force component that is along the muscle line of action is transmitted in this direction:

$$F_{tendon} = F_{fascicle}\ cos\,\theta,$$ [3.5]

where θ is a pennation angle. The second force component $F_{ort} = F_{fascicle}\ sin\,\theta$, which is orthogonal to the first component, always exists and may induce different effects such as a change of the muscle curvature or an increase of the intramuscular pressure.

Fiber shortening is commonly accompanied by fiber rotation (figure 3.11).

The fiber rotation is evidently induced by a moment of force acting on the fiber. The origin of this moment is not obvious. In particular, it is not clear whether this moment of force is active, that is, contributes to the force acting on the tendon, or passive, a consequence of tendon shortening. If

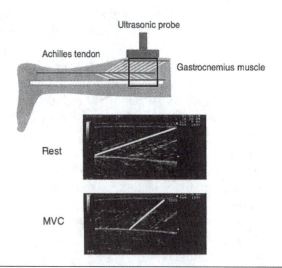

Figure 3.11 The fascicle rotation during maximal isometric action of plantar flexors. Longitudinal ultrasonic images of the gastrocnemius muscle at rest (top) and during maximal isometric plantar flexion (MVC).

Reprinted, by permission, from Y. Kawakami and T. Fukunaga, 2006, "New insights into in vivo human skeletal muscle function," *Exercise and Sport Sciences Reviews* 34(1): 16-21.

the first hypothesis is valid, the moment increases the force acting on the tendon (rotating fiber pulls the tendon). However, if the second hypothesis is valid the situation is opposite: the tendon pulls the fiber end forcing the fiber to rotate.

3.1.5.2.2.2 Speed Transmission: Architectural Gear Ratio
Representative publication: Azizi et al. 2008

This section addresses the relation between fiber and muscle length changes. Both fiber shortening and fiber rotation contribute to aponeurosis movement and muscle shortening. This factor as well as aponeurosis and tendon elasticity may cause the shortening velocity of the whole muscle to exceed the shortening velocity of the fibers. An extreme example is when the active muscle fibers maintain a constant length but rotate to a greater pennation angle so that the whole muscle shortens. The ratio of whole muscle velocity to muscle fiber velocity is called the muscle *architectural gear ratio* (AGR). The fiber rotation (change of fiber pennation angle) often results in AGR greater that 1.0; that is, muscle shortening velocity exceeds fiber shortening velocity.

■ ■ ■ *FROM THE LITERATURE* ■ ■ ■

During Isometric Force Development the Fascicle Length Decreases and Pennation Angle Increases

Source: Ito, M., Y. Kawakami, Y. Ichinose, S. Fukashiro, and T. Fukunaga. 1998. Nonisometric behavior of fascicles during isometric contractions of a human muscle. *J Appl Physiol* 85(4):1230-1235

Fascicle length, pennation angle, and tendon elongation of the human tibialis anterior were measured in vivo by ultrasonography. Subjects (n = 9) were requested to develop isometric dorsiflexion torque gradually up to maximum at the ankle joint angle of 20° plantar flexion from the anatomic position. Fascicle length shortened from 90 ± 7 mm to 76 ± 7 mm (± standard error), pennation angle increased from 10° ± 1° to 12° ± 1°, and tendon elongation increased up to 15 ± 2 mm with graded force development up to maximum. The tendon stiffness increased with increasing tendon force from 10 N/mm at 0 to 20 N to 32 N/mm at

(continued)

> From the Literature   *(continued)*
>
> 240 to 260 N. Young's modulus increased from 157 MPa at 0 to 20 N to 530 MPa at 240 to 260 N. The authors concluded that in isometric actions of a human muscle, the shortening of muscle fibers does mechanical work, some of which is absorbed by the tendinous tissue.

During a contraction the muscle volume remains constant. Therefore, when the length of the muscle decreases, the muscle dimensions in other directions should increase (the muscle bulges). Two major directions of bulging can be distinguished: (a) *muscle thickness*, defined by the distance between two aponeurotic planes (figure 3.12), and (b) *muscle width*, measured in the direction orthogonal to the plane in which muscle thickness is measured (not shown in the figure). In the majority of studies, muscle thickness is assumed constant; however, recent experimental studies demonstrated that during contraction the thickness of some muscles changes.

Fiber rotation decreases a muscle's output force but increases output velocity by allowing the muscle to function at a higher architectural gear ratio (ratio of muscle velocity to fiber velocity). Consider as an example a

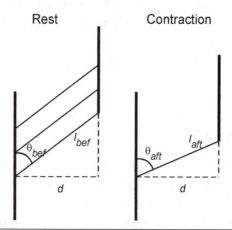

Figure 3.12   Contraction of a unipennate muscle (projection on the thickness plane); the width plane is not shown. The solid vertical lines are aponeuroses. The oblique thin lines are muscle fibers. d is the muscle thickness. The subscripts *bef* and *aft* refer to the values before and after contraction, respectively. The angle of pennation increases during contraction, $\theta_{aft} > \theta_{bef}$.

unipennate muscle with two parallel aponeuroses, similar to one shown in figure 3.12. The pennation angle θ is

$$\theta = \sin^{-1} \frac{d}{l}, \qquad [3.6]$$

where d is the distance between the aponeuroses, the muscle thickness, and l is the fiber length. The muscle width changes are neglected.

The tendon displacement ΔT is related to the fiber length changes as

$$\Delta T = l_{aft} \cos\theta_{aft} - l_{bef} \cos\theta_{bef}, \qquad [3.7]$$

where the subscripts *bef* and *aft* refer to the values recorded at two instances of time, for example, at rest and during a plateau of isometric muscle force, respectively. The ΔT depends on both the changes of the fiber length ($\Delta l = l_{aft} - l_{bef}$) and the pennation angle changes ($\Delta\theta = \theta_{aft} - \theta_{bef}$). The AGR equals $\Delta T / \Delta l$.

If the muscles thickness during contraction stays constant, Δl and $\Delta\theta$ should vary in a coordinated manner, such that the distance between the aponeuroses d does not change. In such a case an alternative equation can be used:

$$\Delta T = d(\cot\theta_{aft} - \cot\theta_{bef}). \qquad [3.8]$$

If the muscle thickness d increases during contraction and Δl is the same, as in the case of constant muscle thickness, the pennation angle should increase more compared with the constant thickness. Such a case would correspond to a larger AGR (because the contribution of the fiber rotation to the ΔT increases).

On the contrary, if the thickness d decreases—the muscle bulges in width—fibers can shorten with minimal rotation or even with no rotation at all. This case is considered less favorable for muscle speed because the fiber rotation no longer contributes to the muscle shortening velocity. In contrast, this situation is considered more favorable for muscle force production because the pennation angle is smaller and hence a component of the fiber forces along the muscle line of action is larger.

■ ■ ■ *FROM THE LITERATURE* ■ ■ ■

Variable Gearing in Pennate Muscles

Source: Azizi, E., E.L. Brainerd, and T.J. Roberts. 2008. Variable gearing in pennate muscles. *Proc Natl Acad Sci U S A* 105(5):1745-1750

(continued)

From the Literature *(continued)*

In pennate muscles, fibers are oriented at an angle to the muscle's line of action and rotate as they shorten, becoming more oblique such that the fraction of force directed along the muscle's line of action decreases throughout a contraction. Fiber rotation decreases a muscle's output force but increases output velocity by allowing the muscle to function at a higher gear ratio (ratio of muscle velocity to fiber velocity). The magnitude of fiber rotation, and therefore gear ratio, depends on how the muscle changes shape in the dimensions orthogonal to the muscle's line of action. The authors showed that gear ratio is not fixed for a given muscle but decreases significantly with the force of contraction. Dynamic changes in muscle shape promote fiber rotation at low forces and resist fiber rotation at high forces. As a result, gearing varies automatically with the load, to favor velocity output during low-load contractions and force output for contractions against high loads. Therefore, changes in muscle shape act as an automatic transmission system, allowing a pennate muscle to shift from high gear during rapid contractions to low gear during forceful contractions. These results suggest that variable gearing in pennate muscles provides a mechanism to modulate muscle performance during mechanically diverse functions.

If each sarcomere shortens by the same amount ΔS, the shortening of the entire fiber will depend on the number of the sarcomeres in the fiber, n. The fiber shortening is then $\Delta l = n\Delta S$ and is proportional to the fiber length. In other words, each fiber shortens by the same relative amount (in percentage of its initial length), which can also be expressed using the so-called *shortening factor*, $f = l / l_0$, actual fiber length divided by its length at rest. From figure 3.12 and equation 3.6, it follows that $\sin\theta_{bef} = d / l_{before}$ and $\sin\theta_{aft} = d / l_{aft}$. Considering that

$$\sin\theta_{aft} = d / fl_{before} = \sin\theta_{bef} / f \qquad [3.9]$$

and

$$\cos\theta_{aft} = \sqrt{1 - \frac{\sin^2\theta_{bef}}{f^2}} = \frac{\sqrt{f^2 - \sin^2\theta_{bef}}}{f}, \qquad [3.10]$$

the relation between the fiber length at rest and the muscle shortening is given by the formula

$$\Delta L = l_{bef}\left[\cos\theta_{bef} - \sqrt{\cos^2\theta_{bef} + f^2 - 1}\right], \qquad [3.11]$$

where ΔL is the muscle shortening or lengthening, l_{bef} is the muscle fiber length at rest, θ_{bef} is the pennation angle at rest, and f is the shortening factor. The equation assumes that all the fibers shorten by an equal percentage of their initial length, for instance, by 20%. The equation is valid when the length of the internal tendons (aponeuroses) remains constant (the aponeuroses are not extended) and the muscle thickness does not change.

3.1.6 Intramuscular Stress and Pressure

3.1.6.1 Specific Muscle Force

Representative publications: Haxton 1944; Kawakami et al. 1994; Fukunaga et al. 1996; Gatton et al. 1999

The value of maximal muscle force per unit of the muscle physiological cross-sectional area (PCSA) is called the *muscle specific force (SF)* or *specific tension* (before World War II, the term *absolute force* was mainly used).

$$SF = Tendon\ Force\ /\ PCSA. \qquad [3.12a]$$

In mammalian muscles, SF is usually between 15 and 40 N/cm^2 with a median value of about 25 N/cm^2 (250 kPa). When SF is determined from individual muscle fibers, values around 350 kPa are commonly reported. If SF for a muscle is known, the maximal muscle force (F_{max}) can be computed as

$$F_{max} = SF \times PCSA. \qquad [3.12b]$$

Equation 3.12b is broadly used in the modeling of muscle function. Specific tension is essentially a unidimensional stress: in parallel-fibered muscles, it is the maximum unidimensional stress. However, in the pennate muscles the tendon force (the muscle force) and fiber forces are in dissimilar directions and the PCSA plane is not orthogonal to the direction of the tendon force. The SF values depend on the muscle pennation angle and hence should vary among the muscles. To avoid the effect of different pennation angles, some authors compute the SF for muscle fascicles $(SF_{fascicle})$ rather than for the muscle. In this case, the muscle force (tendon force) is converted first to fascicle force (the muscle force divided by the cosine of the pennation angle of the fascicles during maximal isometric contraction) and then divided by the PCSA.

■ ■ ■ *FROM THE LITERATURE* ■ ■ ■

Muscle Specific Force, Calculated by Dividing Fascicle Force by PCSA, Increases After Training in Old Age

Source: Reeves, N.D., M.V. Narici, and C.N. Maganaris. 2004. Effect of resistance training on skeletal muscle-specific force in elderly humans. *J Appl Physiol* 96(3):885-892

Leg extension and leg press exercises (2 sets of 10 repetitions at 80% of the 5 repetition maximum) were performed 3 times a week for 14 weeks. Vastus lateralis (VL) muscle fascicle force was calculated from maximal isometric voluntary knee extensor torque with superimposed electric stimuli, accounting for the patellar tendon moment arm length, ultrasound-based measurements of muscle architecture, and antagonist co-contraction estimated from electromyographic activity. Physiological cross-sectional area was calculated from the ratio of muscle volume to fascicle length. Specific force was calculated by dividing fascicle force by PCSA. Fascicle force increased by 11%, from 847.9 ± 365.3 N before to 939.3 ± 347.8 N after training ($p < .05$). Because of a relatively greater increase in fascicle length (11%) than muscle volume (6%), PCSA remained unchanged (pretraining: 30.4 ± 8.9 cm^2; posttraining: 29.1 ± 8.4 cm^2; $p > .05$). Activation capacity and VL muscle root mean square electromyographic activity increased by 5% and 40%, respectively, after training ($p < .05$), indicating increased agonist neural drive, whereas antagonist co-contraction remained unchanged ($p > .05$). The VL muscle-specific tension increased by 19%, from 27 ± 6.3 N/cm^2 before to 32.1 ± 7.4 N/cm^2 after training ($p < .01$), highlighting the effectiveness of strength training for increasing the intrinsic force-producing capacity of skeletal muscle in old age.

Determining *SF* requires knowledge of (a) the maximal isometric force generated by a given muscle and (b) the PCSA of the muscle at the instant of the maximal force production.

For a specific case of a pennate muscle for which the fascicle force per unit area $(SF_{fascicle})$ is determined, the area of the muscle attachment to the tendon *T* (see figure 1.24), and the pennation angle θ are all known, the maximal muscle force (tendon force) can be estimated by from equation 3.13:

$$F_{\max} = SF_{fascicle} \times T \sin\theta \cos\theta \qquad [3.13]$$

The equation can be deduced from equations 3.9, 3. 12b, and 1.11. This task is left to the readers as an exercise.

3.1.6.2 Stress Tensors

So far we have analyzed only the forces acting on the tendons and muscles along a line of muscle shortening or extension. Now, we want to analyze intramuscular forces in three dimensions. Consider an element in a deformed body. The force per unit area of the element surface perpendicular to the acting force is called a *normal stress* or *pressure*. However, forces can also act in other directions, like the forces due to muscle bulging, the crosswise forces on a wraparound tendon, or transverse external forces, such as forces that act on the hip muscles during sitting. In such cases the stresses acting on an element of the body are three-dimensional and can include nonzero normal and tangential components. If the element under consideration is a small hexahedron, a body with six equal square sides, one normal and two tangential force components are acting at each side of the element. It is well known (see *Kinetics of Human Motion,* section **1.1.4**) that any force can be replaced by a parallel force of the same magnitude and a couple. Hence, tangential forces can be moved to a parallel surface of the hexahedron. Because we are considering a very small element, the moment arm of the couple is small and the moment of the couple can be neglected. Therefore, the forces and stresses are limited to the ones acting on the three perpendicular surfaces of the hexahedron, and the entire stress σ_{ij} is represented by a 3×3 matrix known as a *stress tensor* or a *Cauchy stress tensor:*

$$\sigma_{ij} = \begin{bmatrix} \sigma_{xx} & \sigma_{xy} & \sigma_{xz} \\ \sigma_{yx} & \sigma_{yy} & \sigma_{yz} \\ \sigma_{zx} & \sigma_{zy} & \sigma_{zz} \end{bmatrix}, \qquad [3.14]$$

where σ_{xx}, σ_{yy}, and σ_{zz} are the normal stresses, and σ_{xy}, σ_{xz}, σ_{yx}, σ_{yz}, σ_{zx}, and σ_{zy} are the tangential, or shear, stresses. In the symbols σ_{ij}, one of the subscripts indicates the direction of the applied force and the other indicates a plane perpendicular to reference axis x, y, or z. Hence, symbol σ_{xx} designates the normal component of force applied to a unit area in the plane perpendicular to x. The components of the stress tensor depend on the orientation of the coordinate system x, y, and z. When the coordinate system coincides with the eigenvectors of the stress tensor, the stress tensor is represented by a diagonal matrix:

$$\sigma_{ij} = \begin{bmatrix} \sigma_{xx} & 0 & 0 \\ 0 & \sigma_{yy} & 0 \\ 0 & 0 & \sigma_{zz} \end{bmatrix},$$ [3.15]

where σ_{xx}, σ_{yy}, and σ_{zz} are the principal stresses. In research, the x direction is usually assumed to lie along the tendon direction or along a muscle fiber direction. This is valid for long, thin tendons and muscles with parallel muscle fibers. However, in pennate muscles the principal stress can be in other directions, for example, in the tendon direction or somewhere between the muscle fiber direction and the tendon direction. (Unfortunately, reliable methods of local stress determination in tendons and muscles do not exist. For a majority of muscles, principal stress directions are not known.)

In continuum equations of deformable bodies, the stress tensor is commonly expressed as the sum of

- a *volumetric stress tensor,* which tends to change the volume of the stressed body and

- a *deviatoric stress tensor*, which tends to deform the object without changing its volume.

Because volumetric stress tensors tend to change only the volume of the material, their action is similar to that of hydrostatic pressure. Therefore, they are commonly called *hydrostatic stress tensors* although the object under consideration is not liquid and may not contain any liquid at all. However, when the hydrostatic stress tensor is computed for a muscle (this has been done in some studies modeling muscle properties), the term *hydrostatic* is understood in its literal meaning and the hydrostatic stress tensor is believed to represent the pressure of the intramuscular fluid. In the hydrostatic stress tensor, the stress components are equal in all directions. They can be computed from the trace of the diagonal stress matrix (equation 3.15):

$$p = \frac{1}{3}tr\left(\sigma_{ij}\right) = \frac{1}{3}\left(\sigma_{xx} + \sigma_{yy} + \sigma_{zz}\right),$$ [3.16]

where σ_{xx}, σ_{yy}, and σ_{zz} are the magnitudes of the principal stresses. The hydrostatic stress tensor is

$$\sigma_{hs} = \begin{bmatrix} p & 0 & 0 \\ 0 & p & 0 \\ 0 & 0 & p \end{bmatrix}. \qquad [3.17]$$

The deviatoric stress tensor can be computed by subtracting the hydrostatic stress tensor (equation 3.17) from the stress tensor (equation 3.14).

♦ ♦ ♦ DIGRESSION ♦ ♦ ♦

《 Elastographic Methods

Representative publications: Gennisson et al. 2005; Bensamoun et al. 2008

The motivation behind these new methods is to record the local strain and stress, or stress and strain tensors, in muscular tissue. The methods usually involve a low-frequency vibration of the muscle tissue and visualization of the wave propagation with ultrasound or magnetic resonance techniques (*transient elastography* and *magnetic resonance elastography*, respectively). The methods are very promising but are at the early stages of their development. 》

3.1.6.3 Intramuscular Fluid Pressure

Representative publications: Hill 1948; Sejersted et al. 1984; Sejersted and Hargens 1995; Jenkyn et al. 2002

Intramuscular fluid pressure, or simply intramuscular pressure (IMP), arises from combination of the hydrostatic and osmotic pressure.

3.1.6.3.1 Hydrostatic and Osmotic Pressure

Muscles contain large amounts of fluid, such as blood and interstitial fluid. In static conditions, fluid pressure follows *Pascal's law*: pressure is conveyed undiminished to every part of the fluid and to the surfaces of its container. The law is also known as the *law of hydrostatic pressure*. Interstitial fluid pressure is usually called the *intramuscular fluid pressure*. To visualize it, imagine an injection into a patient's muscle. When pressing on the plunger of the syringe, the clinician encounters a certain resistance that increases with the IMP (the resistance also depends on the friction in the syringe and viscosity of the injected liquid). This principle of resistance to injected fluid is used in needle manometers to measure IMP. Needle manometers are fluid-filled devices used to measure the pressure.

Although intramuscular fluid pressure locally follows the law of hydrostatic pressure, the fluid distribution in the whole muscle or muscle group may depend not only on fluid mechanics. The individual muscle cells as well as the entire muscles are separated by membranes that are more permeable to water (solvent) than to the solute. The solvent (water) tends to move through a membrane from a more dilute to a more concentrated solution (such a passage is called *osmosis*). The passage of the solvent through the membrane depends on the permeability of the membrane and the concentration of the solution, that is, the number of moles of solute per kilogram of solvent (the osmolarity of solution), and the *osmotic pressure* that is directly related to the solution osmolarity and represents the minimal amount of pressure required to prevent the inward flow of water across a membrane. When a muscle tissue is swollen, for instance, as a result of a local damage, the pressure in the swollen part of the muscle increases and is not distributed evenly to the other parts of the muscles as the law of hydrostatic pressure requires.

••• *PHYSICS REFRESHER* •••

Laplace's Law

Imagine a child blowing a balloon. When applied to this particular case, Laplace's law (named after Pierre S. de Laplace, French physicist, 1749-1827) establishes a relation between (a) the pressure inside the balloon (or more precisely, the difference between the pressure inside and outside the balloon, Δp) and (b) the tension in the balloon walls (μ) and (c) the curvature of the balloon walls. The Laplace equation is

$$\Delta p = 2\mu C = \mu\left(\frac{1}{R_1} + \frac{1}{R_2}\right), \qquad [3.18]$$

where C is the mean curvature, and R_1 and R_2 are the principal radii of curvature (curvature characteristics are explained in *Kinematics of Human Motion*, section **4.1.1.2**). According to equation 3.18, the pressure in the balloon increases when the tension in the walls is larger (imagine what happens if one blows a balloon with much thicker rubber), the mean curvature *(C)* increases, or both. For another example, consider the human heart: the pressure in the ventricular cavity is generated by the tension in the heart wall muscles and depends on the heart size. The relations follow Laplace's law (*Pressure = Tension / Radius*).

3.1.6.3.2 Factors Affecting Intramuscular Pressure: Application of the Laplace Law

Representative publications: Sejersted et al. 1984; Jenkyn et al. 2002

We would like to apply Laplace's law to a muscle with one distinct fiber direction. We assume that the muscle fibers are arranged in layers from the muscle surface to the muscle center line (like layers of an onion). Then for any layer,

$$\Delta p = \Delta h \mu / R,$$ [3.19]

where Δp is the increment in the pressure, Δh is the thickness of the layer measured in the direction of the radius of curvature R, and μ is the fiber stress or tension. If the fibers run in all directions, there will be two radii of curvature as in equation 3.18. Pressure at any point in the muscle will depend on thickness of the layersΔh, stress μ, curvature $(C = 1/R)$, and the number of externally surrounding muscle layers, n. The layer depth from the muscle surface is $n \cdot \Delta h$. The pressure at a given muscle depth is obtained by summation of equation 3.19:

$$P = P_0 + \Delta h \cdot \mu \sum_{n=0}^{n=i} \frac{1}{R - n\Delta h},$$ [3.20a]

where P_0 is pressure directly underneath the fascia and R is the radius of curvature of the outermost layer. (Equation 3.20 is from Sejersted et al. 1984.) The deeper layers are all curved concentrically in the same plane. Because R is much larger than $n \cdot \Delta h$, equation 3.20a can be written in simplified form:

$$P = P_0 + n\Delta h \cdot \mu / R.$$ [3.20b]

Because actual geometry of human muscles, for instance, the radius of curvature R, is usually not known precisely, equation 3.20 is used for only approximate estimation of the IMP. Several conclusions can be drawn from the equation:

1. For a muscle of given architecture and size, the IMP is proportional to stress μ. Because the force in the muscles is transmitted in part via lateral

pathways during the contraction, the stress μ increases and so does the IMP. Thus, the IMP increases with muscle force. The correlation between the IMP and the muscle force is usually larger than the correlation between the muscle force and its integrated surface EMG. Some authors recommend measuring the IMP to estimate the generated muscle force.

2. The IMP depends on muscle curvature; that is, it is in inverse proportion to the radius R. The IMP during muscle contraction increases because of the inward force generated by the curved muscle fascicles. In the muscles with straight parallel fibers—such as trapezius and rectus abdominis (when the trunk is extended)—the IMP during contraction is low. In such muscles, the muscle force is completely transmitted to the tendons. Another extreme case is circular muscles in which no force is transmitted to any tendon. In isometric conditions, such a muscle exerts force inward to the center of the ring and IMP greatly increases. In pennate muscles, the direction of the force developed by the muscle fibers is not along the tendon. Hence the fiber force has two components, one along the tendon and one perpendicular to it. The perpendicular component tends to curve the fibers and increase the IMP. In such muscles the pressure can be very high; for instance, in human vastus medialis muscle IMP values up to 600 mmHg have been observed.

3. The IMP increases as a function of recording depth $n \cdot \Delta h$.

■ ■ ■ *FROM THE LITERATURE* ■ ■ ■

Monitoring Intramuscular Pressure

Source: Sejersted, O.M., and A.R. Hargens. 1995. Intramuscular pressures for monitoring different tasks and muscle conditions. *Adv Exp Med Biol* 384:339-350

Intramuscular fluid pressure can easily be measured in humans and animals. It follows the law of Laplace, which means that it is determined by the force of the muscle fibers, the recording depth, and fiber geometry (fiber curvature or pennation angle). High resting or postexercise IMPs are indicative of a compartment syndrome due to muscle swelling. Intramuscular pressure increases linearly with force (torque) independent of the mode or speed of muscle action (isometric, eccentric, concentric). Thick, bulging muscles create high IMPs (up to 1,000 mmHg) and force transmission to tendons becomes inefficient (because of muscle bulging, the angle between muscle fibers and the tendon increases and a smaller part of fiber forces is transmitted to the tendon). Intramuscular pressure is a much better predictor of muscle force than is the EMG signal (figure 3.13). During prolonged low-force isometric contractions, cyclic variations in IMP are seen. Because IMP influences muscle blood flow through the muscle pump, alterations in IMP have important implications for muscle function.

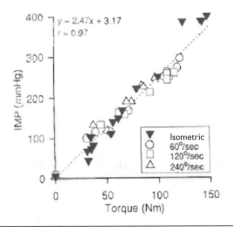

Figure 3.13   Soleus intramuscular pressure (IMP) and joint torque during isometric and concentric contractions in one subject. Contractions were performed at velocities of 0°/s (isometric), 60°/s, 120°/s, and 240°/s. Isometric contractions were performed at 5 different joint angles. Knee and hip joints flexed at 90°. It was concluded that the relation between IMP and torque is independent of velocity, intensity, and mode of contraction.

With kind permission from Springer Science+Business Media: *Advances in Experimental Medicine and Biology,* "Intramuscular pressures for monitoring different tasks and muscle conditions," 384, 339-350, O.M. Sejersted and A.R. Hargens, fig. 2. © Plenum Press 1995.

3.1.6.3.3 Biological Function of Intramuscular Pressure: The Compartment Syndrome

Representative publications: Sejersted and Hargens 1995; Steinberg 2005

Capillary blood pressure is approximately 30 mmHg. If the IMP increases above this level, the blood supply to the muscles is occluded. When occlusion occurs for a short period, for example, during push efforts in running, no adverse effects occur. The periodically changing IMP (known as a "muscle pump") may even help move the blood in veins. However, long-lasting blood occlusion, as seen during compartment syndromes (explained subsequently), can be dangerous.

Muscles of the extremities form fascial compartments, the groups of muscles separated by tough connective tissue called *intermuscular septum*. For instance, the muscles of the upper arm form two compartments separated by the anterior and posterior intermuscular septa (figure 3.14). The muscles of the thigh and lower leg are divided into three fascial compartments each. The muscles in a compartment are often supplied by the same nerve and artery.

After an injury or muscle overuse, high pressure can build up in a compartment that can damage the compartment content (*compartment syndrome*).

▪ ▪ ▪ *From the Literature* ▪ ▪ ▪

Noninvasive Method of Determining the Intramuscular Fluid Pressure by Measuring the Muscle Hardness

Source: Steinberg, B.D. 2005. Evaluation of limb compartments with increased interstitial pressure: An improved noninvasive method for determining quantitative hardness. *J Biomech* 38(8):1629-1635

The purpose of this project was to study a noninvasive method of evaluating limbs at risk of compartment syndrome. The handheld noninvasive compartment syndrome evaluator device was used to measure muscle hardness—force versus indentation depth—by applying a 5.0 mm diameter indenter to a limb muscle compartment. In 18 volunteers (9 men, 9 women), intracompartmental interstitial pressure was elevated. Comparison of interstitial pressure to quantitative hardness measurements in 71 compartments tested showed a statistically significant, strong linear relationship with an average Pearson correlation coefficient of $r = 0.84$ (range 0.78-0.90, $p < .0001$). In addition, for all compartments tested, the mean hardness value for the group with interstitial pressure greater than 50 mmHg (considered compartment syndrome) was statistically different from the mean hardness for the group with interstitial pressure less than 30 mmHg (no compartment syndrome, t-test: $p < .0001$).

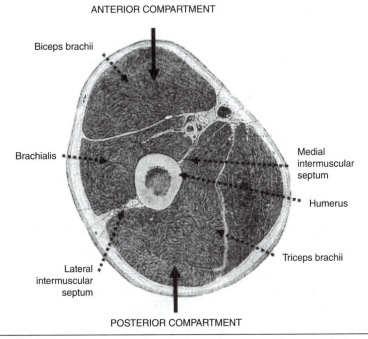

ANTERIOR COMPARTMENT

Biceps brachii

Brachialis

Medial
intermuscular
septum

Humerus

Triceps brachii

Lateral
intermuscular
septum

POSTERIOR COMPARTMENT

Figure 3.14 Fascial compartments of the left upper arm. The permeability of the intermuscular septa to the interstitial fluid is limited. The anterior compartment is supplied by the musculocutaneous nerve, the posterior compartment is supplied by the radial nerve.

Reprinted from A.C. Eycleshymer and D.M. Schoemaker, 1911, *A cross-section anatomy* (New York: D. Appleton and Company).

Because the intermuscular septa that separate the compartments are not very extensible, bleeding into a compartment or muscle swelling can cause the pressure to increase. Increased pressure within the fascial compartment decreases blood supply and compresses the nerves. The patient experiences severe pain. If the steady-state capillary pressure in the compartment becomes much higher than 30 mmHg, the blood supply to the muscles is brought to a halt, permanently damaging the tissues in the compartment. To prevent the loss of the limb, an immediate fasciotomy is required, whereby the fascia is cut to relieve the pressure.

The compartment syndrome can be caused by repetitive overuse of the muscles and occurs in some long-distance runners. This syndrome is known as *chronic compartment syndrome.* The first signs of chronic compartment syndrome are muscle fatigue and pain in the calf region, especially after sustained running. To treat the syndrome, long periods of rest are recommended; as a rule the pain slowly disappears.

3.2 FUNCTIONAL RELATIONS

Representative publications: Wilkie 1956; Close 1972; Gregor 1993; Herzog 2000

This section describes the relations between (a) muscle force and (b) muscle length and speed of shortening.

3.2.1 Force–Length Relations

Representative publications: Kaufman et al. 1989; Rassier et al. 1999

When a joint angle changes, the distance from the muscle origin to insertion (i.e., the length of the muscle–tendon complex) also changes. Muscle length variations have a profound effect on muscle force and other muscle properties. When one is modeling isometric muscle action, muscles are commonly considered to be systems with one input, excitation, and one output, force (single input, single output—SISO). Initial muscle length and velocity are considered initial conditions (parameters) of the system.

3.2.1.1 Force–Length Curves

Representative publications: Blix 1893, 1894; Bobbert et al. 1990; Smeulders et al. 2004

Systematic studies of muscle force–length relations began in the 19th century. In the first experiments, one end of the muscle was fixed and various loads were attached to the other end. Prior to muscle stimulation, muscle length increased due to the suspended load (passive muscle stretching was considered in chapter 2). After stimulation the muscle shortened: the heavier the load the smaller the shortening (figure 3.15).

It was concluded that muscle force decreases with muscle length; that is, when a muscle shortens its force decreases. These observations were confirmed in experiments performed on patients with amputations. When different loads were fixed to a free end of the muscle (the loads were hooked to so-called *cineplastic tunnels* made surgically to apply

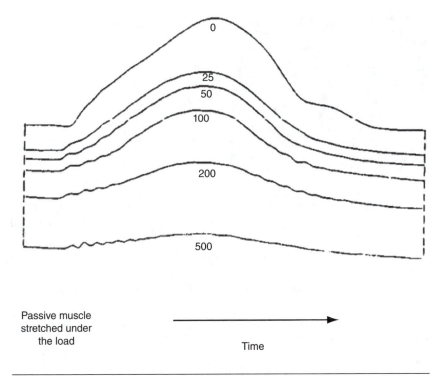

Figure placeholder labels:

0

25
50

100

200

500

Passive muscle
stretched under
the load

Time

Figure 3.15 Muscle length change before and after stimulation (from Blix 1894). The muscle (m. gastrocnemius of the "largest Hungarian frogs that the author was able to find") was fixed at one end. The loads from 25 to 500 g were fixed to another end. Then the muscle was electrically stimulated and the time course of the change of muscle length recorded. *(a)* Before the stimulation the length of the passive muscle increased (the muscle was stretched) and *(b)* the amount of shortening of the active muscle was the larger the smaller the load.

Adapted from M. Blix, 1894, "Die Lange und die Spannung des Muskels," *Scandinavian Archives of Physiology* 5: 149-206.

muscle pulling forces to prosthesis), the maximal height of the load lift was inversely related to the load magnitude (figure 3.16).

With the advent of more sophisticated force recording methods, the muscle force–length relations—or *force–length curves*, as they are commonly called—could be recorded with both ends of the muscle fixed. In these experiments, muscle length is determined prior to stimulation and

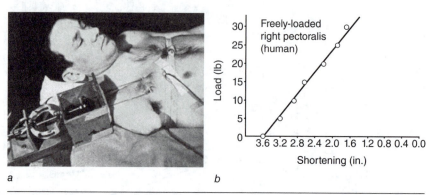

Figure 3.16 Direct measurement of the muscle force in amputees. *(a)* Experimental conditions. The cineplastic tunnel has been made surgically to transmit the muscle forces to the prosthesis. The forces exerted by the pectoralis major were measured with the strain gauge device at different muscle lengths. *(b)* Dependence of the magnitude of the muscle shortening on the loads. The heavier the load, the smaller the muscle shortening. Note that the abscissa axis goes from the right to the left.

Reprinted, by permission, from H.J. Ralston, V. Inman, L.A. Strait, and M.D. Shaffrath, 1947, "Mechanics of human isolated voluntary muscle," *American Journal of Physiology* 151: 612-620.

the muscle develops force in static conditions. In experiments performed on healthy subjects, such an experimental technique corresponds to assuming several body postures at different joint angular positions and exerting maximal force isometrically against an unmovable obstacle (the results of such measurements are presented later in the book).

If passive muscle is stretched beyond its equilibrium length (EL; see figure 2.18 in chapter 2), it resists stretching. The force of resistance of a passive muscle is the *passive force*. *Active force* is the increase in force due to activation of the muscle. At a muscle length that is smaller than the EL, the force recorded at the muscle ends is completely due to the active force. At the muscle length longer than EL, the recorded values of the muscle force—the *total muscle force*—are due to the summation of the passive and active forces (figure 3.17). To visualize the experimental conditions, imagine that a person is performing a stretching exercise in a relaxed state, and when the muscles are stretched—and hence passively resist the external force—the person exerts muscular efforts against the stretching force. In such a situation the exerted force will be due to both the

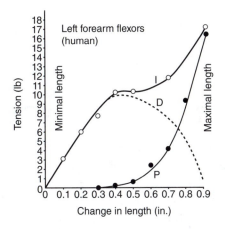

Figure 3.17 Forces recorded from forearm flexors in patients with amputations. The muscles (cineplastic tunnels) were connected to a force-measuring device in a way similar to that presented in the left panel of figure 3.16. The length of the passive muscle was changed in increments, and after that the subjects exerted maximal voluntary contractions. The curves correspond to (1) the total force (marked as I), (2) the passive force (P), and (3) the active force (D). Active force was computed as Total Force – Passive Force. At muscle lengths smaller than the equilibrium length (when the passive force equals zero), the total force equaled the active force. Similar force–length curves were recorded in many muscles in animals. To the best of our knowledge, they were recorded only once on human muscles (in the patients with amputations).

Reprinted, by permission, from H.J. Ralston, V. Inman, L.A. Strait, and M.D. Shaffrath, 1947, "Mechanics of human isolated voluntary muscle," *American Journal of Physiology* 151: 612-620.

passive resistance of the deformed body tissues and active muscle efforts. In this example, the passive force is assumed to be arising solely from the connective tissues acting in parallel to the force-generating (contractile) muscle components and hence is dependent on the muscle length but not on the actively generated muscle force (a more detailed account on the muscle parallel and serial elastic components is given in section **4.4** of chapter 4).

In these experiments, both passive and total forces are recorded and the active force is then computed as Total Force – Passive Force. The total force–length curves have different shapes in different muscles. This is partly due to different content of connective tissue in the muscles and their different passive extensibility (figure 3.18).

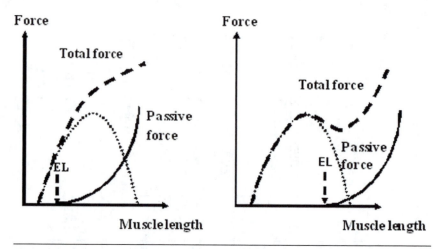

Figure 3.18 Force–length curves in different muscles, schematics. The total force–length relations depend on the relative position of the passive and active force–length curves along the muscle length axis. Left panel: A muscle, when passive, resists extension at small muscle lengths. The equilibrium length of the muscle is at the short muscle length. Right panel: The muscle is highly extensible. When passive, it resists deformation starting at the relatively large muscle length. Note the difference in the total force–length curves, monotonically increasing with the length in the left panel and with a local minimum in the right panel. The difference in the muscle extensibility is supposed to be influenced by the amount of connective tissue in the muscle. Because the leg muscles contain a larger percentage of the connective tissues than the arm muscles, it is accepted by some authors that the left panel represents the behavior of the leg muscles whereas the right panel is closer to the arm muscles. The available experimental evidence is insufficient for definite conclusions.

3.2.1.2 Mechanisms Behind the Active Force–Length Curve

Representative publication: Gordon et al. 1966

In the whole muscle, the active force–length curve is a smooth curve with a single maximum in the medium range of attainable muscle length (see figure 3.17). With the change of the muscle length, the muscle force increases in the ascending limb of the curve and decreases in the descending limb. The muscle length at which maximal force is produced is called the *optimal length* and usually designated as L_0.

In a sarcomere, the force–length curve consists of four straight segments (figure 3.19). The curve, especially its plateau region (region 3-4 in figure 3.19) and descending limb (segment 4-5), is nicely explained by the

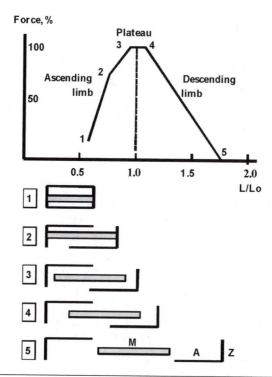

Figure 3.19 Dependence of the force produced by a sarcomere on its length (top) and sarcomeres at the crucial points 1-5 (bottom), schematic. M, myosin filaments; A, actin filaments; Z, membranes separating the sarcomeres (Z-membranes); L_o, optimal length of the sarcomere at which maximal force is exerted. Force is presented as percentage of the maximal force at the optimal length L_o. At length 5, the filaments do not overlap and hence do not produce any force. At lengths 3 and 4, the overlap of the actin and myosin filaments, and hence the force, are maximal. There is no difference between the forces at lengths 3 and 4 because the central zone of the myosin filament does not have crossbridges.

crossbridge theory of muscle contraction as a result of the different extent of overlapping between the actin and myosin filaments at various sarcomere lengths. In the ascending limb, two zones—"steep" (region 1-2) and "shallow" (region 2-3 in figure 3.19)—are distinguished. At these lengths, actin and myosin overlap and the actin filaments from the opposite sides of a sarcomere also overlap, diminishing the crossbridge formation. Several theories addressing the force reduction at lengths shorter than optimal are suggested. Detailed discussion of the theories is beyond the scope of this book.

■ ■ ■ *FROM THE LITERATURE* ■ ■ ■

The Force–Length Curve Explained
by the Crossbridge Theory of Muscle Contraction

Source: Gordon, A.M., A.F. Huxley, and F.J. Julian. 1966. The variation
in isometric tension with sarcomere length in vertebrate muscle fibres. *J
Physiol* 184:170-192

In this paper, the authors overcame serious technical obstacles and
managed to measure (a) the length of the thick (myosin) and thin (actin)
filaments and their overlap at various sarcomere lengths and (b) the
force–sarcomere length curve. The experiments were performed on
small segments of a single muscle fiber that were approximately 3 mm
long and 75 μm in diameter. The authors used a sophisticated electromechanical feedback system to stabilize the length of the experimental
segment and hence the length of the sarcomeres. The schematics
presented in figure 3.19 are based mainly on the results of this classic
paper. In frog muscles, the myosin and actin filaments are 1.65 μm
and 2.0 μm long, respectively. Therefore, the crossbridge theory of
muscle contraction predicts that at the sarcomere length 3.65 μm, the
filaments do not overlap and hence cannot generate any force. This
fact was confirmed experimentally. The plateau region was between
2.0 and 2.2 μm. This paper greatly influenced research in muscle
physiology and biomechanics; as of April 2010 it had been cited in
more than 1,500 research papers (according to Google Scholar).

For human muscles, the available data on the sarcomere lengths are
limited. Still, it can be concluded that the sarcomeres are longer—roughly
between 3.0 and 4.0 μm—than reported for frog muscles.

■ ■ ■ *FROM THE LITERATURE* ■ ■ ■

Sarcomere Length in Humans

Source: Lieber, R.L., and J. Friden. 1997. Intraoperative measurement and
biomechanical modeling of the flexor carpi ulnaris-to-extensor carpi radialis longus tendon transfer. *J Biomech Eng* 119(4):386-391

Sarcomere length was measured intraoperatively using a laser diffraction method in five patients undergoing tendon transfer of the flexor

carpi ulnaris to the extensor carpi radialis longus for radial nerve palsy. Prior to tendon transfer, flexor carpi ulnaris sarcomere length ranged from 2.84 ± 0.12 μm with the wrist flexed to 4.16 ± 0.15 μm with the wrist extended. After transfer into the extensor carpi radialis longus tendon, sarcomere length ranged from 4.82 ± 0.11 μm with the wrist flexed (the new longest position of the flexor carpi ulnaris) to 3.20 ± 0.09 μm with the wrist extended, resulting in a shift in the sarcomere length operating range to significantly longer sarcomere lengths ($p <$ .001). This study demonstrated that intraoperative sarcomere length measurements provide the surgeon with a powerful method for predicting the functional effect of tendon transfer surgery.

The force–length curves discussed here were obtained as a result of interpolation of force measurements performed in static conditions at several preestablished muscle and sarcomere lengths. They do not immediately represent the instantaneous force–length relations that can occur in natural human and animal movements. There are several reasons for this: (a) The relations are recorded for steady-state conditions with the muscles exerting a constant force; that is, the relations neglect transition from rest to activity or from one level of force to another (discussed in section **3.1.2**); (b) they do not account for movement speed (force–velocity relations are described in section **3.2.2**); (c) they neglect an immediate muscle "history," that is, how the muscles arrive at this length—as a result of muscle shortening, or lengthening. During muscle stretch the force is typically larger, and during shortening it is smaller than the force exerted in static conditions (these issues are discussed in more detail in chapter 4).

3.2.1.3 Problem of Muscle Stability

Representative publications: Allinger et al. 1996a, 1996b; Hof and van Zandwijk 2004; Gollapudi and Lin 2009

If the force–length curves obtained in isometric conditions were similar to force–deformation or stress–strain curves recorded in deformable bodies, the descending limb of the curve would signify instability: when the body length increases its resistance to the deformation decreases, thereby tending to cause further lengthening. Such behavior corresponds to "negative stiffness" and is unstable; it leads to material failure. However, muscles behave differently than passive materials. When the muscles on the descending limb of the force–length curve are forcibly stretched, they resist the perturbation and exert a larger force than could be expected from their force–length relation (figure 3.20; see also figure 4.2*b*). In other words, they do not follow

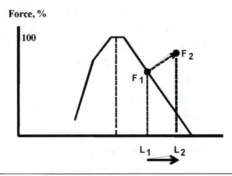

Figure 3.20 Effect of sarcomere extension by an external force on the descending limb of the active force–length curve, schematic. When a sarcomere is stretched from length L_1 to length L_2, the force does not follow the force–length curve; instead it increases from F_1 to F_2.

the instability suggested by the isometric force–length curve but are rather perfectly stable. The behavior of active muscles subjected to stretch is described in more detail in chapter 4.

■ ■ ■ *FROM THE LITERATURE* ■ ■ ■

Mechanical Analogy Explaining Muscle Behavior on the Descending Limb of the Force–Length Curve

Sources: (1) Allinger, T.L., M. Epstein, and W. Herzog. 1996. Stability of muscle fibers on the descending limb of the force–length relation: A theoretical consideration. *J Biomech* 29(5):627-633. (2) Epstein, M., and W. Herzog. 2003. Aspects of skeletal muscle modeling. *Proc R Soc Lond B Biol Sci* 358:1445-1452

The goal of this analogy is to help the reader grasp the possibility of an apparent softening but stable muscle behavior. Imagine two toothbrushes placed in parallel and opposite to each other with their bristles partly penetrating and their handles in opposite directions. The brushes are pulled in opposite directions and are displaced with respect to each other in small steps. Before the force reaches a certain threshold, the system resists the perturbation and is entirely stable. If the threshold is exceeded, the brushes move to another position. The force that is necessary to achieve the next farther displacement is smaller than the previous one because the area of the contact and, hence, the number of the contacting bristles decrease. A force–displacement graph would look like a jagged diagram made of small segments with positive slopes and the negative slope of the entire curve. In such a relation, the instantaneous stiffness is almost everywhere positive whereas the overall behavior represents decreasing resistance (force) to the deformation.

. . . and Its Criticism

Source: Zahalak, G.I. 1997. Can muscle fibers be stable on the descending limbs of their sarcomere length-tension relations? *J Biomech* 30(11-12):1179-1182

The author disagrees with the theory that the muscle fibers can be stable on the descending limb of their static force–length relations (the mathematical analysis provided in the paper to substantiate this view would take too much space and is not shown here). According to the author, the toothbrush analogy is not valid. In a muscle fiber subjected to length perturbation, the crossbridges break and then form new, less-strained bonds. During this time, the force increment rapidly dissipates. In a similar condition, the hairs of the brushes are unlocked and then reengaged with the same strain as before unlocking.

A comment from the book authors: The problem of muscle stability evidently requires further analysis and, especially, direct experimentation.

3.2.1.4 Submaximal Force–Length Curve

Representative publication: Rack and Westbury 1969

The force–length curves discussed previously are typically obtained for maximally activated muscles. If the muscle is not activated maximally, that is, does not exert maximal forces, the dependence of the force on the muscle length can change. During submaximal isometric force development, the tendons are stretched less by the shortening muscle fibers than during maximal force production. Therefore, at the same joint angular configurations, muscle fibers and sarcomeres are longer during submaximal stimulation than during maximal conditions. Hence, the muscle fibers may possibly operate at different sections of the force–length curve under submaximal and maximal conditions.

To obtain a submaximal force–length curve, the muscle should be equally activated at different lengths. The concept of equal activation requires prior agreement on what exactly should be equal, for example, percentage of maximal force attainable at a given length, concentration of intracellular Ca^{2+} ions, EMG magnitude, or something else. If equal activation is understood to be equal intensity of electrical stimulation of the muscle at the same rate of stimulation, there is a reciprocal relation between the stimulus rate and muscle length. At long muscle lengths, low rates yield near-maximal force (i.e., there is weak dependence of the force on the stimulation frequency), whereas at short muscle lengths the maximal force is reached only when the stimulus rate is very high (figure 3.21).

When equal activation is understood to be equal magnitude of the integrated EMG activity, (a) an isometric muscle force–length curve—with the muscle length set before the muscle is activated—is different from the

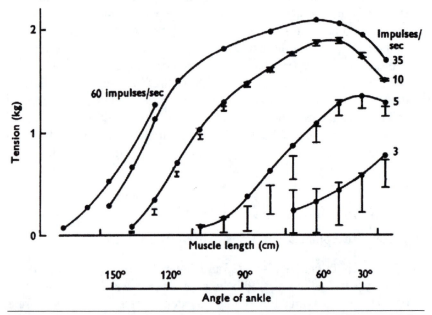

Figure 3.21 The effect of muscle length on active force at different stimulus rates. The vertical lines show the limits of force fluctuations during the stimulation. During the experiment, the ankle joint angle in the cat was changed in increments and fixed. The soleus muscle and sarcomere lengths were determined later on. At high rates of stimulation, the maximal force for a given rate was obtained at an angle of approximately 60° and a mean sarcomere length of about 2.8 μm.

From P.M. Rack and D.R. Westbury, 1969, "The effects of length and stimulus rate on tension in the isometric cat soleus muscle," *Journal of Physiology* 204(2): 443-460. Reprinted by permission of John Wiley & Sons Ltd.

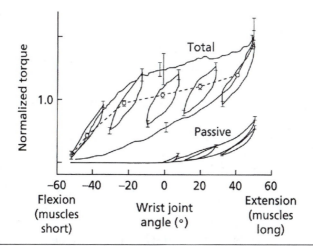

Figure 3.22 Joint angle–joint torque curves for voluntary activation of wrist flexors. These are not the maximal torque data. During the experiment, the subjects were asked to maintain a constant electromyographic (EMG) level of a wrist flexor, the flexor carpi radialis. The EMG level was established at 10% of the maximal voluntary contraction. Joint torque was measured in three contraction conditions: (1) isometric (dotted line), (2) ±10° angular displacements centered at five different angles (small loops), and (3) ±50° angular displacements over full range of motion of the wrist joint (large loop). Narrow loops at bottom of the figure show angle–torque profiles with muscles completely relaxed. The torque values are normalized to mean torques of ±10° angle-torque loops at 0° wrist angle. Note the difference between the values of the torques recorded in the static and dynamic conditions. The difference depends on the amplitude of the joint angular displacement.

Reprinted from *Journal of Biomechanics* 8(2), D.M. Gillard, S. Yakovenko, T. Cameron, and A. Prochazka, "Isometric muscle length-tension curves do not predict angle-torque curves of human wrist in continuous active movements," 1341-1348, copyright 2000, with permission from Elsevier.

curves in continuous active movements, and (b) the force is larger during muscle stretching and smaller during muscle shortening. As a result, during stretch–shortening cycles, hysteresis-like curves are observed. The same is valid for joint angle–moment relations (figure 3.22).

3.2.1.5 Muscle Lengths in the Body: Expressed Sections of the Force–Length Curve

Representative publications: Ichinose et al. 1997; Lieber and Friden 1998; Winter and Challis 2008a, 2008b, 2009

During joint motions, the possible changes in muscle length do not extend over the entire range of the active force–length relation discussed previously (see figure 3.19). In the past, it was commonly accepted that all muscles in the body operate on the ascending limb of the force–length curve; that is, at the same activation level they generate smaller forces at shorter lengths and larger forces at longer lengths. These ideas were later confirmed by experimental evidence; see figure 3.17 as an example. However, recently it has been shown that some muscles (wrist extensors), perhaps only in certain people, operate on the descending limb of the curve. These muscles generate larger forces at shorter lengths and smaller forces at longer lengths. (So far, only static force–length curves were investigated for these muscles. It is not clear what happens with the force of a muscle during concentric action when the muscle shortens on the descending limb of the curve. Does the force decrease, as shown in figures 3.17 and 3.22, or increase? Unfortunately, we were not able to find experimental evidence in the literature on this topic.)

The part of the force–length curve seen in vivo is called the *expressed section of the force–length curve*. Determining the expressed section of the muscle force–length curve is a challenging task for the researchers. In vivo, neither muscle force nor muscle length is precisely known. With the exception of the previously mentioned studies performed on patients with amputations (figures 3.16 and 3.17), direct measurements of force–length relations for human muscles have not been performed. Instead, indirect estimates have been used.

Muscle force is estimated from known values of the joint torques and moment arms at various joint angles. Also, the length of the muscle–tendon unit at these angles is determined. Then, the force–length curves are reconstructed. Because of many assumptions, for example, about the contribution of a given muscle force into the joint moment, the method is not very accurate but can provide qualitative information on whether the muscle is working on the ascending or descending limb of its force–length curve. In addition to the muscle–tendon length, the sarcomere length of muscles can be determined during surgical operations using a laser diffraction method. Then, the force values can be associated with the known sarcomere lengths. Intraoperative measurements of the sarcomere length have been recommended for determining the optimal length of a muscle–tendon unit during reconstructive limb surgery.

■ ■ ■ *From the Literature* ■ ■ ■

Determining the Optimal Length of a Muscle–Tendon Unit During Tendon Transfer

Source: Friden, J., and R.L. Lieber. 2002. Mechanical considerations in the design of surgical reconstructive procedures. *J Biomech* 35(8):1039-1045

Tendon transfers are used to restore arm and hand function after injury to the peripheral nerves or after spinal cord injury. Usually, choosing the length at which the transferred muscle should be attached by a surgeon relies on a muscle's passive resistance. However, this passive force is only significant at relatively long lengths, so using passive force as the major factor in intraoperative decision making may result in overstretch of the muscle–tendon unit and accompanying low active force generation. The authors suggest the use of intraoperative sarcomere length measurements to predict and set the optimal length of the muscle–tendon unit during reconstructive upper-limb surgery.

For two-joint muscles, the force–length curve can be estimated by changing the joint angle at one joint and measuring the maximal joint torque at the other joint. When the joint position varies, the length of the two-joint muscle changes and as a result the values of the maximal torque exerted at another joint may also change. For instance, the maximal extension torque at the knee joint can be measured at two hip angular positions: at the angle of 90° (the subject is sitting) and 180° (the subject is on his or her back in the prone position). When the hip angle is 90°, the length of the m. rectus femoris is shorter than at an angle of 180°. Therefore, if the muscle is working on the ascending limb of its force–length curve, we should expect that with the subject in prone position the maximal knee torque is larger than in the sitting posture. Opposite changes should be expected if the muscle is working on the descending limb of the curve.

Expressed Sections of the Force–Length Curve in Human Muscles

1. Soleus and Tibialis Anterior Muscles Operate Over the Ascending Limb and Plateau Region of the Force–Length Curve

Source: Maganaris, C.N. 2001. Force–length characteristics of in vivo human skeletal muscle. *Acta Physiol Scand* 172(4):279-285

The in vivo force–length relations of the human soleus and tibialis anterior muscles were estimated. Measurements were taken in six men at ankle angles from 30° of dorsiflexion to 45° of plantar flexion in steps of 15° and involved dynamometry, electrical stimulation, ultrasonography, and magnetic resonance imaging (MRI). For each muscle and ankle angle studied, the following three measurements were carried out: (1) dynamometry-based measurement of joint moment at maximal tetanic muscle stimulation, (2) ultrasound-based measurement of pennation angle and fiber length, and (3) MRI-based measurement of tendon moment arm length. Tendon forces were calculated dividing joint moments by moment arm lengths, and muscle forces were calculated dividing tendon forces by the cosine of pennation angles. In the transition from 30° of dorsiflexion to 45° of plantar flexion, the soleus muscle fiber length decreased from 3.8 to 2.4 cm and its force decreased from 3,330 to 290 N. Over the same range of ankle angles, the tibialis anterior muscle fiber length increased from 3.7 to 6 cm and its force increased from 157 to 644 N. The results indicate that the intact human soleus and tibialis anterior muscles operate on the ascending limb and in the plateau region of the force–length relationship.

2. In Different Subjects, Gastrocnemius Operates Over the Ascending or Descending Limb of the Force–Length Curve

Source: Winter, S., and J. Challis. 2008. Reconstruction of the human gastrocnemius force–length curve in vivo: Part 2. Experimental results. *J Appl Biomech* 24(3):207-214

Fourteen male and fourteen female subjects performed maximal isometric plantar flexion with a Biodex dynamometer. Plantar flexion moments were recorded at five ankle angles: −15°, 0°, 15°, 30°, and 40°, with negative angles defined as dorsiflexion. These measurements were repeated for four randomly ordered knee angles over two testing sessions 4 to 10 days apart. The peak force (i.e., the maximal force across the muscle lengths) was determined and its location along the muscle length axis used to assign the force–length curve to one of the three groups. When the peak force occurred at more than 60% of the range of the muscle length variations, the curve was classified as located on the ascending limb; when the peak was below 40% of the muscle length variations, the curve was labeled as located on the descending limb; and when the peak force was between 40% and 60% of the muscle length range, the curve was classified as located in the plateau region. Twenty-four subjects operated on the ascending limb, three operated on the descending limb, and one operated in the plateau region.

3. Wrist Extensors Operate Over the Descending Limb of the Force–Length Curve

Source: Lieber, R.L., and J. Friden. 1998. Musculoskeletal balance of the human wrist elucidated using intraoperative laser diffraction. *J Electromyogr Kinesiol* 8(2):93-100

Sarcomere length measurements were combined with studies on cadaveric extremities to generate biomechanical models of human wrist function and to provide insights into the mechanism by which wrist strength balance is achieved. Intraoperative measurements of the human extensor carpi radialis brevis muscle during wrist joint rotation reveal that this muscle operates on the descending limb of its length–force curve and generates maximum force with the wrist fully extended. The synergistic extensor carpi radialis longus also operates on its descending limb but over a much narrower sarcomere length range. Both muscles generate maximum force with the wrist fully extended. As the wrist flexes, force decreases because the extensors lengthen along the descending limb of their length–force curve and the flexors shorten along the ascending limb of their length–force curve. The net result is a nearly constant ratio of flexor to extensor torque over the wrist range of motion and a wrist that is most stable in full extension. The wrist torque motors appear to be designed for balance and control rather than maximum torque-generating capacity.

The force–length curve in humans is not constant. It shifts to longer lengths with fatigue. Also, muscle length can change because of an increasing number of sarcomeres (during growth and perhaps also as a result of systematic stretches) or a decreasing number of sarcomeres (e.g., as a result of joint immobilization at short muscle lengths).

3.2.2 Force–Velocity Relations

Representative publications: Fenn and Marsh 1935; Hill 1938; Wilkie 1949; Pertuzon and Bouisset 1973; Komi 1990; Yeadon et al. 2006

Everyone intuitively understands the relation between force and velocity in human movements: we cannot lift heavy objects at the same speed as lighter objects. Force–velocity muscle properties have attracted great interest among researchers and have been a source of intensive discussion and, unfortunately, misunderstanding over several decades. These properties are fundamental for understanding muscle function.

3.2.2.1 A Piece of History: Muscle Viscosity Theory and Heat Production

Representative publications: Hill 1922, 1938; Gasser and Hill 1924; Fenn 1924; Mommaerts 1970; Needham 1971; Zatsiorsky 1997

In the beginning of the 1920s, A.V. Hill and his coworkers performed several fundamental studies on muscle mechanics and heat production. (Archibald Vivian Hill, 1886-1977, was a British scientist who, with O.F. Meyerhof, won the Nobel Prize in Physiology or Medicine in 1922 for their discovery relating to the production of heat in the muscle.) In some of the experiments, a stimulated frog muscle was allowed to shorten against different loads. The results are schematically presented in figure 3.23.

Four facts attracted attention: with the increasing load magnitude, (1) the latent period increased while (2) the amount of muscle shortening and (3) the shortening speed decreased. Although the precise mechanisms responsible for such changes were not clear, on an intuitive level they were expected: muscles need more time to develop larger force, the amplitude of lifting decreases with the load as it should if muscles work on the ascending limb of their force–length curve, and muscles shorten at a smaller speed when they overcome larger resistance.

The fourth fact was not expected: over periods of time the muscles shorten at a constant speed. During these periods, the muscle length–time relations were essentially straight lines. This fact did not immediately agree with the

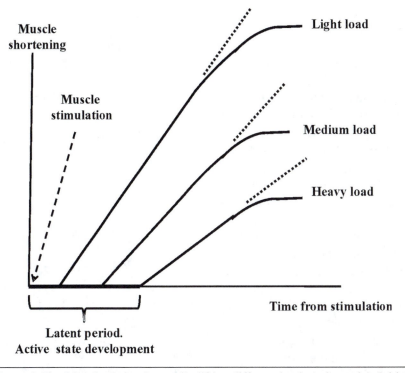

Figure 3.23 Muscle shortening while lifting different loads (schematic). Initial length of the muscle is fixed; that is, the muscle cannot shorten or be stretched beyond a given length. The muscle is stimulated and then released and thus allowed to shorten while lifting various loads (below developed force at a given length). The slope of the broken lines is indicative of the speed of shortening.

second law of Newton: if muscles were exerting constant force the loads should accelerate according to the equation $F = ma$ (this equation is so well known that we do not bother to explain the meaning of its symbols).

The constant muscle speed during shortening has been explained by the *viscosity hypothesis* (for the definition of viscosity, see section **2.1.2.1** and figure 2.9). According to the hypothesis, under standard stimulation and different loads a muscle generates similar internal forces but these forces are not manifested externally because of the viscous resistance within the muscle. As a result the muscle shortens at a constant speed. The effect is similar to what happens in skydiving. During freefall, the jumper's speed increases to the level at which air resistance equals the gravity force acting on the jumper's body; from this moment on, the speed of falling is constant, the "terminal velocity."

The viscosity hypothesis, also known as the *viscoelastic hypothesis*, assumes that the active muscles behave as systems consisting of elastic (springlike) and damping elements (viscous dashpots). Thus, a muscle could be approximated by the mechanical analogy of a spring working in a viscous medium. The theory is based on a postulate that when a muscle progresses from rest to activity, new elastic bonds are formed and the muscle becomes stiffer. For that reason, (a) an excited muscle has a shorter equilibrium length (i.e., the length at which it exerts zero force at its ends) compared with a relaxed muscle, (b) an active muscle provides larger resistance to stretch than a relaxed muscle, and (c) when a muscle is activated it stores potential energy that is liberated during contraction. The latter postulate is based on the fact that when springs are deformed, they store a certain amount of elastic potential energy. When they are allowed to shorten this energy is converted into mechanical work and heat. The percentage of the energy converted into work and heat depends on the shortening conditions. However, for a given deformation, the total amount of the stored and liberated energy (Work + Heat) remains constant (figure 3.24). In other words, according to the viscoelastic theory, muscle activation adds a fixed amount of energy to the muscle.

According to the viscosity hypothesis, the greater the speed of shortening, the larger the energy losses due to internal friction (viscosity) and hence the larger the percentage of energy that is converted into heat and the less

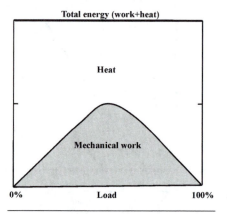

Figure 3.24   Mechanical work and heat production according to the viscoelastic theory of muscle contraction. In this schematic, mechanical work and heat are represented as the shaded and white areas, respectively. At the zero loads, the muscle end is free and when the muscle shortens the muscle force does not do any mechanical work (because the force is zero). Hence, all the energy goes into overcoming internal friction (viscosity) and converts into heat. In this case the speed of muscle shortening is maximal. When the load is maximal (100%), the muscle contracts isometrically and the mechanical work is again zero. Remember that in the simplest case when the force acting on the object *(F)* and the displacement of the point of force application *(d)* are in the same direction, work *(W)* equals the product $W = F \times d$, and if one of the factors equals zero the product is also zero. No work is performed in this case.

mechanical work done. In the past it seemed that human experiments agreed well with this theory. Studying elbow flexion with maximal effort against different inertial resistance, A.V. Hill (1922) found that mechanical work decreased with movement speed and explained this finding as a result of energy loss due to viscosity resistance proportional to the velocity. The viscosity hypothesis became a dominant theory of muscle biomechanics. The theory was simple, elegant, and unfortunately wrong.

It took considerable time and effort to recognize the weaknesses of the theory. Between 1922 and 1938, dozens of papers were published on the muscle viscosity phenomenon. The theory required that the total amount of energy generated by an active muscle (work plus heat) be constant and independent of the speed of shortening (as shown in figure 3.24). However, as early as 1924, W.O. Fenn, who recorded both muscle mechanical work and heat production, demonstrated that the total energy released by the muscle was not constant when it shortened against different loads and at different speeds. (Wallace O. Fenn, 1893-1971, was an American scientist who worked in the laboratory of A.V. Hill in Britain from 1922 to 1924.) When a stimulated muscle was allowed to shorten against a load, that is, when it was allowed to do mechanical work, the amount of liberated heat increased in proportion to the muscle force and the distance of shortening. The heat production (in excess of the so-called activation heat) was found to be proportional to the work done by the muscle (figure 3.25). Hence, it was acknowledged that the total energy liberated during muscle contraction is not constant. The amount of energy is not determined solely by the muscle activation: the energy rather increases with the work done. This fact became known as the *Fenn effect*. In the concluding remarks of his paper, Fenn (1924, p. 395) mentioned that "the existence of an excess heat liberation is to some extent inconsistent with the idea that a stimulated muscle is a new elastic body. It is suggested that contraction is analogous rather

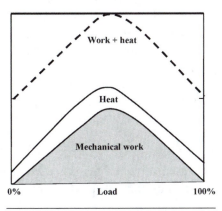

Figure 3.25 Mechanical work and heat production according to the Fenn effect. In this schematic, the mechanical work is represented as the shaded area and the heat production is represented as the area below the heat curve. The total amount of energy (Work + Heat) equals the sum of the two curves (broken line). The figure is not drawn to scale.

to the winding up of an anchor chain by a windlass than to the lifting of a weight by the energy of a stretched spring." To use another metaphor, muscle behaviors more closely resemble electric motors that increase both electricity (energy) consumption and heat production when they are heavily loaded. Mechanical springs, in contrast, possess a fixed amount of elastic potential energy when deformed.

The effect was largely overlooked until 1938, when Hill published his famous paper on the heat of shortening (as of July 2011 this paper was cited in 2,706 research publications, according to Google Scholar) . In this study, Hill discovered—among other things—that during eccentric muscle action, the total energy produced by the muscle is less than in concentric contraction, a fact that is incompatible with the idea of the decisive role of muscle viscosity. This is because viscous resistance depends only on movement speed, not on the movement direction. Imagine that you move your arm in a barrel of oil. Whether you move the arm to the right or to the left, you experience similar viscous resistance at equal speeds. For muscles, this direction independence does not hold: when active muscles shorten they generate more energy than when they are forcibly stretched (eccentric muscle actions are described in more detail in chapters 4 and 7). A.V. Hill concluded that "the viscosity hypothesis must be dismissed" (1938, p. 193). In biomechanics literature, this theory is not used anymore. However, in some motor control studies the viscoelastic hypothesis is still in use and the muscles are modeled as springs with changeable stiffness and variable rest (zero deformation) length. Such models are used probably because the muscle properties that contradict these theories are not important for the models under discussion. The reader is advised to regard these models as phenomenological rather than as representative of true muscle mechanics.

3.2.2.2 Hill's Force–Velocity Curve

Representative publications: Fenn and Marsh 1935; Hill 1938; Katz 1939; Caiozzo et al. 1981; Winters 1990

If a muscle contracts against constant loads that vary from trial to trial, with increasing load the speed of shortening decreases. The relation between the load (muscle force) and speed of shortening is known as the *force–velocity curve*, or *Hill's force–velocity curve* (in appreciation of the 1938 seminal paper by A.V. Hill). Experimentally the curve was first recorded by Fenn and Marsh (1935) (figure 3.26). These force–velocity curves are (a) descending and (b) concave upward. Note that the viscosity hypothesis predicted

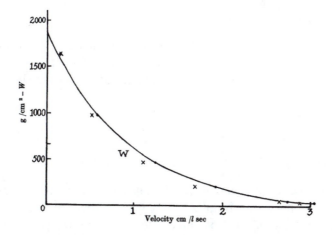

Figure 3.26 Force–velocity curve obtained from a frog sartorius muscle. Ordinate: muscle force (W) in grams per square centimeter cross-section (we left in this graph the designation W used by Fenn and Marsh instead of the usual force symbol *F* used throughout the book). The curve was well-fitted by the equation

$$V = V_0 e^{-F/b} - kF,$$ [3.21]

where *F* = force (or load), *V* = velocity of shortening, V_0 = shortening velocity of unloaded muscle at *F* = 0, and *b* and *k* are empirical constants.

From W.O. Fenn and B.S. Marsh, 1935, "Muscular force at different speeds of shortening," *Journal of Physiology* 85(3): 277-297. Adapted by permission of John Wiley & Sons Ltd.

that the force–velocity relation should be a straight line (provided that the spring is linear and that the viscous behavior does not depend on the speed) and a relation between the muscle force *F* at the velocity of muscle shortening *v* should be $F = F_0 - kv$, where F_0 is the force at zero speed and *k* is a coefficient.

Hill's force–velocity curves are obtained by interpolating the force and velocity data recorded in several trials (i.e., by fitting the data to an equation of an assumed form) while one of the parameters of the task (e.g., the load) was systematically changed. Because of that these relations are at times called *parametric relations.*

Force–velocity relations analogous to Hill's curves can also be recorded in human movements when a certain parameter of a task changes among the trials. The relations are obtained in the following way:

1. Subjects perform several trials with maximal efforts (e.g., flexion or extension at a joint).

2. A certain parameter of the task is varied in a systematic way from trial to trial (e.g., mass of an implement in throwing or velocity of an isokinetic device in which the speed of muscle shortening is controlled).

3. In each trial, both force and velocity are measured, F_m and V_m. Each F_m–V_m pair corresponds to a certain value of the parameter.

4. The values of F_m and V_m are plotted versus each other.

Some examples of the parametric relations encountered in several human activities are presented in table 3.1. Note, however, that the examples presented in table 3.1 do not immediately characterize Hill's curves for individual muscles. Rather the examples represent the combined action of several muscles that may have different force–velocity properties and different joint moment arms. Still, the curves as a rule appear similar: they are descending and concave upward. Certain exceptions are noted; for example, during in vivo human knee extension at low velocity of movement the force is lower than can be expected from the Hill's force–velocity relation, which can partly be explained by inhibition of the knee extensors to prevent injury (figure 3.27).

Table 3.1 Parametric Force–Velocity Relations in Athletic Movements

Activity	Variable factor (parameter)	Force	Velocity
Cycling	Gear ratio	Applied to the pedal	Pedaling frequency (or velocity)
Rowing, kayaking, canoeing	Blade area of an oar or a paddle	Applied to the oar or to the paddle	The blade with respect to the water
Throwing	Mass of the implement	Exerted upon the implement	Implement at the release
Standing vertical jump	Modified body weight: in different trials, the weight added (e.g., by wearing a waist belt) or reduced (e.g., via a suspension system)	Exerted on the ground	Body at the beginning of the flight

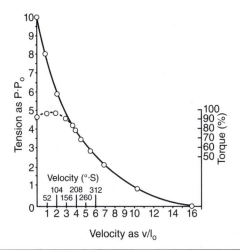

Figure 3.27 Experimental force–velocity relations of in vivo human muscles (a broken curve drawn through open circles) and isolated animal muscles (a solid hyperbolic-like curve) as determined in two separate experiments under similar loading conditions. Note the difference between the curves in the high force range. The relation for human muscles is obtained in isokinetic knee extension with maximal efforts.

Reprinted, by permission, from J.J. Perrine and V.R. Edgerton, 1978, "Muscle force-velocity and power-velocity relationships under isokinetic loading," *Medicine & Science in Sports & Exercise* 10(3): 159-166.

3.2.2.3 Other Types of the Force–Velocity Curves

Representative publication: Zatsiorsky 2003

Parametric force–velocity curves are obtained by interpolating the force–velocity data recorded in several trials in which one of the task parameters varied among the trials. Other types of force–velocity relations also exist.

3.2.2.3.1 Force–Velocity Relations in Single Movement

Representative publications: Komi 1990; Gregor 1993

If force and velocity are recorded continuously in a single trial and plotted versus each other, the obtained force–velocity curves can be quite different from the parametric curves discussed above. An example is presented in figure 3.28.

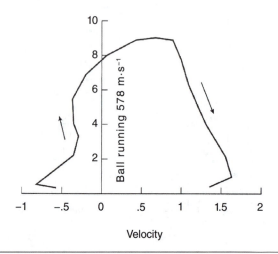

Velocity

Figure 3.28 The force–velocity profile of the human triceps surae muscle during ball-of-the-foot-contact running at 5.78 m/s. Force (in kN) represents the Achilles tendon force recorded with a buckle transducer, whereas velocities represent the rate of the muscle–tendon unit changes (m/s) for the gastrocnemius muscle.

Adapted from *Journal of Biomechanics* 23(Suppl 1), P.V. Komi, "Relevance of in vivo force measurements to human biomechanics," 23-34, copyright 1990, with permission from Elsevier.

In some cases, force–velocity curves recorded in a single trial may appear similar to Hill curves. This happens if the velocity increases monotonically to a certain saturation level and, consequently, the acceleration monotonically decreases (figure 3.29). When the resistance is provided by inertia, the relation between the force and acceleration is given by Newton's second law: $F = ma$. If the mass is constant, the force–velocity relation is in essence the acceleration–velocity relation. In such a case, the observed force–velocity relation is a consequence of velocity saturation: when velocity increases the acceleration decreases. This relation is not a parametric relation (all parameters of the task are constant). The corresponding curve can be called a *pseudo-Hill curve*, indicating simply that the curve was obtained in a different way than the typical Hill curve. Pseudo-Hill curves should be distinguished from the real Hill curves described previously. The pseudo-Hill curves are usually straight lines.

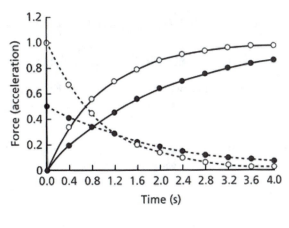

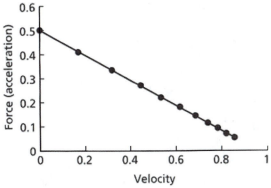

Figure 3.29 Pseudo-Hill curves result from the inverse relation between the speed and acceleration in all-out sprints from the rest. Top panel: Normalized velocity (solid lines) and acceleration (broken lines) in a single trial. If intracycle velocity fluctuations are neglected or filtered, such curves are registered at the beginning of the sprint running and cycling. The velocity curve $V(t)$ can be described by an equation:

$$V(t) = V_{max}(1-e^{-kt}),$$ [3.22]

where V_{max} is the maximal velocity and k is an acceleration constant (the magnitude of acceleration at the start, $t = 0$). The acceleration equals $a(t) = V_{max}ke^{-kt}$. The velocity and acceleration curves are normalized with respect to the V_{max} and correspond to $k = 0.5$ (circles) and $k = 1.0$ (triangles). When velocity increases, the acceleration decreases. When velocity is maximal, the acceleration is zero. Bottom panel: The plot of acceleration (force) versus velocity for $k = 0.5$.

■ ■ ■ *FROM THE LITERATURE* ■ ■ ■

Pseudo-Hill Curves in All-Out Cycling Sprints

Source: Butelli, O., H. Vandewalle, and G. Peres. 1996. The relationship between maximal power and maximal torque-velocity using an electronic ergometer. *Eur J Appl Physiol* 73(5):479-483

In the experiment, subjects performed a single all-out sprint on a cycle ergometer. The crank angle–torque curves were recorded during the first 10 revolutions. Each point corresponds to the average velocity and torque during 1 revolution (figure 3.30).

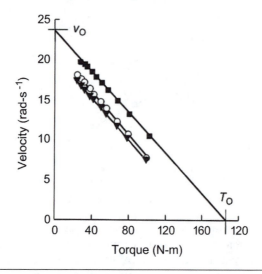

Figure 3.30 Relations between the pedal velocity and torque in three cyclists. Note that the relations are straight lines.

With kind permission from Springer Science+Business Media: *European Journal of Applied Physiology*, "The relationship between maximal power and maximal torque-velocity using an electronic ergometer," 73(5), 479-483, O. Butelli, H. Vandewalle, and G. Pérès, fig. 1. © Springer-Verlag 1996.

3.2.2.3.2 Nonparametric Force–Velocity Relations

Representative publication: Zatsiorsky 2003

Nonparametric force–velocity relations occur between the maximal force generated by a subject in a muscular strength test (F_{mm}) and the maximal movement speed V_m. Such relations help to address questions regarding the role of F_{mm} in improving movement speed. An example is the relation between the F_{mm} in a leg extension and the takeoff velocity (or the height) of a squat vertical jump. The question is whether stronger athletes jump higher. In contrast to parametric relations, the slopes at nonparametric relations are either positive (the larger the F_{mm} the larger the V_m) or zero. The relation depends on the amount of resistance: the larger the resistance (e.g., mass of an implement), the higher the correlation between F_{mm} and V_m (figure 3.31).

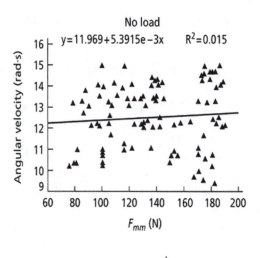

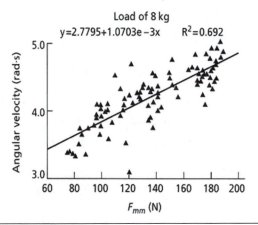

Figure 3.31 Nonparametric force–velocity relations. The maximal isometric force of shoulder flexion recorded with the arm in anatomic posture (muscular strength, F_{mm}) versus the maximal velocity of shoulder flexion V_m with the arm extended. $n = 100$ subjects. Upper panel. No load; there is no significant correlation between F_{mm} and V_m. Bottom panel. Load (a dumbbell) of 8 kg in the hand; there is a high correlation between F_{mm} and V_m.

Adapted, by permission, from V.M. Zatsiorsky, 1995, *Science and practice of strength training* (Champaign, IL: Human Kinetics), 41.

3.2.2.4 Mathematical Description of the Force–Velocity Curve: The Hill Characteristic Equation

Representative publications: Hill 1938; Epstein and Herzog 1998

We return now to the discussion of parametric force–velocity relations. In the literature, the relations have been successfully fitted using several empirical equations; see as an example the equation in the caption to figure 3.26. However, the most popular and broadly used is the equation suggested by A.V. Hill (1938). This seminal paper was mainly devoted to measuring the heat and work (and hence the energy) production during muscle shortening. At that time, as compared with the Fenn study of 1923-1924, the measurement technique was greatly improved: Hill was able to measure temperature changes as small as 0.003 °C over periods of only hundredths of a second. He found that when an active muscle was allowed to shorten, it released—in excess of the activation heat (the heat corresponding to bringing the muscle into readiness)—the shortening heat. The shortening heat was proportional to the shortening distance *(x)* and independent of load. The total amount of liberated heat is then equal to *ax,* where *a* is the so-called *heat constant* (*a* = shortening heat/cm) or the coefficient of shortening heat. Because during shortening the muscle force produces Work = $F \times x$, where F and x are scalars, the energy liberation (Work + Shortening Heat) equals

$$\text{Energy Liberation} = (Fx) + (ax) = (F + a)x. \qquad [3.23]$$

Taking the time derivative yields

$$\text{Rate of Energy Liberation} = (F+a)\frac{dx}{dt} = (F+a)V, \qquad [3.24]$$

where V is the velocity of shortening (expressed as cm/s). The rate was found to be a negative linear function of the load

$$\text{Rate of Energy Liberation} = b(F_0 - F), \qquad [3.25]$$

where F_0 is the force at $V = 0$ and b is a constant, the increase in energy rate per unit decrease in load (or force). Combining equations 3.24 and 3.25 yields

$$(F + a)V = b(F_0 - F) \qquad [3.26a]$$

which can be also written as

$$(F + a)(V + b) = b(F_0 + a) = \text{Constant}. \qquad [3.26b]$$

Equation 3.26b is known as the *characteristic equation* or *the Hill equation*. The equation relates muscle force and velocity during concentric muscle actions at different values of a task parameter, such as for instance the lifted load. The equation is not valid for eccentric muscle actions.

Equation 3.26 can be investigated by purely mechanical methods. In the mechanics domain, *a* has the dimensionality of force and *b* the dimensionality of velocity. The equation is a hyperbolic one; the force–velocity curve can be considered a part of a hyperbola with the coordinate axes shown in figure 3.32. Mathematically, the equation represents a rectangular hyperbola with asymptotes $F = -a$ and $V = -b$.

When solved for F, the equation is

$$F = \frac{F_0 b - aV}{V + b}.$$
[3.26c]

Setting $F = 0$ (the muscle is unloaded and its velocity of shortening is maximal, V_0), the following relations can be determined

$$F_0 b = aV_0$$
[3.27a]

or

$$a/F_0 = b/V_0.$$
[3.27b]

The ratios a/F_0 and b/V_0 are dimensionless. The ratio a/F_0 is on average equal to 0.25. The smaller the ratio, the more curved is the line. For slow-twitch

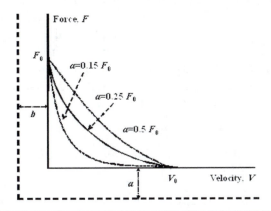

Figure 3.32 Force–velocity relation as a part of a hyperbola. Note the dependence of the line curvature on the value of coefficient *a* (discussed later in the text) and the geometric representation of constants *a* and *b*.

muscle fibers in humans the ratio was found to be around 0.15, whereas for fast-twitch muscle fibers the ratio was larger, and hence the force–velocity relations were less curved. For the joint torque–joint angular velocity curves, ratios as high as 0.5 have been found in elite sprinters and power athletes (javelin throwers), whereas in elite endurance athletes the ratios are as low as 0.1.

In equation 3.26 the maximal muscle force F_0 is determined for the optimal muscle length. To use it for all muscle lengths l, equation 3.26c is modified as

$$F = \left(\frac{F_0 b - aV}{V + b} \right) f(l), \qquad [3.28]$$

where $f(l)$ represents the dependence of the isometric muscle force on the muscle length ($\theta \le f(l) \le 1.0$).

In 1938 A.V. Hill reported that the values of coefficient a were the same whether they were determined from thermal measurements or from the force–velocity curve. This finding allowed the characteristic curve to be considered a fundamental property of a muscle. However, later studies performed with the improved methods showed that the coefficients a determined from the heat measurements and from pure mechanical analysis of the force–velocity relations are not exactly the same. Therefore, there is no compelling evidence to consider equation 3.26 as more fundamental than other empirical equations describing the force–velocity relations. Still, this equation remains the most popular and is used much more often than other force–velocity equations.

The popularity of this equation stems from its simplicity and familiarity. Its parameters can be determined from simple standard tests. The equation is convenient for the analysis of muscle–tendon units and multiple muscle systems. Unfortunately, the equation has no immediate connection with the underlying physiological mechanisms of muscle contraction. Still, this is the best force–velocity equation that we have.

3.2.2.5 Power–Velocity Relations

Representative publications: Faulkner et al. 1986; Lichtwark and Wilson 2005a, 2005b

Mechanical power generated by a muscle equals the product $F \times V$, where both force (F) and velocity (V) are scalars. At the ends of the force–velocity curve, when either F or V is zero the power is also zero. In the force–velocity graphs, the power at force and velocity values F_i or V_i, respectively, equals the area of a rectangle with a vertex F_i, V_i (figure 3.33).

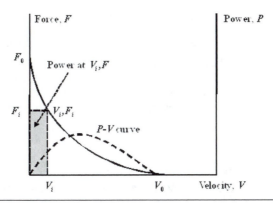

Figure 3.33   Dependence of mechanical muscle power *(P)* on shortening velocity, *V* (broken line). In the force–velocity graph, the power at the force and velocity values V_i and F_i equals the area of the shaded rectangle.

From equation 3.26c it follows that the mechanical power *P* generated by a muscle is

$$P = FV = \frac{F_0 b - aV}{V + b} V.$$ [3.29]

The power is maximal when the derivative *dP/dV* equals zero. Taking the derivative we obtain

$$\frac{dP}{dV} = \frac{dF}{dV} v + F = \frac{(F_0 + a)b^2 - a(V + b)^2}{(V + b)^2} = 0.$$ [3.30]

Hence, velocity V_m at which maximal power is generated is

$$V_m = b\left(\sqrt{F_0/a + 1} - 1 \right).$$ [3.31]

Velocity V_m is larger for larger ratios F_0/a or smaller ratios a/F_0. Hence, we can expect that endurance athletes with low a/F_0 ratio can achieve the V_m at higher relative velocity than can sprinters (do not forget that we discuss here the velocity normalized with respect to individual maximum). For a/F_0 = 0.25, the maximal power occurs at approximately 33% of the maximal velocity of the unloaded muscle V_0.

The power–velocity relations are important for practical applications, for instance, when selecting the optimal gear ratio of a bicycle at which maximal muscle power can be generated.

3.2.3 Force–Length–Velocity Relations

Representative publications: Abbott and Wilkie 1953; Baratta et al. 1993; van den Bogert et al. 1998; Chow et al. 1999

In muscle modeling studies, when force–length and Hill-type force–velocity relations for a muscle are known, they are frequently combined in one expression

$$F = F_{max} f(L) f(V) A,$$ [3.32]

where F_{max} is the maximal isometric force at maximal activation A_{max} and optimal length L_0; $f(L)$ and $f(V)$ are normalized force–length and force–velocity relations, respectively; and A is muscle activation. In the relations $f(L)$ and $f(V)$, the force is expressed in units of F_{max}, and the force–velocity relation $f(V)$ is considered for the optimal muscle length L_0; that is, at L_0 the isometric force $F_{max} = 1$. The equation implicitly assumes that the functions $f(L)$ and $f(V)$ are independent. Equation 3.32 constitutes a three-dimensional surface that can be visualized graphically for convenience of interpretation (figure 3.34).

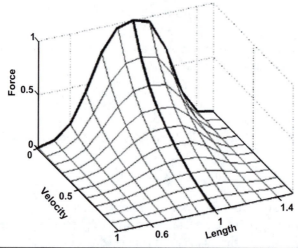

Figure 3.34 The normalized force–length–velocity relation for a muscle. Values of force, length, and velocity are normalized to F_{max}, L_0, and V_0, respectively. Activation is considered maximal, $A = A_{max}$. Function $f(L)$ can be described by the equation

$$f(L) = \exp\left(-\left|\frac{(L/L_0)^\beta - 1}{\omega}\right|^\rho\right),$$ [3.33]

where β, ρ, and ω are constants that describe the shape and breadth of the $f(L)$ relation. Function $f(V)$ is described by equation 3.27c, where $a/F_0 = 0.439$.

Because most probably the force–length relation *f(L)* does not scale precisely with the activation *A*, equation 3.33 can be considered only a first approximation of the real force–length–velocity relation.

3.3 HISTORY EFFECTS IN MUSCLE MECHANICS

Representative publications: Herzog and Leonard 2000; Rassier and Herzog 2004

So far the muscle functional relations, force–length (section **3.2.1**) and force–velocity (**3.2.2**), have been considered with an implicit assumption that they do not depend on preceding history of muscle action. Here and in chapters 4 and 7 the reader will see that the ability of muscle to develop force and do mechanical work depends on whether the active muscle was shortening or lengthened immediately prior to force development.

3.3.1 Force Depression After Muscle Shortening

Representative publications: De Ruiter et al. 1998; Lee and Herzog 2009

When a muscle develops isometric force immediately after its active short-ening, the force magnitude is reduced. The following experimental protocol is typically used to investigate this phenomenon. The muscle is maximally stimulated in isometric conditions at a given length (e.g., $0.9L_o$), and a steady-state isometric force at this length is recorded. After a rest period, the relaxed muscle is placed at a longer length, maximally stimulated, and then allowed to shorten with a constant speed below V_{max} to the same length of $0.9L_o$. During the shortening phase, the muscle force sharply decreases (faster than expected from the force–velocity relation, section **3.2.2**), and after redevelopment of isometric force its steady-state magnitude is sub-stantially reduced compared with isometric force developed at the same length without preliminary shortening (figure 3.35). This force reduction is called *shortening-induced force depression*.

As seen in figure 3.35, force depression in isometric conditions after shortening is directly related to the magnitude of shortening and lasts until offset of the stimulation. In human muscle, the force depression appears greater (37% of the maximum isometric force) than in isolated muscles or fibers (up to 25%). Force depression increases with the magnitude of isometric force before and during shortening (or with stimulation level). If stimulation is interrupted by a short period (several hundred milliseconds for human muscles) so that the force drops to zero, the force deficit is

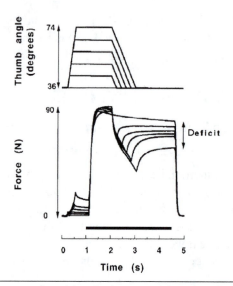

Figure 3.35 Force produced by the human adductor pollicis muscle (bottom panel) during maximal stimulation (50 Hz, a thick horizontal line above the time axis) and 5 isovelocity muscle shortening actions with different magnitudes (top panel, thumb angle is proportional to muscle length). The muscle was passively stretched to different thumb angles in each contraction. The difference between the control isometric force at 36° (top force trace at 4.5 s) and the redeveloped isometric force was defined as the force deficit.

From C.J. De Ruiter, A. De Haan, D.A. Jones, and A.J. Sargeant, 1998, "Shortening-induced force depression in human adductor pollicis muscle," *Journal of Physiology* 507(2): 583-591. Reprinted by permission of John Wiley & Sons Ltd.

substantially or completely eliminated after stimulation resumes. No force deficit is observed after unloaded shortening (i.e., with maximum velocity) or when nonuniform length distribution along individual fiber segments has been eliminated.

Although the exact mechanisms responsible for the shortening-induced force depression are still largely unknown, the facts presented here allow for the hypothesis that nonuniformity of sarcomeres along muscle fibers may be responsible for force depression after shortening. Sarcomere non-uniformity is larger during greater magnitudes of shortening and when muscle shortens against greater loads. Short relaxation of the muscle from the force-depressed state allows reestablishment of uniform distribution of sarcomere lengths and thus elimination of force depression after resumption of stimulation.

3.3.2 Effects of Muscle Release: Quick-Release and Controlled-Release Methods: Series Muscle Components

Representative publications: Gasser and Hill 1924; Hill 1950; Jewell and Wilkie 1958

If a fully activated muscle is suddenly released and allowed to shorten against different constant loads (figure 3.36), the muscle shortening after release consists of two distinct phases. This procedure is usually referred to as the *quick-release method*.

The first phase has a relatively fast speed of shortening (slope of the shortening trace) that does not depend on load; that is, the slope is essentially constant. The magnitude of this shortening is greater for smaller loads. The behavior of the muscle during the initial shortening resembles the behavior of a spring; the element of the muscle responsible for this

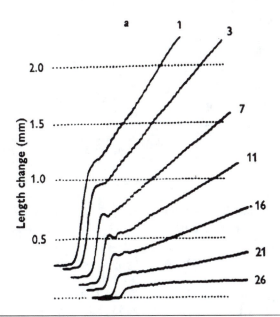

Figure 3.36 Experimental records of muscle shortening that occurs when the load on the muscle is suddenly reduced (quick release) from the full isometric tension to the value (in grams of weight) shown alongside each trace (frog sartorius, $L_o = 31$ mm, weight 76 mg, temperature 2 °C; muscle is released 0.7 s after the start of a 1 s tetanus [30 shocks per second]; time interval between dots 1 ms).

From B.R. Jewell and D.R. Wilkie, 1958, "An analysis of the mechanical components in frog's striated muscle," *Journal of Physiology* 143(3): 515-540. Reprinted by permission of John Wiley & Sons Ltd.

behavior is called a *series elastic component* (SEC). Other parts of SEC—in addition to the elements in the muscle belly—are the tendinous structures like aponeurosis and tendon. Functionally, the SEC behaves as a spring connected in series to the force generator in the muscle (the *contractile component*). The muscle SEC plays an important role in absorbing energy during muscle stretch and releasing it during shortening, thus providing additional mechanical work and power during muscle shortening.

By plotting the amount of muscle length change during the initial shortening versus the corresponding load, we can obtain the force–length relation for the SEC (figure 3.37).

The stress–strain relation obtained for the SEC is nonlinear and monotonic (figure 3.37): its instantaneous slope (or stiffness of the SEC) is relatively low at low muscle forces and increases with increasing muscle forces. The area under the stress–strain curve is elastic strain energy stored in the SEC during isometric development of a force. The elastic energy is released during release of the muscle. The amount of stored strain energy depends on the SEC stiffness, the maximum force the muscle is able to develop at a given length, and the maximum SEC elongation. The SEC elongation in isolated skeletal muscles at maximal isometric force is, on average, 2% to 4% of L_0.

Besides the quick-release method, the so-called *controlled-release method* is also used to determine the SEC properties. In this method, an activated muscle is suddenly released and allowed to shorten over a given distance, say

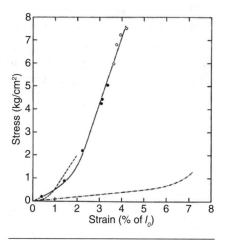

Figure 3.37 Stress–strain curves of the series elastic component of frog gastrocnemius (continuous line), of sartorius (interrupted line) (data from Jewel and Wilkie 1958), and of rat gracilis anticus (stippled line) (data from Bahler 1967). The stress is expressed in kg/cm² of muscle cross-section and the strain as a percentage of muscle resting length l_0. Open circles correspond to data obtained during releasing muscle immediately after stretch; black circles indicate muscle release from isometric contraction.

From G.A. Cavagna, 1970, "The series elastic component of frog gastrocnemius," *Journal of Physiology* 206(2): 257-262. Reprinted by permission of John Wiley & Sons Ltd.

2 mm. Then, at the new length, the muscle exerts isometric tension. The starting value of the newly developed force is, however, smaller than the value prior to the release. By comparing the decrease in muscle tension (ΔF) with the amount of the length change (Δl), we can determine the stiffness of the SEC, or its compliance. The quick-release and controlled-release methods are illustrated in figure 3.38. The methods yield similar results.

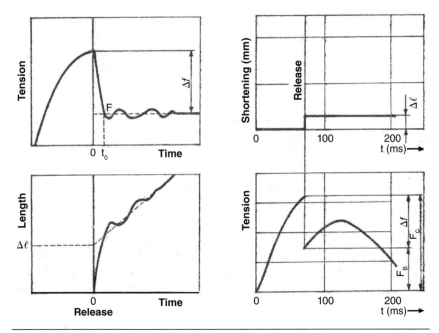

Figure 3.38   Quick-release (left panels) and controlled-release (right panels) methods, schematics. An active muscle that is exerting an isometric tension against an unmovable obstacle is suddenly released and is allowed to shorten. In the quick-release method the muscle shortens against a load; in the controlled-release method it shortens a given distance. When the SEC properties are an object of interest, the changes in the muscle tension (ΔF) are compared with the changes in the muscle length (Δl). To study the force–velocity relation, which represents the properties of the contractile muscle component, the rate of muscle length change during the second phase of the muscle shortening in the quick-release method (a left panel on the bottom) is compared with the load lifted (see figure 3.36).

Adapted from V.M. Zatsiorsky, A.S. Aruin, and S.N. Seluyanov, 1981, *Biomechanics of human musculoskeletal system* (in Russian) (Moscow: FiS Publishers), 92.

■ ■ ■ *FROM THE LITERATURE* ■ ■ ■

Elastic Energy Stored in the SEC
of Human Triceps Surae Muscle

Source: Hof, A.L. 1998. *In vivo* measurement of the series elasticity release curve of human triceps surae muscle. *J Biomech* 31(9):793-800

The force–extension characteristic of the series-elastic component of the human triceps surae muscle was measured in vivo by means of a hydraulic controlled-release ergometer. At moments of 150 and 180 Nm, which correspond to the peak moments in walking and running, respectively, the elastic energy stored in the SEC was 23 to 37 J and 31 to 57 J, respectively. This energy can be used in a subsequent phase of positive (concentric) work in both walking and running.

Because the SEC is connected in series with the contractile component, its resistance to stretch (stiffness) and hence the amount of elastic energy that can be accumulated in the SEC depend on the muscle tension and not the muscle length. This prediction is confirmed in the quick-release experiments performed on human subjects (figure 3.39).

The second, slower phase of shortening seen in figure 3.36 depends on the load. When the shortening speed during this phase (the slope of the length trace) is plotted against the load, the obtained relation coincides with the Hill force–velocity curve (see figures 3.26, 3.27, and 3.32 and equation 3.26b). Thus, muscle shortening during the second phase is caused by muscle contractile component. The serial elastic and contractile components represent mechanical properties of the muscles, not their structural elements. The components are parts of a three-component Hill muscle model that is described in more detail in chapter 4, section **4.4**.

3.4 SUMMARY

In vivo muscle excitation is a function of the number of recruited motor units and their firing frequency. During the latent period (from stimulation onset to an observable force increase), the muscle progresses from rest to an active state. The active state develops rapidly whereas the force at the

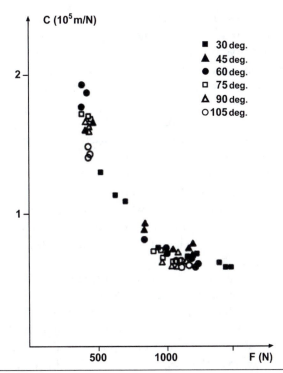

Figure 3.39 The relation between the estimates of the SEC compliance *(C)* and the muscle tension *(F)* for a group of elbow flexor muscles: a quick-release method. The subjects exerted an elbow flexion torque of different magnitudes at different joint angles from 30° to 105°. Then the trigger was suddenly released. The lumped muscle force was estimated for an equivalent elbow flexor assuming that the moment arm of it equals the moment arm of the biceps brachii. The angular values in the figure are for the initial angles before release. Note that the compliance (and stiffness) values do not depend on the joint angle or, hence, on the muscle length.

Adapted from J.C. Cnockaert, E. Pertuzon, F. Goubel, and F. Lestienne, 1978, Series-elastic component in normal human muscle. In *Biomechanics VI-A: International Series on Biomechanics,* vol. 2A, edited by E. Asmussen and K. Jorgensen (Baltimore: University Park Press), 73-78.

muscle–tendon junction develops more slowly. In many athletic events, the time available for force development is smaller than the time necessary to produce maximal force.

Relaxation from maximal isometric contraction involves two phases. The first phase is characterized by slow linear force decay. The second phase is exponential with a rapid force decline at the beginning and a slower decline at the end. In situ, the first part of muscle relaxation is governed by reduction of motor unit firing rate whereas the second part is governed by derecruitment of active motor units.

Muscles contract without substantially changing their volume. The same is true of individual muscle fibers. Muscle volume constancy is a fundamental feature of muscle mechanics.

The forces generated in sarcomeres are transmitted not only along the line connecting the tendons of origin and insertion but also in the lateral direction to adjacent structures (lateral force transmission, myofascial force transmission, epimuscular, myofascial force transmission). Although lateral force transmission is a well-established fact, the magnitude of the laterally transmitted force is unknown in many cases. During muscle contraction, the lengths of the individual sarcomeres can change differently (sarcomere nonuniformity). Some of the sarcomeres may contract whereas others (weak sarcomeres) may be stretched.

Muscle architecture—the arrangement of fibers and their bundles (fascicles) in a muscle—has a profound effect on muscle performance and force transmission. In parallel-fibered muscles, the fibers are arranged either in series or in parallel. When two fibers are connected in parallel, the force of the multifiber set is twice the force of the individual fibers but the speed of shortening is the same as the speed of an individual fiber. When the fibers are connected in series, the force at the end of the series equals the force of a single fiber but the speed of muscle shortening is twice the speed of shortening of the individual fiber. In the series-fibered muscles, the fibers can be connected in two ways: end-to-end (serial fibers with serial connections) or side-by-side (serial fibers with lateral connections).

In fusiform muscles (e.g., biceps brachii), individual fibers have different lengths. The fibers located on the periphery of the muscle are longer than the fibers at the muscle center. When the muscle as a whole shortens, the peripheral fibers shorten more than the central fibers.

The total force from two or more motor units can be different from the sum of the forces from separately activated motor units (nonlinear summation of motor unit forces). The difference can be as large as 20%. When the forces from simultaneously activated units are smaller than the sum of separately activated unit forces, the summation is called subadditive; when the forces are larger the summation is called superadditive.

In a pennate muscle, only the fiber force component that acts along the muscle line of action is transmitted to the tendon. A second force component, one that acts orthogonal to the first component, is also present and can induce a variety of effects including changing muscle curvature and increasing intramuscular pressure. In pennate muscles, fiber shortening is commonly accompanied by fiber rotation. Both fiber shortening and fiber rotation contribute to muscle shortening. The shortening velocity of the whole muscle can exceed the shortening velocity of the fibers. The ratio of whole muscle velocity to muscle fiber velocity is called the muscle architectural gear ratio.

The maximal muscle force per unit of the muscle physiological area (PCSA) is called the muscle specific force *(SF)*. In mammalian muscles it is usually between 15 and 40 N/cm^2 with a median value of about 25 N/cm^2. In pennate muscles, the tendon forces and the fiber forces are in dissimilar directions. To avoid effects of different pennation angles, some authors compute the *SF* for the muscle fascicles *($SF_{fascicle}$)* rather than for the muscle. In this case, the muscle force (tendon force) is converted first to the fascicle force and then divided by the PCSA.

Intramuscular fluid pressure, or simply intramuscular pressure (IMP), arises from the combination of hydrostatic and osmotic pressures. The IMP increases linearly with muscle force. If the IMP increases above the level of blood pressure in the capillaries (approximately 30 mmHg), the blood supply to the muscles is occluded. Long-lasting blood occlusion, as seen during compartment syndromes, can be dangerous.

Muscle force depends on both the muscle length and the speed of muscle length changes. Force–length and force–velocity relations are known as functional relations or characteristic relations (functions).

When a passive muscle is stretched beyond its equilibrium length (EL), it resists the stretching. This resistance is called passive force. Active force, in contrast, results from muscle activation. At muscle lengths shorter than the EL, muscle forces are completely attributable to active force. At muscle lengths longer than the EL, the recorded values of the muscle force—the total muscle force—are attributable to a summation of the passive and active forces. The active force–length curve determined in a whole muscle or muscle group is a smooth curve with a single maximum in the medium range of attainable muscle length. With the change of muscle length, the muscle force increases on the ascending limb of the curve and decreases on the descending limb. The muscle length at which the maximal force is produced is called the optimal length. In a sarcomere, the force–length curve consists of four linear segments: two ascending segments, shallow and steep; a plateau region; and the descending limb. This characteristic curve results from the properties of actin–myosin overlap at various sarcomere lengths. The part of the force–length curve seen in vivo is called the expressed section of the force–length curve.

Muscles behave stably when on the descending limb. When forcibly stretched, they resist the perturbation and exert a larger force than could be expected from their static force–length relation.

During submaximal contractions, the tendons are stretched less than during maximal contractions. Therefore, at the same joint angular configurations muscle fibers and sarcomeres are longer during submaximal contractions than during maximal contractions. Hence, the muscle fibers may possibly operate at different sections of the force–length curve under submaximal and maximal contractions.

If a muscle contracts against constant loads that vary from trial to trial, with greater loads the speed of shortening decreases. The mechanisms behind this fact attracted great interest among researchers and were a source of intensive discussion, and misunderstanding, over several decades. According to the viscosity hypothesis suggested in the 1920s, the muscles work as prestretched springs submerged in a viscous substance. The springs possess a certain amount of potential energy. During contraction, the potential energy transforms into mechanical work and heat. The sum Work + Heat equals the potential energy and is the same at different conditions of shortening. Later it was shown that this analogy (theory) is incorrect. The total energy liberated during muscle contraction is not constant; it increases with the work done by the muscle (the Fenn effect).

During eccentric muscle action, the total energy produced by the muscle is less than in concentric action, a fact that is incompatible with the idea of the decisive role of muscle viscosity. The viscous resistance depends on the movement speed but not on the movement direction. Imagine that you move your arm in a barrel of oil. Whether you move the arm to the right or to the left, you should experience similar viscous resistance at equal speeds. For muscles, this does not hold: when active muscles shorten they generate more energy than when they resist the stretching.

The relation between the load (muscle force) and speed of shortening is known as the force–velocity curve or the Hill force–velocity curve (in appreciation of the seminal paper by A.V. Hill, 1938). The Hill force–velocity curves are (a) descending and (b) concave upward. They are conveniently approximated by a parabolic equation:

$$(F + a)(V + b) = b(F_0 + a) = \text{Constant},$$

where F is force, V is velocity, F_0 is the force at $V = 0$, and a and b are constants. The equation is known as the Hill equation. The equation is not valid for eccentric muscle action or reversible muscle action, when the muscles have been prestretched immediately prior to concentric muscle action. The Hill's force–velocity curves are obtained by interpolating the force and velocity data recorded in several trials in which one of the parameters of the task (e.g., the load) was systematically changed. Because of that these relations are sometimes called the parametric relations.

Mechanical power generated by a muscle equals the product $F \times V$, where both force (F) and velocity (V) are scalars. When either F or V is zero, the power is also zero. The maximal power is produced when the velocity is around one third of the maximal velocity attainable at zero resistance. The power–velocity relations are important for practical applications, for instance, when selecting the optimal gear ratio of a bicycle at which maximal muscle power can be generated.

Besides parametric relations, other force–velocity relations attract interest:

1. Force–velocity relations in single movement: force and velocity are recorded continuously in a single trial and plotted versus each other.

2. Nonparametric force–velocity relations occur between the maximal force generated by a subject in a muscular strength test (F_{mm}) and the maximal movement speed V_m. In contrast to the parametric relations, which are negative, the nonparametric relations are either positive (the larger F_{mm}, the larger V_m) or zero.

Isometric force developed by muscle immediately after shortening against a load is smaller than isometric force developed at the same length without preliminary shortening. This force deficit is called shortening-induced force depression.

If an activated muscle is suddenly released and allowed to shorten against a load, the muscle shortening consists of two phases: a fast speed phase and a slow speed phase (the quick-release method). The first phase is due to shortening of the SEC, and the second is due to shortening of the contractile component. The SEC behaves as a spring connected in series to the force generator in the muscle (the contractile component). In the controlled-release method, an activated muscle is suddenly released and allowed to shorten over a given distance. Then, the muscle exerts isometric tension at a new length.

3.5 QUESTIONS FOR REVIEW

1. Explain the meaning of the terms *in vivo, in situ,* and *in vitro.*

2. Discuss excitation–activation coupling.

3. Discuss the active state of the muscle during transition from rest to activity. How can the active state be determined using only mechanical means (i.e., without monitoring motoneurons)?

4. How long does it take to develop maximal force in humans? Compare this time with the time available for force production in such activities as sprint running and throwing.

5. Discuss the mechanical events that occur during the transition from maximal isometric force to rest.

6. How does the muscle volume change during the transition from rest to activity?

7. Discuss lateral force transmission in muscle fibers.

8. It has been shown experimentally that single muscle-fiber forces and the fiber perimeter (radius) are linearly related. However, at the whole-muscle level force is a function of the physiological cross-sectional area (radius squared). Do these facts contradict each other? Why or why not?

9. Consider two identical muscle fibers connected either (a) in series or (b) in parallel. Both fibers are activated simultaneously in a similar way. Describe the expected patterns of force and velocity of the (a) and (b) fiber arrangements.

10. Define muscle *architectural gear ratio* (AGR). Compare muscle and fiber shortening velocities at AGR > 1.0.

11. In a pennate muscle, the angle of pennation is 30° and the force produced by a muscle fascicle is 10 N. Compute the force transmitted along the tendon.

12. During muscle contraction, the muscle length decreases but its volume remains constant. Hence, the size of the muscle in a direction(s) orthogonal to the direction of the muscle length change increases. In other words, the muscle deforms. Analyze the cases when either the muscle thickness or the muscle width increases. How do these different patterns of muscle deformation affect the force and velocity transmissions from the muscle fibers to the tendon?

13. Define *specific force* of a muscle. What are the typical magnitudes of the specific force in mammalian muscles?

14. Prove that the maximal muscle force (tendon force) generated by a pennate muscle is

$$F_{max} = SF_{fascicle} \times T sin\theta cos\theta$$

where $SF_{fascicle}$ is the fascicle force per unit area (specific tension), T is the area of the muscle attachment to the tendon, and θ is the pennation angle.

15. What is a *stress tensor*? Is a muscle *specific force* a stress tensor?

16. Discuss the factors affecting intramuscular fluid pressure.

17. What are the contributions of the passive and active forces to the muscle force–length curve at different muscle lengths?

18. How is the active force–length curve explained by the cross-bridge theory of muscle contraction?

19. Discuss muscle stability on the descending limb of the force–length curve.

20. What are the *expressed sections* of the muscle force–length curves in the human body?

21. Explain how the viscosity hypothesis of muscle contraction was proven incorrect.

22. Discuss Hill's force–velocity curve (parametric force–velocity relation). Write Hill's equation and explain its meaning.

23. Besides parametric force–velocity relations, other force–velocity relations exist. Name and discuss them.

24. How does mechanical power generated by a muscle depend on the muscle shortening velocity?

25. Define shortening-induced force depression.

26. Explain the quick-release method and the controlled-release method.

27. Discuss the series elastic component (SEC) and the contractile component (CC) of a muscle. How can existence of these components be shown experimentally?

3.6 LITERATURE LIST

Abbott, B.C., and D.R. Wilkie. 1953. The relation between velocity of shortening and the tension-length curve of skeletal muscle. *J Physiol* 120:214-223.

Allinger, T.L., M. Epstein, and W. Herzog. 1996a. Stability of muscle fibers on the descending limb of the force–length relation: A theoretical consideration. *J Biomech* 29(5):627-633.

Allinger, T.L., W. Herzog, and M. Epstein. 1996b. Force–length properties in stable skeletal muscle fibers-theoretical considerations. *J Biomech* 29(9):1235-1240.

Azizi, E., E.L. Brainerd, and T.J. Roberts. 2008. Variable gearing in pennate muscles. *Proc Natl Acad Sci U S A* 105(5):1745-1750.

Bahler, A.S. 1967. Series elastic component of mammalian skeletal muscle. *Am J Physiol* 213:1560-1564.

Baratta, R.V., M. Solomonow, R. Best, and R. D'Ambrosia. 1993. Isotonic length/force models of nine different skeletal muscles. *Med Biol Eng Comput* 31(5):449-458.

Barclay, C.J., and G.A. Lichtwark. 2007. The mechanics of mouse skeletal muscle when shortening during relaxation. *J Biomech* 40(14):3121-3129.

Baskin, R., and P. Paolini. 1965. Muscle volume changes: Relation to the active state. *Science* 148:971-972.

Baskin, R.J., and P.J. Paolini. 1966. Muscle volume changes. *J Gen Physiol* 49(3):387-404.

Baskin, R.J., and P.J. Paolini. 1967. Volume change and pressure development in muscle during contraction. *Am J Physiol* 213(4):1025-1030.

Benninghoff, A., and H. Rollhauser. 1952. Zur inneren Mechanik des gefiederten Muskels [Interior mechanics of pennate muscles] (in German). *Pflugers Arch* 254(6): 527-548.

Bensamoun, S.F., K.J. Glaser, S.I. Ringleb, Q. Chen, R.L. Ehman, and K.N. An 2008. Rapid magnetic resonance elastography of muscle using one-dimensional projection. *J Magn Reson Imaging* 27(5):1083-1088.

Blemker, S.S., P.M. Pinsky, and S.L. Delp. 2005. A 3D model of muscle reveals the causes of nonuniform strains in the biceps brachii. *J Biomech* 38(4):657-665.

Blix, M. 1893. Die Lange und die Spannung des Muskels. *Scand Arch Physiol* 4:399-409.

Blix, M. 1894. Die Lange und die Spannung des Muskels. *Scand Arch Physiol* 5:149-206.

Bloch, R.J., and H. Gonzalez-Serratos. 2003. Lateral force transmission across costameres in skeletal muscle. *Exerc Sport Sci Rev* 31(2):73-78.

Bobbert, M.F., G.C. Ettema, and P.A. Huijing. 1990. The force–length relationship of a muscle-tendon complex: Experimental results and model calculations. *Eur J Appl Physiol Occup Physiol* 61:323-329.

Buller, A., and D. Lewis. 1965. The rate of tension development in isometric tetanic contractions of mammalian fast and slow skeletal muscle. *J Physiol* 176:337-354.

Butelli, O., H. Vandewalle, and G. Peres. 1996. The relationship between maximal power and maximal torque-velocity using an electronic ergometer. *Eur J Appl Physiol* 73(5):479-483.

Caiozzo, V.J., J.J. Perrine, and V.R. Edgerton. 1981. Training-induced alterations of the in vivo force–velocity relationship of human muscle. *J Appl Physiol* 51(3):750-754.

Cavagna, G.A. 1970. The series elastic component of frog gastrocnemius. *J Physiol Lond* 206(2):257-262.

Chow, J.W., W.G. Darling, J.G. Hay, and J.G. Andrews. 1999. Determining the force–length-velocity relations of the quadriceps muscles: III. A pilot study. *J Appl Biomech* 15(2):200-209.

Clamann, H.P., and T.B. Schelhorn. 1988. Nonlinear force addition of newly recruited motor units in the cat hindlimb. *Muscle Nerve* 11(10):1079-1089.

Close, R.I. 1972. Dynamic properties of mammalian skeletal muscles. *Physiol Rev* 52(1):129-197.

Cnockaert, J.C., E. Pertuzon, F. Goubel, and F. Lestienne. 1978. Series-elastic component in normal human muscle. In *Biomechanics VI-A. International Series on Biomechanics*, Vol. 2A, edited by E. Asmussen and K. Jorgensen. Baltimore: University Park Press, pp. 73-78.

De Ruiter, C.J., A. De Haan, D.A. Jones, and A.J. Sargeant. 1998. Shortening-induced force depression in human adductor pollicis muscle. *J Physiol* 507(Pt 2):583-591.

Drost, M.R., M. Maenhout, P.J. Willems, C.W. Oomens, F.P. Baaijens, and M.K. Hesselink. 2003. Spatial and temporal heterogeneity of superficial muscle strain during in situ fixed-end contractions. *J Biomech* 36(7):1055-1063.

Epstein, M., and W. Herzog. 1998. *Theoretical Models of Skeletal Muscle*. Chichester, UK: Wiley.

Epstein, M., and W. Herzog. 2003. Aspects of skeletal muscle modelling. *Philos Trans R Soc Lond B Biol Sci* 358(1437):1445-1452.

Eycleshymer, A.C., and D.M. Schoemaker. 1911. *A Cross-Section Anatomy*. New York: Appleton. (http://web.mac.com/rlivingston/Eycleshymer/Welcome.html)

Faulkner, J.A., D. Claflin, and K.K. McCully. 1986. Power output of fast and slow fibers from human skeletal muscles. In *Human Muscle Power*, edited by N.L. Jones, N. McCartney and A.J. McComas. Champaign, IL: Human Kinetics, pp. 81-94.

Fenn, W.O. 1924. The relation between the work performed and the energy liberated in muscular contraction. *J Physiol Lond* 58:373-395.

Fenn, W.O., and B.S. Marsh. 1935. Muscular force at different speeds of shortening. *J Physiol Lond* 85:277-297.

Friden, J., and R.L. Lieber. 2002. Mechanical considerations in the design of surgical reconstructive procedures. *J Biomech* 35(8):1039-1045.

Fukunaga, T., R.R. Roy, F.G. Shellock, J.A. Hodgson, and V.R. Edgerton. 1996. Specific tension of human plantar flexors and dorsiflexors. *J Appl Physiol* 80(1):158-165.

Gasser, H.S., and A. Hill. 1924. The dynamics of muscle contraction. *Proc R Soc Lond B Biol Sci* 96:398-437.

Gatton, M L., M.J. Pearcy, and G.J. Pettet. 1999. Difficulties in estimating muscle forces from muscle cross-sectional area: An example using the psoas major muscle. *Spine* 24(14):1487-1493.

Gennisson, J.L., C. Cornu, S. Catheline, M. Fink, and P. Portero. 2005. Human muscle hardness assessment during incremental isometric contraction using transient elastography. *J Biomech* 38(7):1543-1550.

Gillard, D.M., S. Yakovenko, T. Cameron, and A. Prochazka. 2000. Isometric muscle length-tension curves do not predict angle-torque curves of human wrist in continuous active movements. *J Biomech* 33(11):1341-1348.

Gillis, J.M. 1985. Relaxation of vertebrate skeletal muscle. A synthesis of the biochemical and physiological approaches. *Biochim Biophys Acta* 811(2):97-145.

Gollapudi, S.K., and D.C. Lin. 2009. Experimental determination of sarcomere force–length relationship in type-I human skeletal muscle fibers. *J Biomech* 42(13):2011-2016.

Gordon, A.M., A.F. Huxley, and F.J. Julian. 1966. The variation in isometric tension with sarcomere length in vertebrate muscle fibres. *J Physiol* 184:170-192.

Gregor, R.J. 1993. Skeletal muscle mechanics and movement. In *Current Issues in Biomechanics*, edited by M.D. Grabiner. Champaign, IL: Human Kinetics, pp. 171-212.

Haxton, H.A. 1944. Absolute muscle force in the ankle flexors of man. *J Physiol* 103:267-273.

Herzog, W. 2000. Mechanical properties and performance in skeletal muscles. In *Biomechanics in Sport*, edited by V.M. Zatsiorsky. Oxford, UK: Blackwell Science, pp. 21-32.

Herzog, W., and T.R. Leonard. 2000. The history dependence of force production in mammalian skeletal muscle following stretch-shortening and shortening-stretch cycles. *J Biomech* 33(5):531-542.

Hill, A.V. 1922.The maximum work and mechanical efficiency of human muscles, and their most economical speed. *J Physiol* 56:19-41.

Hill, A.V. 1938. The heat of shortening and the dynamic constants of muscle. *Proc R Soc Lond B Biol Sci* 126-B:136-195.

Hill, A. 1948. The pressure developed in muscle during contraction. *J Physiol* 107:518-526.

Hill, A.V. 1949a. The abrupt transition from rest to activity in muscle. *Proc R Soc Lond B Biol Sci* 136(884):399-420.

Hill, A.V.1949b. The onset of contraction. *Proc R Soc Lond B Biol Sci* 136(883):242-254.

Hill, A.V. 1949c. Is relaxation an active process? *Proc R Soc Lond B Biol Sci* 136(884):420-435.

Hill, A.V. 1950. The development of the active state of muscle during the latent period. *Proc R Soc Lond B Biol Sci.* 137(888):320-329.

Hof, A.L. 1998. In vivo measurement of the series elasticity release curve of human triceps surae muscle. *J Biomech* 31(9):793-800.

Hof, A.L., and J.P. van Zandwijk. 2004. Is the descending limb of the force–length relation in the human gastrocnemius muscle ever used? *Clin Anat* 17(6):528-529; author reply 530.

Huijing, P.A. 2009. Epimuscular myofascial force transmission: A historical review and implications for new research. International Society of Biomechanics Muybridge Award lecture, Taipei, 2007. *J Biomech* 42(1):9-21.

Ichinose, Y., Y. Kawakami, M. Ito, and T. Fukunaga. 1997. Estimation of active force–length characteristics of human vastus lateralis muscle. *Acta Anat Basel* 159(2-3):78-83.

Ito, M., Y. Kawakami, Y. Ichinose, S. Fukashiro, and T. Fukunaga. 1998. Nonisometric behavior of fascicles during isometric contractions of a human muscle. *J Appl Physiol* 85(4):1230-1235.

Jenkyn, T.R., B. Koopman, P. Huijing, R.L. Lieber, and K.R. Kaufman. 2002. Finite element model of intramuscular pressure during isometric contraction of skeletal muscle. *Phys Med Biol* 47(22):4043-4061.

Jewell, B.R., and D.R. Wilkie. 1958. An analysis of the mechanical components in frog's striated muscle. *J Physiol* 143:515-540.

Julian, F., and R. Moss. 1976. The concept of active state in striated muscle. *Circ Res* 38(1):53-59.

Katz, B. 1939. The relation between force and speed in muscular contraction. *J Physiol* 98:45-64.

Kaufman, K.R., K.N. An, and E.Y. Chao. 1989. Incorporation of muscle architecture into the muscle length-tension relationship. *J Biomech* 22(8-9):943-948.

Kawakami, Y., and T. Fukunaga. 2006. New insights into in vivo human skeletal muscle function. *Exerc Sport Sci Rev* 34(1):16-21.

Kawakami, Y., K. Nakazawa, T. Fujimoto, D. Nozaki, M. Miyashita, and T. Fukunaga. 1994. Specific tension of elbow flexor and extensor muscles based on magnetic resonance imaging. *Eur J Appl Physiol Occup Physiol* 68(2):139-147.

Kisiel-Sajewicz, K., A. Jaskólski, and A. Jaskólska. 2005. Current knowledge in studies on relaxation from voluntary contraction. *Hum Mov* 6(2):136-148.

Komi, P.V. 1990. Relevance of in vivo force measurements to human biomechanics. *J Biomech* 23(Suppl 1):23-34.

Lee, E.J., and W. Herzog. 2009. Shortening-induced force depression is primarily caused by cross-bridges in strongly bound states. *J Biomech* 42(14):2336-2340.

Lichtwark, G.A., and A.M. Wilson. 2005a. Effects of series elasticity and activation conditions on muscle power output and efficiency. *J Exp Biol* 208(Pt 15):2845-2853.

Lichtwark, G.A., and A.M. Wilson. 2005b. A modified Hill muscle model that predicts muscle power output and efficiency during sinusoidal length changes. *J Exp Biol* 208(Pt 15):2831-2843.

Lieber, R.L., and J. Friden. 1997. Intraoperative measurement and biomechanical modeling of the flexor carpi ulnaris-to-extensor carpi radialis longus tendon transfer. *J Biomech Eng* 119(4):386-391.

Lieber, R.L., and J. Friden. 1998. Musculoskeletal balance of the human wrist elucidated using intraoperative laser diffraction. *J Electromyogr Kinesiol* 8(2):93-100.

Lyons, M.F., and A. Aggarwal. 2001. Relaxation rate in the assessment of masseter muscle fatigue. *J Oral Rehabil* 28(2):174-179.

Maganaris, C.N. 2001. Force–length characteristics of in vivo human skeletal muscle. *Acta Physiol Scand* 172(4):279-285.

Mommaerts, W. 1970. What is the Fenn-Effect? Muscle is a regulatory engine the energy output of which is governed by the load. *Naturwissenshaften* 57(7):326-330.

Monti, R.J., R.R. Roy, J.A. Hodgson, and V.R. Edgerton. 1999. Transmission of forces within mammalian skeletal muscles. *J Biomech* 32(4):371-380.

Needham, D. 1971. *Machina Carnis*. Cambridge, UK: University Press.

Otten, E. 1988. Concepts and models of functional architecture in skeletal muscle. *Exerc Sport Sci Rev* 16:89-137.

Pappas, G.P., D.S. Asakawa, S.L. Delp, F.E. Zajac, and J.E. Drace. 2002. Nonuniform shortening in the biceps brachii during elbow flexion. *J Appl Physiol* 92(6):2381-2389.

Patel, T.J., and R.L. Lieber. 1997. Force transmission in skeletal muscle: From actomyosin to external tendons. *Exerc Sport Sci Rev* 25:321-363.

Perrine, J.J. and V.R. Edgerton. 1978. Muscle force–velocity and power-velocity relationships under isokinetic loading. *Med Sci Sports Exerc* 10(3):159-166.

Pertuzon, E., and S. Bouisset. 1973. Instantaneous force–velocity relationship in human muscle. In *Biomechanics III, Medicine and Sport*, Vol. 8, edited by S. Cerquiglini, A. Venerando and J. Wartenweiler. Basel, Switzerland: Karger, 230-234.

Rack, P.M., and D.R. Westbury. 1969. The effects of length and stimulus rate on tension in the isometric cat soleus muscle. *J Physiol* 204(2):443-460.

Ralston, H., V. Inman, L. Strait, and M. Shaffrath. 1947. Mechanics of human isolated voluntary muscle. *Am J Physiol* 151:612-620.

Rassier, D.E., and W. Herzog. 2004. Considerations on the history dependence of muscle contraction. *J Appl Physiol* 96(2):419-427.

Rassier, D.E., B.R. MacIntosh, and W. Herzog. 1999. Length dependence of active force production in skeletal muscle. *J Appl Physiol* 86(5):1445-1457.

Reeves, N.D., M.V. Narici, and C.N. Maganaris. 2004b. Effect of resistance training on skeletal muscle-specific force in elderly humans. *J Appl Physiol* 96(3):885-892.

Riewald, S.A., and S.L. Delp. (1997) The action of the rectus femoris muscle following distal tendon transfer: Does it generate knee flexion moment? *Dev Med Child Neurol* 39(2):99-105.

Ritchie, J.M., and D.R. Wilkie. 1958. The dynamics of muscular contraction. *J Physiol* 143(1):104-113.

Sandercock, T.G. 2005. Summation of motor unit force in passive and active muscle. *Exerc Sport Sci Rev* 33(2):76-83.

Sandercock, T.G., and H. Maas. 2009. Force summation between muscles: Are muscles independent actuators? *Med Sci Sports Exerc* 41(1):184-190.

Savelberg, H. 2000. Rise and relaxation times of twitches and tetani in submaximally recruited, mixed muscle: A computer model. In *Skeletal Muscle Mechanics: From Mechanisms to Function*, edited by W. Herzog. Chichester, UK: Wiley, pp. 225-240.

Sejersted, O.M., A.R. Hargens, K.R. Kardel, P. Blom, O. Jensen, and L. Hermansen. 1984. Intramuscular fluid pressure during isometric contraction of human skeletal muscle. *J Appl Physiol* 56(2):287-295.

Sejersted, O.M., and A.R. Hargens. 1995. Intramuscular pressures for monitoring different tasks and muscle conditions. *Adv Exp Med Biol* 384:339-350.

Smeulders, M.J., M. Kreulen, J.J. Hage, P.A. Huijing, and C.M. van der Horst. 2004. Intraoperative measurement of force–length relationship of human forearm muscle. *Clin Orthop Relat Res* (418):237-241.

Steinberg, B.D. 2005. Evaluation of limb compartments with increased interstitial pressure. An improved noninvasive method for determining quantitative hardness. *J Biomech* 38(8):1629-1635.

Swammerdam, J. 1758. *The Book of Nature*. London: Seyffert.

Tax, A.A., and J.J. Denier van der Gon. 1991. A model for neural control of gradation of muscle force. *Biol Cybern* 65(4):227-234.

van den Bogert, A.J., K.G. Gerritsen, and G.K. Cole. 1998. Human muscle modelling from a user's perspective. *J Electromyogr Kinesiol* 8(2):119-124.

van der Linden, B.J., H.F. Koopman, H.J. Grootenboer, and P.A. Huijing. 1998. Modelling functional effects of muscle geometry. *J Electromyogr Kinesiol* 8(2):101-109.

Van Leeuwen, J.L., and C.W. Spoor. 1992. Modelling mechanically stable muscle architectures. *Philos Trans R Soc Lond B Biol Sci* 336(1277):275-292.

Van Leeuwen, J.L., and C.W. Spoor. 1993. Modelling the pressure and force equilibrium in unipennate muscles with in-line tendons. *Philos Trans R Soc Lond B Biol Sci* 342(1302):321-333.

Wilkie, D. 1956. The mechanical properties of muscle. *Br Med Bull* 12:177-182.

Wilkie, D.R. 1949. The relation between force and velocity in human muscle. *J Physiol* 110:249-280.

Winter, S.L., and J.H. Challis. 2008a. Reconstruction of the human gastrocnemius force–length curve in vivo: Part 1. Model-based validation of method. *J Appl Biomech* 24(3):197-206.

Winter, S.L., and J.H. Challis. 2008b. Reconstruction of the human gastrocnemius force–length curve in vivo: Part 2. Experimental results. *J Appl Biomech* 24(3):207-214.

Winter, S.L., and J.H. Challis. 2009. The expression of the skeletal muscle force–length relationship in vivo: A simulation study. *J Theor Biol* 262(4):634-643.

Winters, J.M. 1990. Hill-based models: A systems engineering approach. In *Multiple Muscle Systems. Biomechanics and Movement Organization*, edited by J.M. Winters and S.L.-Y. Woo. New York: Springer-Verlag, pp. 69-93.

Yeadon, M.R., M.A. King, and C. Wilson. 2006. Modelling the maximum voluntary joint torque/angular velocity relationship in human movement. *J Biomech* 39(3):476-482.

Yucesoy, C.A. 2010. Epimuscular myofascial force transmission implies novel principles for muscular mechanics. *Exerc Sport Sci Rev* 38(3):128-134.

Zahalak, G.I. 1997. Can muscle fibers be stable on the descending limbs of their sarcomere length-tension relations? *J Biomech* 30(11-12):1179-1182.

Zatsiorsky, V.M. 1995. *Science and Practice of Strength Training.* Champaign, IL: Human Kinetics.

Zatsiorsky, V.M. 1997. On muscle and joint viscosity. *Muscle Control* 1:299-309.

Zatsiorsky, V.M. 2003. Biomechanics of strength and strength training. In *Strength and Power in Sport,* edited by P.V. Komi. Oxford, UK, Blackwell Science, pp. 439-487.

Zatsiorsky, V.M., A.S. Aruin, and V.N. Seluyanov.1981. *Biomechanics of Musculoskeletal System* (in Russian). Moscow: FiS Publishers. [The book is also available in German: Saziorski, W.M., A.S. Aruin, and W.N. Selujanow. 1984. *Biomechanik des Menschlichen Bewegungsapparates.* Berlin: Sportverlag, 1984.]

CHAPTER

4

MUSCLES AS FORCE AND ENERGY ABSORBERS
Muscle Models

The chapter addresses eccentric muscle action, also known as plyometric action, and describes the most popular muscle models. A muscle is said to act eccentrically when it is active and being forcibly stretched; muscle length increases while resisting external forces (e.g., weight of load, force produced by other muscles). Such situations commonly occur; for instance, in the early support period during running, the muscles that are leg extensors act eccentrically.

Section **4.1** addresses muscle behavior during stretch, section **4.2** is concerned with muscle behavior after stretch, and section **4.3** deals with energy dissipation. Section **4.4** considers muscle models.

After defining major terms, the chapter describes basic facts of mechanical behavior of electrically stimulated isolated muscle under stretch (i.e., during eccentric action). Stretch of an active muscle modifies (often increases) muscle force. If a muscle is allowed to shorten immediately after stretch, it can do more work and produce higher power without a significant increase in metabolic energy expenditure. Animals often take advantage of these effects by allowing muscles to stretch before they shorten and thereby generate mechanical energy for motion. This sequence of muscle stretch and shortening is called the stretch–shortening cycle, and it is described in detail in chapter 7.

To characterize eccentric action quantitatively, let us consider an isolated muscle that is maximally excited by electrical stimulation and that has one end fixed and the other attached to a load (figure 4.1). If the load exceeds the force developed by the muscle, the muscle will yield; that is, it will be stretched while being active. The rate of eccentric action can be conveniently defined as the product of muscle force (F_m) and the velocity of the point of muscle force application to the load (or the rate of muscle length change; V_m): $P_m = F_m \cdot V_m$. Because F_m and V_m are collinear, in the present discussion they are considered scalars. The product P_m was defined as muscle power (see for details chapter 3 in this book and chapter 6 in *Kinetics of Human Motion*). Because eccentric muscle actions involve opposite directions of muscle force (which tends to shorten the muscle) and the velocity of its application point, P_m is negative, indicating that during eccentric action the muscle absorbs energy. The amount of absorbed energy, or the negative work done by the muscle, can be determined by integrating the negative muscle power over the time period of muscle stretch or by integrating the force generated by the muscle over the muscle length change.

Our definition of eccentric action is valid if we consider the muscle as a whole without paying attention to behavior of its components: muscle fascicles and the aponeurosis and tendon (passive tissue connected to the contractile component in series). As you will see in chapter 7, there may be situations when, while the whole active muscle–tendon unit is being stretched, the muscle fascicles maintain a constant length or even shorten because of an increase in the length of the tendon. Can we call this type of muscle action eccentric? Probably not. We have to distinguish between forcibly stretching the muscle–tendon unit as a whole and forcibly stretching only the muscle belly.

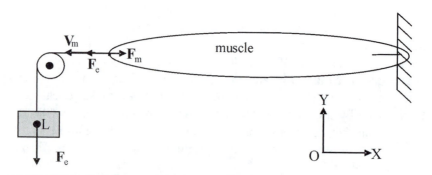

Figure 4.1 Definition of eccentric muscle action. Eccentric muscle action occurs when the tension developed by the muscle, F_m, is smaller than the external force F_e. In this example, weight of load L ($|F_e| > |F_m|$), and the direction of velocity of the point of muscle force application is opposite the direction of muscle force action. For further explanations, see text.

4.1 MUSCLE MECHANICAL BEHAVIOR DURING STRETCH

Representative publications: Lewin and Wyman 1927; Katz 1939; Abbott and Aubert 1952; Edman et al. 1978; Rassier 2009

When a fully activated muscle or a fiber is stretched with a moderate speed from one constant length to another, the force recorded on its end exceeds the maximum isometric force at the same muscle length (figure 4.2). At the end of stretch, the force can be 2 times larger than the maximum

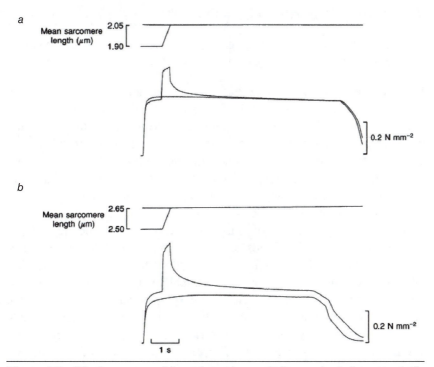

Figure 4.2   Displacement and force (stress) records from a single frog muscle fiber during tetanic stimulation at two different sarcomere lengths. *(a)* Stretch from 1.9 to 2.05 μm sarcomere length compared with ordinary isometric tetanus at 2.05 μm. *(b)* Comparison of stretch from 2.50 to 2.65 μm sarcomere length with isometric tetanus at 2.65 μm. In the figure, *(a)* denotes the velocity-dependent force enhancement during stretch, and *(b)* indicates the residual force enhancement after stretch; the latter appears above optimal sarcomere length 2.2 μm.

From K.A. Edman and T. Tsuchiya, 1996, "Strain of passive elements during force enhancement by stretch in frog fibers," *Journal of Physiology* 490(Pt 1): 191-205. Reprinted by permission of John Wiley & Sons Ltd.

isometric force at the same length, a phenomenon known as dynamic force enhancement. This force enhancement is velocity dependent—the peak force typically increases with the stretch velocity (the force–velocity relation in the eccentric muscle actions is described later in the text, see figure 4.3 and section **4.1.1.1.**) When the stretch is completed and the muscle length is kept constant at a new level, muscle force starts decreasing and reaches a value that is still larger than the force of isometric action at the same muscle length. This residual force enhancement after stretch lasts as long as the muscle is active.

4.1.1 Dynamic Force Enhancement

Representative publications: Sugi 1972; Flitney and Hirst 1978; van Atteveldt and Crowe 1980; Pinniger et al. 2006

Dynamic force enhancement during stretch is thought to be associated with increased strain of attached crossbridges between actin and myosin myofilaments (see figure 1.2). In this situation, crossbridges act like springs and develop force that increases with their strain. The time history of muscle force during stretch depends on several factors: rate of stretch, stretch magnitude, initial muscle length, stimulation intensity, muscle fatigue, and temperature. The rate of stretch is usually expressed in the units of optimal muscle length L_0 per second, L_0/s.

When the (constant) rate of muscle stretch is low to moderate (e.g., up to 10 L_0/s for a bundle of muscle fibers from semitendinosus muscle of the frog at 20 °C, which roughly corresponds to the maximum shortening velocity of this muscle), the force of maximally stimulated muscle starts to increase at the beginning of stretch, initially faster and then more slowly, and reaches a maximum at the end of stretch. Then the force starts to decay to a level that often exceeds the isometric force at the same muscle length insofar as stimulation continues. The peak force developed during such stretches increases with stretch rate (figure 4.3).

As mentioned, this behavior of muscle force during stretch with low to moderate speeds is explained by deformation of crossbridges, which causes an increase in force. The initial higher rate of force increase results presumably from deformation of all attached crossbridges. The rate of force increase in this phase does not markedly depend on stretch velocity. A subsequent slowdown in the force increase (figure 4.2) can be explained by a partial detachment of some crossbridges and by formation of new crossbridges that are not as strongly deformed. The rate of force increase during this phase is stretch-velocity dependent.

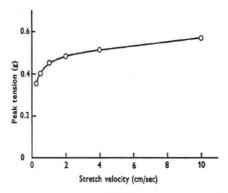

Figure 4.3 Force developed by fully activated bundles of muscle fibers from semitendinosus muscle of the frog above the initial isometric level as a function of stretch velocity. Initial isometric level of force 0.6 to 0.65 g; initial sarcomere length 1.9 to 2.3 μm; amount of stretch 5%.

From H. Sugi, 1972, "Tension changes during and after stretch in frog muscle fibres," *Journal of Physiology* 25(1): 237-253. Reprinted by permission of John Wiley & Sons Ltd.

4.1.1.1 Force–Velocity Relation for Lengthening Muscle

Representative publications: Katz 1939; Joyce et al. 1969; Otten 1987; Cole et al. 1996; Krylow and Sandercock 1974; Brown et al. 1999

The dependence of peak force on stretch velocity of maximally stimulated muscle during dynamic force enhancement does not follow the Hill force–velocity equation (equation 3.26b). When speed V becomes negative (see equation 3.26c), the predicted force increase is much smaller than observed in experiments. The predicted force (from the Hill equation) increases indefinitely when speed of muscle lengthening approaches a constant b, whereas experimentally observed peak forces converge to a constant force value of about $2F_0$ (see figure 4.3).

The force–velocity relation during lengthening of maximally stimulated muscles is not as well established as the force–velocity relation during muscle shortening (see section **3.2.2**). There is no unique equation describing eccentric force–velocity relation that is accepted by all researchers. This could be explained by the fact that when the Hill force–velocity equation for shortening was proposed, it was thought that constants a and b in this equation had a fundamental physiological meaning related to energy production in the muscle, which later, after more accurate measurements, was determined to be incorrect.

Table 4.1   Empirical Force–Velocity Relations for Lengthening Muscle

Equation	Notes	Source
$F/F_0 = 1.8 - \dfrac{0.8(1+V/V_{max})}{1-7.56V/kV_{max}}$	F and F_0 are peak muscle force during stretch at constant velocity ($V < 0$) and maximum isometric force at optimal muscle length, respectively; V_{max} is maximum shortening velocity, $k = 0.25$ for fast twitch muscles, and $k = 0.17$ for slow-twitch muscles.	Otten 1987
$V = \dfrac{b(F_0 - F)}{2F_0 - F + a}$	$V < 0$ is velocity of stretch (mm/s); F and F_0 are peak eccentric force and maximum isometric force (g wt), respectively $(F_0 < F)$; a and b are Hill equation constants measured at optimal length $(a/F_0 = 0.284, b = 11.51$ mm/s).	Krylow and Sandercock 1997
$F/F_0 = \dfrac{b - (a_{v_o} + a_{v1}L + a_{v2}L^2)V}{b+V}$	$V > 0$ is velocity of stretch (L_0/s); L is fascicle length (in units of L_0); $a_{V0} = -1.53, a_{V1} = 0, a_{V2} = 0, b = 1.05$ are constants for fast-twitch muscles; $a_{V0} = -4.70, a_{V1} = 8.41, a_{V2} = -5.34, b = 0.18$ are constants for slow-twitch muscles.	Brown et al. 1999

Several equations have been proposed in the literature to describe the eccentric force–velocity relation (table 4.1).

4.1.1.2 Give Effects

Representative publications: Sugi 1972; Flitney and Hirst 1978; Colombini et al. 2007

When stretch velocity is higher than 20 L_o/s, muscle fibers initially develop a very high force, which before the end of stretch is sharply decreasing to values close to or below the prestretch isometric force at the corresponding fiber length. The sudden reduction of force during fast stretches is called *slip* or *give* and is explained by forced detachments of the crossbridges due to very high forces arising from large (≥ 12 nm) and fast deformation of the crossbridges. After the detachment, the crossbridges start reattaching to thin actin filaments, which partially restores force. The slip magnitude depends on stretch velocity: the higher the velocity, the greater decrease in force.

4.1.2 Residual Force Enhancement

Representative publications: Abbott and Aubert 1952; Edman et al. 1978; Morgan 1994; Telley et al. 2006; Joumaa et al. 2008

After reaching peak force at the end of muscle stretch with moderate velocity (figure 4.2*b*), the force starts to decay and reaches a steady-state level that exceeds that of the isometric force at the same muscle length. The mechanism for this residual force enhancement is not clear, but several explanations have been proposed. According to one, the residual force enhancement is a result of sarcomere inhomogeneity during stretch. When an active muscle fiber is stretched on the descending limb of the force–length curve, shorter sarcomeres are able to develop higher forces than longer sarcomeres. As a result, the stronger sarcomeres shorten and stretch the weaker sarcomeres until the latter rapidly extend (pop) and their passive structural components equilibrate the force of shorter and stronger sarcomeres. This explanation is consistent with the fact that the residual force enhancement is greater when a muscle is stretched from the descending limb of the force–length relation than when it is stretched from the plateau or the ascending limb (see figure 4.2).

According to the preceding explanation, during residual force production, sarcomere lengths should be nonuniform: there should be short and very long popped sarcomeres. Recent measurements of half-sarcomere lengths in individual myofibrils at isometric and eccentric conditions demonstrated highly nonuniform sarcomere length distributions but failed to find popped sarcomeres. These observations lead to another explanation of the residual force enhancement, according to which the residual force after stretch originates mostly from noncrossbridge structural elastic elements that might reside within the myofibril but outside crossbridges. One likely candidate is titin, whose resistance to stretch is much higher in active than nonactive muscle because of calcium-dependent titin–actin interactions. The mechanisms of residual force enhancement are still an object of active research and discussion.

■ ■ ■ *FROM THE LITERATURE* ■ ■ ■

Noncrossbridge Contribution to Residual Force Enhancement

Source: Bagni, M.A., G. Cecchi , B. Colombini, and F. Colomo. 2002. A non-cross-bridge stiffness in activated frog muscle fibers. *Biophys J* 82(6):3118-3127

(continued)

Force responses to fast ramp stretches of various amplitude and velocity, applied during tetanic contractions, were measured in single intact fibers from frog tibialis anterior muscle. The fast tension transient produced by the stretch was followed by a period, lasting until relaxation, during which the tension remained constant at a value that greatly exceeded the isometric tension. The excess of tension was termed *static tension,* and the ratio between the force and the accompanying sarcomere length change was termed *static stiffness.* The static stiffness was independent of the active tension developed by the fiber and independent of stretch amplitude and stretching velocity in the whole range tested; it increased with sarcomere length in the range 2.1 to 2.8 μm, to decrease again at longer lengths. These results suggest that activation increased the stiffness of some sarcomere structures outside the crossbridges. The authors suggested that titin could constitute one such structure because its stiffness could increase upon fiber stimulation and thus lead to increased sarcomere stiffness.

Digression

On Muscle Stiffness

Some researchers in physiology and motor control use the term *muscle stiffness* in models and explanations. In mechanics, stiffness is defined as the ratio of the applied force to the amount of deformation of a body ($F/\Delta L$); it is measured in newtons per meter (N/m). The concept of stiffness is properly applied to the deformation of passive elastic bodies for which the deformation depends only on the force, not on the deformation speed or time history. However, during and after forcible muscle stretch, muscle force is not constant at the same muscle length (see preceding sections on the dynamic and residual force enhancement). Hence, the ratio of the force to the amount of deformation (stiffness) is not constant; it depends on the time of recording and is a time-dependent function. Also, the speed of the stretch has to be known. Without this information, the data cannot be interpreted correctly. For a more detailed discussion on muscle and joint stiffness, see section **3.5** in *Kinetics of Human Motion.*

4.2 MUSCLE SHORTENING AFTER STRETCH

As demonstrated in the previous sections, a muscle stretched at slow to moderate speeds is able to develop much higher forces than during isometric contractions. Therefore, when a muscle is released during the dynamic or residual force enhancement phases and allowed to shorten against a load, one should expect a greater mechanical output from the muscle compared with release from an unstretched isometric condition. Because the dynamic and residual force enhancements are likely to be based on the crossbridge and noncrossbridge mechanisms, respectively, it is possible that energy expenditure during shortening after muscle release from the dynamic and residual force enhancements will be different. Let us first consider consequences of muscle stretch on mechanical output during subsequent shortening.

4.2.1 Work and Power During Shortening After Stretch

Representative publications: Cavagna 1970; Cavagna and Citterio 1974; Sugi and Tsuchiya 1981; Edman 1999; Edman and Tsuchiya 1996; Cavagna et al. 1994

Because the muscle contractile component (CC) and series elastic component (SEC) are connected in series, when muscle contracts isometrically (i.e., the total muscle–tendon complex length is constant), the contractile element shortens (i.e., the actin and myosin myofilaments slide with respect to each other) and the SEC stretches. The amount of elastic energy that the SEC stores during isometric contraction can be determined from the force–length (or stress–strain) relation of the SEC (see figure 3.37).

As demonstrated previously (see figure 4.3), a stretched muscle can develop force levels up to 2 times greater than an unstretched muscle. Correspondingly, strain energy stored in the SEC can also increase by a factor of 2 (area under the stress–strain curve in figure 3.37). This, in turn, may contribute to the ability of the muscle to either shorten against heavier loads at a given shortening velocity or to shorten faster at a given load.

When a fully activated muscle is released immediately after stretch, positive work can also increase by a factor of 2. The ability of the muscle to do more work after it is stretched is expressed in the shift of the force–velocity curve to the right (figure 4.4). In other words, at the same shortening velocity the muscle produces a greater force after stretch, and at the same exerted force a prestretched muscle shortens faster. If a fully activated muscle is allowed to shorten immediately or soon after the stretch, the muscle can shorten against a load equal to the maximum isometric force of the muscle at this length; that is, a stretched muscle can lift a weight that without the stretch it cannot raise. The influence of stretch on muscle performance is more pronounced at slow shortening velocities than at fast ones (see figure 4.4).

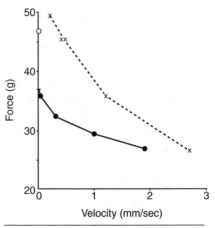

Figure 4.4 Force–velocity relations of fully activated frog semitendinosus obtained after release and shortening with different loads. Full points and continuous line, release from a state of isometric contraction; crosses and interrupted line, release at the end of stretching. The release lengths are the same for both conditions.

From G.A. Cavagna and G. Citterio, 1974, "Effect of stretching on the elastic characteristics and the contractile component of frog striated muscle," *Journal of Physiology* 239(1): 1-14. Reprinted by permission of John Wiley & Sons Ltd.

Work enhancement after stretch increases with stretch velocity, initial muscle length, and temperature and decreases if there is a pause between the stretch and shortening. The mechanisms responsible for work enhancement are not clear, because the elastic energy stored in the strained crossbridges is fully discharged by a small muscle release. If nonuniformity in the sarcomere lengths is substantial during muscle release (which is more probable when a muscle is released from the descending limb of the force–length curve), work and power enhancement can be explained by recoil of the popped sarcomeres that developed high force in their passive structural elements. This explanation would not hold, however, if the muscle is released at the plateau of the force–length relation where little sarcomere nonuniformity is observed. To explain stretch-induced work enhancement during release at the plateau of the force–length relation, Italian physiologist Giovanni Cavagna and his colleagues (1994) suggested that the crossbridges strained by the stretch are brought to a new mechanical state, become stronger and stiffer, and thus are capable of producing greater forces. This hypothesis awaits thorough testing.

> ■ ■ ■ *FROM THE LITERATURE* ■ ■ ■
>
> ## Work Enhancement Occurs When Muscle Shortens Immediately or Soon After Stretch (i.e., During Decay of Dynamic Force Enhancement)
>
> Source: Cavagna, G.A., B. Dusman, and R. Margaria. 1968. Positive work done by a previously stretched muscle. *J Appl Physiol* 24:21-32
>
> The positive work done by a muscle that shortens immediately after being stretched in the contracted state, W', was found to be greater than the positive work done by the same muscle during shortening from a state of isometric contraction, W, the speed, the length, and the extent of shortening being the same. The experiments were made on isolated toad sartorius and frog gastrocnemius and in human forearm flexors. W' and W were measured at different speeds of stretching and shortening and, on isolated muscles only, at different average lengths of the muscle: W' and W increased both with speed and length up to 2.5 times. The greater amount of work done after stretching is not entirely accounted for by the elastic energy stored during stretching. The CC itself is responsible in part for this increase. The force developed by the CC, when muscle shortens after being stretched, is greater than the force developed, at the same speed and length, when the muscle shortens starting from a state of isometric contraction.

4.2.2 Energy Consumption During Stretch and Efficiency of the Muscle Shortening After Stretch

Representative publications: Fenn 1923, 1924; Abbott et al. 1951; Hill and Howarth 1959; Curtin and Woledge 1979; Constable et al. 1997; Linari et al. 2003

Metabolic energy expenditure of isolated skeletal muscles (measured, e.g., as the sum of muscle heat production and mechanical work done by the muscle) is lower during stretch of active muscle than during shortening or isometric development of force at the same stimulation level. Some authors, including Nobel Prize laureate Sir A.V. Hill, reported that a substantial portion of a muscle's negative work (energy absorbed by the muscle) does not appear in total muscle heat production. Hill and his colleagues suggested three possibilities to explain this fact: "(a) the work is absorbed in driving backwards chemical processes which have actually occurred as a normal part of contraction; (b) . . . the work is absorbed in some other chemical or physical process at present unknown; and (c) . . . the work is wholly degraded into heat, but that chemical processes normally occurring in contraction are prevented by the stretch" (Abbott et al. 1951, p. 102).

Evidence for the first and second possibilities has not been found. It is more likely that the rate of adenosine triphosphate (ATP, energy source for crossbridges) splitting is reduced during stretching of active muscle and that negative work is not used in muscles' chemical reactions. The rate of ATP splitting is especially small at low velocities of stretch ($\sim 0.2\ L_0$/s) compared with isometric force development (up to 4 times).

A low-rate ATP-splitting phenomenon also occurs during residual force enhancement after stretch despite the enhanced forces. This phenomenon can be partly explained by the increased length of sarcomeres beyond the region of overlap between actin and myosin filaments, which allows passive structural element in the fibers to develop high forces without energy consumption.

If mechanical efficiency of work production by an isolated muscle or a fiber is defined as the ratio of work done and total energy expenditure,

$$e = \dot{W}/(\dot{Q} + \dot{W}),\qquad\qquad [4.1]$$

(where e is mechanical efficiency, $\dot{W}$ is positive mechanical power, and $\dot{Q}$ is rate of heat production), muscle released at the state of the residual force enhancement (see figure 4.2b) should have a higher efficiency in moving the load compared with a case where muscle is released from an isometric contraction at the same length without a preceding stretch. This is because following stretch the muscle develops higher forces, but this force excess does not require additional energy expenditure. As discussed in chapter 7, this assertion may be true.

■ ■ ■ *FROM THE LITERATURE* ■ ■ ■

Energy Storage During Stretch of Active Single Fibers

Source: Linari, M., R.C. Woledge, and N.A. Curtin. 2003. Energy storage during stretch of active single fibres from frog skeletal muscle. *J Physiol* 548(Pt 2):461-474

Heat production and force were measured during tetanic contractions of single muscle fibers from anterior tibialis of frog. During stimulation, fibers were kept under isometric conditions, stretched, or allowed to shorten (at constant velocity) after isometric force had reached a plateau. The energy change was evaluated as the sum of heat and work (work = integral of force with respect to length change). Net energy absorption occurred during stretch at velocities greater than about 0.35 L_0/s (L_0 is fiber length at resting sarcomere length 2.10 μm). Heat produced by 1 mm segments of the fiber as well as in the whole fiber was measured. The authors concluded that energy absorption is not an artifact of patchy heat production. The maximum energy absorption, 0.092 $\pm$ 0.002 P_0L_0 (where P_0 is isometric force at L_0, which is proportional to the fiber cross-sectional area, and thus the product P_0L_0 was an estimate of fiber volume), occurred during the fastest stretches (1.64 L_0/s) and amounted to more than half of the work done on the fiber. Energy absorption occurred in two phases. The amount in the first phase, 0.027 $\pm$ 0.003 P_0L_0, was independent of velocity beyond 0.18 L_0/s. The quantity absorbed in the second phase increased with velocity and did not reach a limiting value in the range of velocities used. After stretch, energy was produced in excess of the isometric rate, probably from dissipation of the stored energy. About 34% (0.031 P_0L_0/0.092 P_0L_0) of the maximum absorbed energy could be stored elastically (in crossbridges, tendons, and thick, thin, and titin filaments) and by redistribution of crossbridge states. The remaining energy could have been stored in stretching transverse, elastic connections between myofibrils.

4.3 DISSIPATION OF ENERGY

Representative publications: Gasser and Hill 1924; Tidball 1986; Lin and Rymer 1998

A muscle subjected to periodic stretching and shortening damps the imposed oscillations or, in other words, dissipates the oscillation energy. The ability of the muscle to dissipate energy increases with activation level and with the magnitude of the length change. For example, oscillation damping is approximately 40 times greater in an active muscle compared with a passive one (figure 4.5).

The ability of muscles to dissipate the body's mechanical energy seems to have important implications for many human movements. For example, in

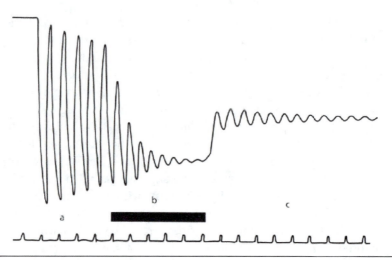

Figure 4.5   Damping oscillations in a spring connected to a muscle *(a)* unexcited, *(b)* excited, and *(c)* unexcited again. The damping becomes greatly amplified when the muscle is excited. Time marks, 0.2 s.

Reprinted, by permission, from H.S. Gasser and A.V. Hill, 1924, "The dynamics of muscle contraction," *Proceedings of the Royal Society of London, Series B: Biological Sciences* 96: p. 417, fig. 11.

athletic activities such as landing in gymnastics, muscles acting eccentrically have to dissipate the body's energy in a short period. The ability to dissipate energy also acts as an important protective mechanism by decelerating limbs as they reach the limits of their range of motion.

■ ■ ■ *FROM THE LITERATURE* ■ ■ ■

Energy Recoil and Dissipation in Muscle–Tendon Units

Source: Morgan, D.L., U. Proske, and D. Warren. 1978. Measurements of muscle stiffness and the mechanisms of elastic energy storage in hopping kangaroos. *J Physiol* 282:253-261

A kangaroo hopping above a certain speed consumes less oxygen than do quadruped mammals. This is explained by elastic energy storage in tendons and ligaments. Energy can be stored in a tendon by stretching it but only when in-series muscle fibers are stiff enough to resist most of the length change. The authors measured length and tension changes in the contracting gastrocnemius muscle of the wallaby *Thylogale* during rapid, controlled stretches and from this determined the amount of movement in muscle fibers and tendon. When the muscle developed close to maximum isometric tension, up to 8 times as much movement occurred in the tendon as in the muscle fibers. This is possible because of the long and compliant tendon. Measurement of work absorption by the muscle with a full length of free tendon and when the tendon had been shortened showed that with the shortened tendon, a larger proportion of movement occurred in the muscle fibers, producing a steep increase in work absorption by the muscle and a consequent increase in energy loss.

4.4 MECHANICAL MUSCLE MODELS

Representative publications: Hatze 1978; Otten 1987; Zajac 1989; Sandercock and Heckman 1997

Several muscle models have been suggested in the literature. Some are very detailed and include modeling internal mechanics in the sarcomeres, such as titin involvement and interaction of the actin–myosin filaments. This level of detail is beyond the scope of this book. We limit the present discussion to the so-called Hill-type model that is intended to represent a muscle's gross mechanical behavior; that is, a class of phenomenological models with only input (activation) and output (force) as the physiologically realistic components. Such models typically are as follows:

• A single input–single output model for which the input is muscle excitation and the output is the muscle force. Limiting the input to a single excitation variable means that the model neglects the facts that the muscle consists of many fibers and motor units and that the force level control is realized via two mechanisms, recruitment of motor units and their rate control.

• The model is a lumped-parameter model in which such mechanical features as force generation, muscle elasticity, and damping are assigned to one or two of the muscle elements. Although these elements can be associated with certain morphological structures, the model is not intended to realistically replicate muscle architecture or morphology. Instead, the model represents only the muscle behavior at a gross phenomenological level.

After describing the model, we address its scaling to specific muscles in individual subjects.

4.4.1 Hill-Type Model

Representative publications: Zajac 1989; Winters 1990

In section **3.3.2**, such concepts as series elastic component (SEC) and contractile component (CC), also called contractile element, of a muscle were introduced. The concepts were introduced to explain the results of experiments performed using the quick-release method (see figure 3.36). During these experiments, a muscle at optimal length L_0 with fixed ends is stimulated maximally until it develops a steady isometric force, F_0. Then one end of the muscle is released and the muscle is allowed to shorten, lifting a load, $F < F_0$ (afterload). The muscle length changes occur in two phases: (1) near-instantaneous muscle shortening Δl, whereby the shortening amplitude increases monotonically with decreasing the difference $(F_0 - F)$ whereas the shortening speed does not depend on the $(F_0 - F)$ difference; and (2) relatively slow muscle shortening at an almost-constant speed V. During this phase, the speed is higher for smaller values of the $(F_0 - F)$ difference. The first stage of muscle shortening was assigned to the SEC shortening and the second to the CC shortening. The SEC and CC are deemed to be connected in series. To account for the passive resistance of an unstimulated muscle (illustrated in figure 3.17), a second elastic element, connected in parallel to the CC, was introduced into the model (parallel elastic component, PEC). The CC possesses two fundamental properties discussed previously: an active force–length relation (described in section **3.2.1**) and a force–velocity relation (section **3.2.2**). Because the fundamental facts on which the model is based were obtained by A.V. Hill and his co-workers, these models are commonly called *Hill-type models* or *three-element muscle models*. These models can be represented by two structures with similar mechanical properties (figure 4.6).

The structures, especially model A, resemble the Kelvin model of viscoelastic materials (standard linear model, see figure 2.8c) with one important difference. Instead of a dashpot (damping component), as in the standard linear model, the Hill-type model includes a contractile element that possesses more complex properties. The dashpot has only one property: it provides resistance proportional to the rate of strain change (the speed). In the Hill-type model, in addition to damping, the CC possesses the properties of force generation and the force–length and force–velocity relations. The damping in the model is also nonlinear (to explain the nonlinear force–velocity curve).

The elastic elements in the model, PEC and SEC, represent mainly the properties of the connective tissues (albeit it was shown that crossbridges also possess elastic properties and that some mechanical properties of

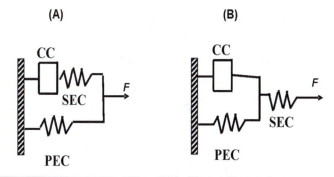

Figure 4.6 Element arrangements in the Hill-type muscle models.

structural proteins that constitute PEC, i.e., titin, are activation-dependent). The action of the CC depends on its current length and velocity as well as on the activation level. Effects of other factors, such as temperature, are not considered.

The model predicts the following phenomena of muscle behavior:

- With both muscle ends fixed, muscle activation results in CC shortening and SEC lengthening. The muscle–tendon unit as a whole exerts force.

- At lengths longer than rest length, muscle force is the sum of the passive force (caused by PEC resistance to forceful stretching) and the active force generated by the CC (as illustrated in figure 3.17).

- When a muscle, fixed at both ends, is activated and then allowed to shorten against a smaller load (a quick release), the muscle length–time history consists of two parts: a rapid length decrease due to SEC shortening and slow length changes due to CC contraction.

The reader is invited to find the corresponding examples in chapters 3 and 4. The model evidently has some limitations; it does not describe all muscle properties. In particular, it does not predict time-dependent muscle behavior such as force depression due to muscle shortening (described in section **3.3.1**), force enhancement due to muscle lengthening (described in section **4.1**), and the effects of muscle fatigue. The model in its basic form also does not explain the time delay between the CC-induced force generation onset and the force recorded at the muscle end (the existence of such a delay follows from experiments with sudden stretch of a muscle during a latent period between muscle stimulation and onset of a muscle force increase, see figure 3.2). To incorporate such a delay in Hill-type

■ ■ ■ *FROM THE LITERATURE* ■ ■ ■

Tendon Slack Delays the Force Development

Source: Muraoka, T., T. Muramatsu, T. Fukunaga, and H. Kanehisa. 2004. Influence of tendon slack on electromechanical delay in the human medial gastrocnemius in vivo. *J Appl Physiol* 96(2):540-544

The purpose of this study was to clarify the influence of tendon slack on electromechanical delay in the human medial gastrocnemius. Electromechanical delay and medial gastrocnemius tendon length were measured at each of five ankle joint angles (−30°, −20°, −10°, 0°, and 5°; positive values for dorsiflexion) using percutaneous electrical stimulation and ultrasonography, respectively. The relative electromechanical delay, normalized with the maximal electromechanical delay for each subject, decreased dependent on the extent of decrease in medial gastrocnemius tendon slack. There were no significant differences in electromechanical delay among the joint angles (−10°, 0°, and 5°) where medial gastrocnemius tendon slack was taken up. These results suggest that the extent of tendon slack is an important factor for determining electromechanical delay.

models, the activation input is shifted in time by a delay that accounts for the excitation–contraction time or an additional assumption that the SEC at rest is slack is made.

Nevertheless, Hill-type models are still the most popular models in muscle biomechanics and are broadly used to explore muscle action in human and animal movement.

4.4.2 Model Scaling

Representative publications: Zajac 1989; Challis 2000

As follows from chapters 2 and 3, the main functional relations for a muscle are these:

- Force–length curve for a passive muscle
- Force–length curve for a tendon with aponeurosis
- Force–length curve for an active muscle
- Force–velocity curve for an active muscle

By expressing force as a percentage of maximal force, expressing velocity as muscle lengths per second, and normalizing the muscle length to a reference length (e.g., to the muscle length at the center of the studied range of motion), we can present Hill-type equations in a dimensionless form. The muscle shortening velocity also can be normalized using the so-called time-scaling parameter, which is a ratio of the optimal muscle length L_0 (i.e., the length at which maximal isometric force is generated), to the maximal shortening velocity expressed in the units of muscle length per second:

$$\tau_0 = L_0 / V_m. \qquad [4.2]$$

With such a convention, the normalized maximum shortening velocity is

$$V_{norm} \equiv \frac{V_m}{L_0} \equiv \frac{1}{\tau_c}. \qquad [4.3]$$

The dimensionless form of the relations serves as a normalization scheme that permits comparisons between different muscles. However, to apply a dimensionless model to a specific muscle in a given subject, we must determine the model parameters. Examples include maximum isometric force F_0, optimal muscle (or fiber) length L_0, and the time-scaling parameter. If a muscle–tendon unit is modeled, the following parameters should also be specified:

1. The tendon slack length normalized by the optimal muscle fiber length L_0, that is, by the length at which the maximal fiber force can be exerted $L_{T,Slack} / L_0$ (described in section **1.5.2**)
2. The ratio of the maximal muscle force generated at the optimal muscle length and the corresponding tendon extension

The preceding two parameters are necessary to characterize the sharing of the muscle–tendon unit length changes between the muscle belly and the tendon. All of the preceding parameters are features of the Hill-type muscle models.

▪ ▪ ▪ *FROM THE* **LITERATURE** ▪ ▪ ▪

Effect of Varying Muscle Parameters on the Expressed Section of the Force–Length Curve

Source: Winter, S.L., and J.H. Challis. 2008. Reconstruction of the human gastrocnemius force–length curve in vivo: Part 1. Model-based validation of the method. *J Appl Biomech* 24(3):197-206

(continued)

From the Literature   *(continued)*

The triceps surae muscle group was modeled. The following muscle parameters were varied until the muscle was operating entirely on the ascending limb, on the plateau region, or over the descending limb of the force–length curve: (1) the ratio $L_{T,slack}$ / $L_{F,Opt}$ (for an explanation of symbols, see the preceding text); (2) the muscle fiber length in the reference position $L_{F,REF}$ which was 0° of plantar flexion and 0° of knee flexion; (3) maximal isometric force at the optimal fiber length; (4) the tendon extension at the maximal muscle force generated at the optimal length of the muscle fiber; (5) the muscle fiber pennation angle; and (6) the width parameter of the force–length relation. By manipulating these parameters the investigators could change the expressed region of the force–length curve.

4.5 SUMMARY

A muscle acts eccentrically if it is active and its length is increasing under an external force. Muscle power (P_m) equals the product of muscle force (F_m) and the velocity of the point of muscle force application to the load (or the rate of muscle length change, V_m; $P_m = F_m \cdot V_m$). In eccentric muscle actions, muscle power is negative. The amount of absorbed energy, negative work done by the muscle, can be determined by integrating the negative muscle power over the time period of muscle stretch or by integrating the force generated by the muscle over the muscle length change.

If fully activated muscle is stretched with a constant and relatively low speed (<10 L_o/s), the force developed by the muscle initially increases quickly with subsequent slowing until the end of stretch. The force developed during the stretch is called dynamic force enhancement; it exceeds the force developed by the muscle isometrically at the same length. The dynamic force depends on speed of stretch; the enhancement increases with speed if it is below about 10 L_o/s. Several equations have been proposed to describe the force–velocity relation for lengthening muscle. At slow stretch speeds (<1 L_o/s), the increase in force is sharp, and as speed increases the force increase slows and approaches values up to 2 F_o at fast speeds. The mechanism of dynamic force enhancement is not fully understood. A likely explanation is that during the stretch, the formed crossbridges are stretched and brought to a new mechanical state in which they become stronger and stiffer and thus generate greater forces in response to stretch. A decrease in the rate of force increase during stretch is explained by a detachment of some crossbridges that reach their maximum extension and by formation of new crossbridges that are not as strongly deformed.

When fully activated muscle is stretched at relatively high velocities (e.g., 20 L_o/s), muscle fibers initially develop a very high force, which starts decreasing sharply before the end of stretch to values close to or below the isometric force at the corresponding fiber length. The sudden reduction of force during fast stretches is called *slip* or *give* and is explained by forced mechanical detachments of the crossbridges by very high forces arising from large and fast deformation of the crossbridges. After the detachment, the crossbridges start reattaching to thin actin filaments, which partially restores force. The magnitude of the slip effect depends on stretch velocity: the higher the velocity, the greater the decrease in force.

After reaching peak force at the end of muscle stretch with moderate velocity (<10 L_o/s), the force starts to decay and reaches a steady-state level exceeding that of the isometric force at the same muscle length. This force level is called residual force enhancement. It is more pronounced if, after stretch, muscle length lies on the descending limb of the force–length relationship. The residual force enhancement is larger at greater magnitudes of stretch. One possible explanation for this observation is that sarcomeres stretched beyond the plateau of the force–length relation are not homogeneous in length. As a result, sarcomeres that happen to be shorter are able to develop higher forces than sarcomeres at longer lengths. Stronger sarcomeres continue to shorten and stretch the weaker sarcomeres until the latter rapidly extend (pop) and their passive structural components equilibrate the force of the shorter and stronger sarcomeres. Additionally, part of the residual force enhancement is explained by stretch of activation-dependent but noncrossbridge structures (i.e., structural protein titin).

If following a stretch a fully activated muscle is allowed to shorten against a load, it is capable of doing more work and producing more power compared with a muscle released from isometric contraction without prestretch. If a muscle is allowed to shorten immediately after a stretch, it is capable of moving a load corresponding to the maximum force of the muscle at the same length, but the muscle's maximum shortening velocity does not change. The ability of muscles to do more work after a stretch can be explained by two distinct mechanisms. The crossbridges strained by the stretch are brought to a new mechanical state, become stronger and stiffer, and thus are capable to produce greater forces. A second explanation is that the popped sarcomeres that developed high passive forces are thus able to move a greater load while shortening from the descending limb of the isometric force–length relation.

During stretch of an active muscle, the muscle consumes less metabolic energy compared with isometric and especially concentric contractions. The higher force developed by the fully activated muscle during and after stretch does not require extra energy consumption. As a result, muscle released following stretch is capable of doing more work with the same energy expenditure; that is, muscle efficiency in this situation is higher than when muscle shortens from isometric contraction.

The ability of a muscle to dissipate mechanical energy during periodic length changes is much higher (40 times greater) when it is fully active compared with when it is passive.

Hill-type muscle models are single input–single output models with lumped parameters. They consist of three elements, or components: contractile component, series elastic component, and parallel elastic component. The components do not replicate the muscle architecture or morphology; they represent mechanical properties rather than the structures. To apply the model to a specific muscle, the model parameters should be properly scaled.

4.6 QUESTIONS FOR REVIEW

1. Define concentric, isometric, and eccentric muscle action.

2. Define a type of muscle action in a situation when the muscle–tendon complex is active and stretched by an external force but muscle fascicles do not change their length.

3. Define a dynamic force enhancement. Under what mechanical conditions can it be produced?

4. How can dynamic force enhancement be explained in terms of known muscle properties?

5. Describe the force–velocity relation for muscle lengthening. How does it differ from the Hill force–velocity relation for muscle shortening?

6. Explain reasons for the sudden decrease in muscle force during very fast muscle lengthening.

7. Define residual force enhancement. Under what mechanical conditions it is produced?

8. How can residual force enhancement be explained using the notion of popped sarcomeres?

9. How can the structural protein titin contribute to residual force enhancement?

10. What happens to the force–velocity relation if fully activated muscle is allowed to shorten against different loads following stretch? How would parameters of the force–velocity relationship (maximum load, maximum shortening velocity) change?

11. Discuss possible reasons for the changes in the force–velocity relation after stretch.

12. Explain why mechanical efficiency of muscle released after stretch is higher than the efficiency of muscle released from isometric contraction at the same muscle length.

13. How does muscle activity affect energy dissipation in the muscle?

14. What are the main elements of the Hill-type muscle models and how are these elements arranged with respect to each other?

15. What features of muscle behavior do Hill-type models explain?

16. What features of muscle behavior do Hill-type models not explain?

17. Discuss muscle model scaling.

4.7 LITERATURE LIST

Abbott, B.C., and X.M. Aubert. 1952. The force exerted by active striated muscle during and after change of length. *J Physiol* 117(1):77-86.

Abbott, B.C., X.M. Aubert, and A.V. Hill. 1951. The absorption of work by a muscle stretched during a single twitch or a short tetanus. *Proc R Soc Lond B Biol Sci* 139(894):86-104.

Bagni, M.A., G. Cecchi, B. Colombini, and F. Colomo. 2002. A non-cross-bridge stiffness in activated frog muscle fibers. *Biophys J* 82(6):3118-3127.

Brown, I.E., E.J. Cheng, and G.E. Loeb. 1999. Measured and modeled properties of mammalian skeletal muscle: II. The effects of stimulus frequency on force-length and force-velocity relationships. *J Muscle Res Cell Motil* 20(7):627-643.

Cavagna, G.A. 1970. The series elastic component of frog gastrocnemius. *J Physiol* 206(2):257-262.

Cavagna, G.A., and G. Citterio. 1974. Effect of stretching on the elastic characteristics and the contractile component of frog striated muscle. *J Physiol* 239(1):1-14.

Cavagna, G.A., B. Dusman, and R. Margaria. 1968. Positive work done by a previously stretched muscle. *J Appl Physiol* 24:21-32.

Cavagna, G.A., N.C. Heglund, J.D. Harry, and M. Mantovani. 1994. Storage and release of mechanical energy by contracting frog muscle fibres. *J Physiol* 481(Pt 3):689-708.

Challis, J.H. 2000. Muscle-tendon architecture and athletic performance. In *Encyclopedia of Sports Medicine. Biomechanics in Sport,* edited by V.M. Zatsiorsky. Oxford, UK: Blackwell Science, pp. 33-55.

Cole, G.K., A.J. van den Bogert, W. Herzog, and K.G. Gerritsen. 1996. Modelling of force production in skeletal muscle undergoing stretch. *J Biomech* 29(8):1091-1104.

Colombini, B., M. Nocella, G. Benelli, G. Cecchi, and M.A. Bagni. 2007. Crossbridge properties during force enhancement by slow stretching in single intact frog muscle fibres. *J Physiol* 585(Pt 2):607-615.

Constable, J.K., C.J. Barclay, and C.L. Gibbs. 1997. Energetics of lengthening in mouse and toad skeletal muscles. *J Physiol* 505(Pt 1):205-215.

Curtin, N.A., and R.C. Woledge. 1979. Chemical change, production of tension and energy following stretch of active muscle of frog. *J Physiol* 297:539-550.

Edman, K.A. 1999. The force bearing capacity of frog muscle fibres during stretch: its relation to sarcomere length and fibre width. *J Physiol* 519(Pt 2):515-526.

Edman, K.A., G. Elzinga, and M.I. Noble. 1978. Enhancement of mechanical performance by stretch during tetanic contractions of vertebrate skeletal muscle fibres. *J Physiol* 281:139-155.

Edman, K.A., and T. Tsuchiya. 1996. Strain of passive elements during force enhancement by stretch in frog muscle fibres. *J Physiol* 490(Pt 1):191-205.

Fenn, W.O. 1923. A quantitative comparison between the energy liberated and the work performed by the isolated sartorius muscle of the frog. *J Physiol* 58(2-3):175-203.

Fenn, W.O. 1924. The relation between the work performed and the energy liberated in muscular contraction. *J Physiol* 58(6):373-395.

Flitney, F.W., and D.G. Hirst. 1978. Cross-bridge detachment and sarcomere "give" during stretch of active frog's muscle. *J Physiol* 276:449-465.

Gasser, H.S., and A.V. Hill. 1924. The dynamics of muscular contraction. *Proc Royal Soc Lond B Biol Sci* 96:398-437.

Hatze, H. 1978. A general myocybernetic control model of skeletal muscle. *Biol Cybern* 28(3):143-157.

Hill, A.V., and J.V. Howarth. 1959. The reversal of chemical reactions in contracting muscle during an applied stretch. *Proc R Soc Lond B Biol Sci* 151:169-193.

Joumaa, V., T.R. Leonard, and W. Herzog. 2008. Residual force enhancement in myofibrils and sarcomeres. *Proc Biol Sci* 275(1641):1411-1419.

Joyce, G.C., P.M. Rack, and D.R. Westbury. 1969. The mechanical properties of cat soleus muscle during controlled lengthening and shortening movements. *J Physiol* 204(2):461-474.

Katz, B. 1939. The relation between force and speed in muscular contraction. *J Physiol* 96(1):45-64.

Krylow, A.M., and T.G. Sandercock. 1997. Dynamic force responses of muscle involving eccentric contraction. *J Biomech* 30(1):27-33.

Lewin, A., and J. Wyman. 1927. The viscous elastic properties of muscle. *Proc R Soc Lond B Biol Sci* 101:219-243.

Lin, D.C., and W.Z. Rymer. 1998. Damping in reflexively active and areflexive lengthening muscle evaluated with inertial loads. *J Neurophysiol* 80(6):3369-3372.

Linari, M., R.C. Woledge, and N.A. Curtin. 2003. Energy storage during stretch of active single fibres from frog skeletal muscle. *J Physiol* 548(Pt 2):461-474.

Morgan, D.L. 1994. An explanation for residual increased tension in striated muscle after stretch during contraction. *Exp Physiol* 79(5):831-838.

Morgan, D.L., U. Proske, and D. Warren. 1978. Measurements of muscle stiffness and the mechanism of elastic storage of energy in hopping kangaroos. *J Physiol* 282:253-261.

Muraoka, T., T. Muramatsu, T. Fukunaga, and H. Kanehisa. 2004. Influence of tendon slack on electromechanical delay in the human medial gastrocnemius in vivo. *J Appl Physiol* 96(2):540-544.

Otten, E. 1987. A myocybernetic model of the jaw system of the rat. *J Neurosci Meth* 21(2-4):287-302.

Pinniger, G.J., K.W. Ranatunga, and G.W. Offer. 2006. Crossbridge and non-crossbridge contributions to tension in lengthening rat muscle: force-induced reversal of the power stroke. *J Physiol* 573(Pt 3):627-643.

Rassier, D.E. 2009. Molecular basis of force development by skeletal muscles during and after stretch. *Mol Cell Biomech* 6(4):229-241.

Sandercock, T.G., and C.J. Heckman. 1997. Force from cat soleus muscle during imposed locomotor-like movements: experimental data versus Hill-type model predictions. *J Neurophysiol* 77(3):1538-1552.

Sugi, H. 1972. Tension changes during and after stretch in frog muscle fibres. *J Physiol* 225(1):237-253.

Sugi, H., and T. Tsuchiya. 1981. Enhancement of mechanical performance in frog muscle fibres after quick increases in load. *J Physiol* 319:239-252.

Tidball, J.G. 1986. Energy stored and dissipated in skeletal muscle basement membranes during sinusoidal oscillations. *Biophys J* 50(6):1127-1138.

Telley, I.A., R. Stehle, K.W. Ranatunga, G. Pfitzer, E. Stussi, and J. Denoth. 2006. Dynamic behaviour of half-sarcomeres during and after stretch in activated rabbit psoas myofibrils: sarcomere asymmetry but no "sarcomere popping." *J Physiol* 573(Pt 1):173-185.

van Atteveldt, H., and A. Crowe. 1980. Active tension changes in frog skeletal muscle during and after mechanical extension. *J Biomech* 13(4):323-331.

Winter, S.L, and J.H. Challis. 2008. Reconstruction of the human gastrocnemius force-length curve in vivo: part 1-model-based validation of method. *J Appl Biomech* 24(3):197-206.

Winters, J.M. 1990. Hill-based muscle models: A systems engineering perspective. In *Multiple Muscle Systems: Biomechanics and Movement Organization,* edited by J.M. Winters and S.Y. Woo. New York: Springer-Verlag, pp. 69-93.

Zajac, F.E. 1989. Muscle and tendon: properties, models, scaling, and application to biomechanics and motor control. *Crit Rev Biomed Eng* 17(4):359-411.

II

MUSCLES IN THE BODY

Part II addresses biomechanical aspects of muscle functioning in the body. Chapter 5 deals with muscle force transmission to bones and muscle force transformation to the moments of force at joints. The biomechanics of two-joint muscles are discussed in chapter 6. Eccentric muscle action and the stretch–shortening cycle (reversible muscle action) are addressed in chapter 7. Chapter 8 deals with biomechanical aspects of muscle coordination.

FROM MUSCLE FORCES TO JOINT MOMENTS

This chapter describes the transformation of force generated by a skeletal muscle to a moment of force at a joint and the way that mechanical properties of tendons and aponeuroses as well as geometry of joints and bones affect the transformation. As described previously (chapter 1 in *Kinetics of Human Motion*), a moment of force quantifies a turning effect of a force acting on a body; it depends on the force and the distance from the line of force action to a point or axis of rotation (the moment arm). Thus, transformation from muscle force to a moment of force depends on both muscle force and the position of the line of muscle force action with respect to the joint axis or center; the latter is affected by various factors.

Force generated by a muscle is transmitted to the bones through multiple mechanical pathways that include (1) elastic tissues in series with the muscle (aponeurosis and tendon), whose mechanical behavior modifies muscle force pattern; (2) so-called *soft tissue skeleton* (*ectoskeleton*, from the Greek *ecto-* or *ect-*, meaning "outside, external, beyond") that includes fascia and tendons with aponeuroses of neighboring muscles, which makes it possible to transmit generated force through tissues parallel to the muscle and causes the line of muscle force action to depend on generated forces; (3) bony prominences of different shapes over which muscle and tendon wrap around; and (4) tendon–bone junction (enthesis). Complexity of the muscle force–joint moment transformation comes also from the fact that the lines of action of many skeletal muscles pass multiple joints through several anatomical planes, thus creating moments of force with respect to two or more joints and axes of rotation at each joint.

The description of the muscle force–moment of force transformation starts with considering the sites where muscle meets its aponeurosis and tendon (muscle–tendon junction, section **5.1.1**) and then proceeds to bone–tendon attachments, section **5.1.2**. Section **5.2** describes structure and functions of the ectoskeleton in human movements. Several examples of how fascia in the human body contributes to movement are given. Muscle moment arms are considered in detail in section **5.3**. Specifically, muscle moment arm definitions are given in section **5.3.1** for muscles whose line of action is modeled as a vector or as a curved path. Methods of muscle moment arm measurements in humans are addressed in section **5.3.2**, including geometric and functional methods. Factors that affect the moment arm magnitude (i.e., joint angles, muscle force, and bone size) are discussed in section **5.3.3**. Finally, the muscle moment arm matrix, called the muscle Jacobian, which describes the transformation of forces of multiple muscles to joint moments, is derived and explained in section **5.3.4**.

5.1 FORCE TRANSMISSION: FROM MUSCLE TO BONE

Mechanical properties of tendons, aponeuroses, and passive muscles were considered in chapter 2. Here attention is paid to the mechanical interactions among these structures in active and passive conditions as well as to how the tendon and aponeurosis affect the functional relations, such as force–length and force–velocity of the muscle–tendon unit (MTU).

5.1.1 From Muscle to Tendon

Representative publications: Tidball 1991; Trotter 2002; Monti et al. 2003; Finni et al. 2003

The connection between a muscle and its tendon is called a *muscle–tendon junction* (MTJ). At the sites where muscle fibers meet collagen fibers, the force is mainly transmitted through shear stresses exerted on neighboring fibers. The mechanism is similar to the mechanism of force transmission between the neighboring nonspanning muscle fibers (see figure 1.12). Force transmission through the shear forces allows for decreasing the local stresses. Such a force transmission is possible because (a) the muscle fibers and the tendon fibers densely interweave with each other and (b) cell membranes at the MTJ are extensively folded in a complicated manner (figure 5.1). The strength of the MTJ, that is, its resistance to mechanical damage, is provided by the dense interweaving of the fibrillar component of the basal membrane about the collagen fibrils of the tendon.

The folding allows for increasing contact area between muscle and collagen fibers by approximately 10 to 20 times. The membrane folding and the corresponding increase in contact area between the muscle and tendon

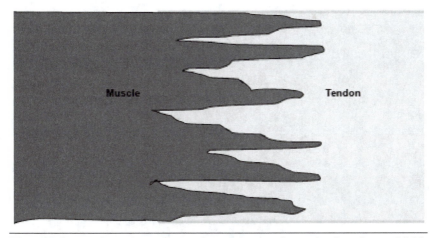

Figure 5.1 A schematic representation of the muscle–tendon junction and folding. Dark area on the left depicts a muscle fiber with characteristic invaginations at its end filled with extracellular matrix material and collagen fibers of the tendon (lighter area on the right).

fibers result in a corresponding stress reduction at the MTJ during muscle–tendon force transmission, which otherwise would be higher by an order of magnitude. Membrane folding also improves adhesion between muscle and collagen fibers at the MTJ.

The force-transmitting and stress-bearing functions of the MTJ are reflected in membrane-folding reduction following muscle disuse or injury. After an injury (e.g., an MTJ tear), the folding may not restore fully, increasing the risk of a secondary tear. The extent of membrane folding depends on the muscle fiber type. Slow-twitch fibers that are regularly recruited during everyday activities have approximately 2 times greater membrane folding than the fast-twitch fibers, which are used sporadically. This difference in the MTJ design between slow- and fast-twitch fibers is linked to the higher proportion of damaged fast-twitch fibers after eccentric exercise (see section **7.5.3**).

Even though the MTJs are highly specialized for load transmission, muscle strain injuries commonly occur near or at the MTJ (a manifestation of the rule of the weakest link). One of the most frequent injuries of this type is observed at the proximal and distal MTJs of hamstrings (figure 5.2). This injury often happens in the late swing of running when the hamstrings are acting eccentrically; that is, they are active, produce high force, and elongate because of large external forces (see chapters 4 and 7 for a discussion of eccentric muscle action).

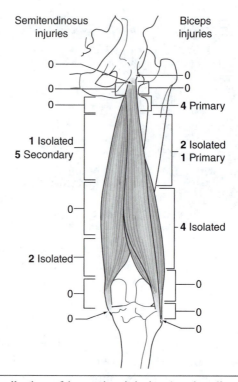

Figure 5.2 Distribution of hamstring injuries (semitendinosus and long head of biceps muscles) in a group of 15 athletes. The injuries (indicated by numbers) mainly occurred at or near muscle–tendon junctions. "Primary" and "Secondary" denote injury sites with the greatest and smaller abnormalities within a muscle group. "Isolated" denotes injury in a single muscle.

From A.A. De Smet and T.M. Best, 2000, "MR imaging of the distribution and location of acute hamstring injuries in athletes," *American Journal of Roentgenology (AJR)* 174(2): 393-399. Reprinted with permission from the *American Journal of Roentgenology.*

Strain-related injuries at the MTJ could result from differences in mechanical properties among the muscle, aponeurosis, and tendon. Although there are substantial discrepancies in reported values for strain and stiffness of these structures during standard muscle activities (see section **2.1.1.3** in chapter 2), it is likely that stiffness differences between active muscle and collagen fibers at the MTJ contribute to strain-related failures.

Border areas between muscle and its tendinous tissues undergo complex deformations during muscle activity. During isometric muscle efforts, muscle fibers shorten and stretch aponeurosis (internal tendon) and tendon as discussed in chapters 2 and 3. Deformation of tendinous tissues is not uniform over the whole tissue area. The aponeurosis is stretched more in some areas than in other locations and can even be shortened in certain

---

■ ■ ■ *FROM THE LITERATURE* ■ ■ ■

Strain Injuries Occur Near or at MTJ

Source: Tidball, J.G., G. Salem, and R. Zernicke. 1993. Site and mechanical conditions for failure of skeletal muscle in experimental strain injuries. *J Appl Physiol* 74:1280-1286

Using whole frog semitendinosus muscles with intact muscle–tendon and tendon–bone junctions, the authors investigated the rate and site of injury following imposed strain at physiological rates to failure to determine the site of lesion in both electrically stimulated and nonstimulated muscles. Using electron microscopic examination, the investigators found that all failures occurred at or near the proximal myotendinous junction in both stimulated and nonstimulated muscles.

---

areas. Such nonuniform behavior of tendinous tissues may result from a nonuniform recruitment of motor units, especially during submaximal activation. In addition, the orientation of muscle fibers along the aponeurosis may vary, creating different deforming forces in the aponeurosis plane even if all muscle fibers generate equal forces. Furthermore, properties of tendinous tissue (e.g., thickness) vary across regions, and thus different regions provide dissimilar resistance to deforming forces. These details of tendinous tissue behavior in vivo may explain the discrepancy in the literature regarding strain and stiffness values of aponeuroses and tendons (see section **2.1.1.3**).

5.1.2 From Tendon to Bone

Representative publications: Maganaris et al. 2004; Doschak and Zernicke 2005; Benjamin et al. 2006

Tendon–bone junction, the site of insertion of tendon or fascia into bone, is called *enthesis*. The insertions may be fibrous or fibrocartilaginous. Fibrous insertions are formed by dense fibrous connective tissue that links the tendon to the periosteum (membrane covering the outer surface of bone) and to bone through a calcified anchorage created by collagen fibers (Sharpey's fibers). Fibrocartilaginous insertions are those entheses in which cartilage formation has occurred. In these tendon–bone junctions, the tendon tissue gradually transforms from pure fibrous connective tissue to uncalcified fibrocartilage, then to calcified fibrocartilage, and then to bone.

Gradual continuous changes in tissue properties from compliant tendons to stiffer tendon–bone junctions and bones provide better conditions for force transmission and reduction of stress concentration. Figure 5.3 shows how, at similar stresses in different tendon regions, the strain gradually reduces from the MTJ, to midtendon, and then to the tendon–bone junction.

In many entheses, the tendons fan out at the attachment sites. This helps to distribute load over a larger bone area and thus contributes to stress reduction. The fanning out is often accompanied by the reorganization of tendon fibers into a network of fiber bundles that either provides a uniform load transmission across the entire enthesis or creates conditions for load transmission by different parts of the enthesis at different joint angles.

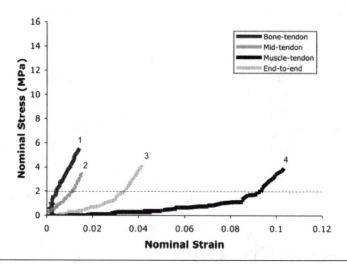

Figure 5.3 Regional differences in stress–strain properties of the tibialis anterior tendon unit of a rat tested in vitro. The tendon was stretched at a rate of 0.01/s, and the deformation and tendon force were recorded. Stress was obtained by dividing measured force by mean regional tendon cross-sectional area. The nominal strain was calculated as the ratio of tendon deformation and tendon slack length. Curve 1, tendon–bone junction region; curve 2, tendon midsection; curve 3, the entire tendon; 4, region close to muscle (muscle–tendon junction).

Reprinted, by permission, from E.M. Arruda, S. Calve, R.G. Dennis, K. Mundy, and K. Baar, 2006, "Regional variation of tibialis anterior tendon mechanics is lost following denervation," *Journal of Applied Physiology* 101(2): 1113-1117.

In the areas around the tendon attachment, the trabecular bone architecture corresponds to the direction of transmitted forces (*Wolff's law*; after German anatomist Julius Wolff, 1836-1902). Trabecular alignment in the direction of applied load ensures bone strength and safe force transmission in areas of high stress concentration. Trabecular bone architecture in the vicinity of the enthesis is highly adaptable. Muscle disuse decreases trabecular thickness, alignment, and number.

Another common feature of entheses is deposits of adipose tissue (fat) around the tendon–bone attachment sites. Fat deposits are typical between fascicles, where they fan out at the attachment sites, and under tendons, where they attach to the bone at an angle. The function of these fat deposits is thought to contribute to the increase in surface area at the tendon–bone junction, promote movement between tendon and bone, and dissipate stress. Nerve endings in the fat tissue can also play a mechanosensory role.

Tissue inflammation, tear, or rupture at or in the vicinity of enthesis (*insertional tendinopathy*) is a common pathological condition that occurs in tendons of many human muscles, especially in the supraspinatus (rotator cuff tendinopathy), common wrist extensor (lateral epicondylitis), quadriceps, and patellar tendon (jumper's knee). The cause of insertional tendinopathy is not clear. Traditionally, tendinopathies have been considered overuse injuries that result from excessive tendon loading. This may cause initial microinjury that progresses with time to clinically apparent symptoms (inflammation, pain) if healing processes cannot keep up with overuse-dependent tendon tear. Additional factors, such as unskilled task performance or movement errors, could contribute to increased loading of the tendon and tendinopathies.

Some studies of insertional tendinopathy do not support the traditional point of view concerning tendon overuse, because the pathological changes are primarily found not at the tendon itself but at the insertion site of the tendon to bone (at the enthesis) or very close to it. The pathological condition is often seen on the side of the enthesis that is subjected to lower tensile strain or even to compressive loads, which is inconsistent with the overuse hypothesis. According to an alternative explanation of insertional tendinopathy, the injuries might be related to adaptive cartilage-like or atrophic changes in parts of the enthesis that experience either compressive loads or a lack of tensile loads. Such tissue weakening may predispose the enthesis to injury when occasional high tensile loads are placed on it. These two hypotheses addressing insertional tendinopathy have been studied intensively, but no conclusions have been reached.

5.1.3 Tendon Elasticity and Isometric Force–Length Relation

Representative publications: Rack and Ross 1984; Lieber and Brown 1992; Delp and Zajac 1992; Loren and Lieber 1995; Alexander 2002; Arampatzis et al. 2005

The force generated by muscle fibers at a given level of muscle activation is a function of fiber length and velocity (see chapter 3) as well as history-dependent effects such as preceding shortening (see section **3.3.1**) or stretch (see chapters 4 and 7). If muscle fibers are considered together with their tendons, another important factor that affects force development is the elasticity of the muscle–tendon–bone connection. The force patterns recorded at the ends of an isolated MTU are different from force patterns generated by muscle fibers alone because of the series elastic component of MTU (see, e.g., figure 3.2). When muscles are activated in a fixed joint position (isometric action), muscle fascicles shorten and the in-series tendinous tissues are stretched, although not uniformly.

What effect does this series elasticity have on the basic force–length relation of a sarcomere and MTU? A short answer is that the effect is substantial. To consider the effect of series elasticity on the force–length relation in more detail, recall first the classic force–length relation for a sarcomere obtained by Gordon, Huxley, and Julian in 1966 (figure 5.4*a;* see also figure 3.19).

The relation in figure 5.4*a* was obtained on muscle fibers isolated from in-series connective tissues and with sarcomere length maintained at a constant length by a servo control system. Thus, sarcomeres for which the classic force–length relation was obtained by Gordon, Huxley, and Julian can be considered to be sarcomeres that are rigidly connected to external environment. Attaching the sarcomeres to the environment through serial elastic elements distorts the force–length relation (figure 5.4*b*). If the initial length of such a sarcomere in a passive condition is on the ascending limb, then postactivation force developed at the sarcomere ends will pull and stretch the in-series element, and thus the sarcomere will shorten until its force becomes the same as the force developed by the stretched series elastic component. At this point, because of the shorter sarcomere length, the steady-state isometric sarcomere force will be lower than the force developed by the same sarcomere activated at the same initial length but rigidly attached to environment (figure 5.4*b*). If the initial length of passive sarcomere in-series with elastic elements is on the descending limb of the force–length relation, then postactivation sarcomere shortening will increase the steady-state force. As a result of in-series elasticity, the force plateau will be shifted to longer sarcomere lengths (figure 5.4*b*). Note that the force values at lengths 1 and 5 are zero and do not change. This is because

A **B**

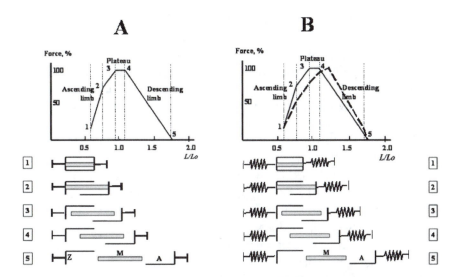

Figure 5.4 Force–length relation for a sarcomere and relative positions of myosin (M) and actin (A) myofilaments in a sarcomere (separated by Z-membrane) at specific sarcomere lengths 1 through 5 (vertical dotted lines; schematic). *(a)* Classic force–length relation for a sarcomere (cf. figure 3.19). This relation was obtained on individual myofibers isolated from series elastic tissues. This fact is reflected in the diagram—Z-membranes are rigidly fixed. As a result, the sarcomere length and overlap between myosin and actin filaments are the same in passive and fully activated fibers. *(b)* Force–length relation for a sarcomere attached to a series elastic element. Passive lengths of the sarcomere before activation (1-5) are the same as in part *(a)*. When the sarcomere is fully activated, the force developed at the Z-membranes (at lengths above 1 and below 5) pulls on the elastic element, which stretches the element and shortens the sarcomere until their forces become equal. Because of the sarcomere shortening, overlap between myosin and actin becomes different than that occurring with a sarcomere that has rigidly fixed ends, and the force–length relation is distorted (dashed line) from the classic force–length relation (solid line). On the ascending limb, the force developed in the sarcomere with series elasticity is less than the force predicted by the classic relation; on the descending limb this force is higher, and the peak forces occur at longer lengths.

at these lengths, the sarcomere is unable to develop force and therefore no sarcomere shortening can occur.

Now consider, instead of a single sarcomere or many sarcomeres in series, an MTU (i.e., a muscle with in-series tendinous tissues); its force–length relation will also change in a similar manner as described for a sarcomere. Specifically, for an MTU compared with the muscle belly, the force will be lower (higher) on the ascending (descending) limb and the peak force will be shifted to a longer MTU length.

There is another functionally important modification of the force–length relation that is present at the MTU level but not at the sarcomere level: elastic tendon causes the operational range of an MTU to increase. In other words, the range of MTU lengths at which the MTU can develop active force (lengths between the minimal and maximal muscle lengths, 1 and 5 in figure 5.4) increases. In contrast, an elastic tendon causes the slope of the ascending limb of the force–length curve to decrease in an MTU (compare continued and dashed lines in figure 5.4*b*); that is, the rate of MTU force development will be slower.

The extent to which the MTU force–length relation is modified by tendon elasticity depends on the ratio $L_{T,Slack} / L_o$ (described in sections **1.5.2** and **4.4.2**), where $L_{T,Slack}$ and L_o are tendon slack length and optimal fiber length, respectively. In general, the longer the tendon compared with muscle fiber, the more muscle fiber shortening and tendon stretch occur during isometric muscle action. In human leg and arm muscles, this ratio varies between ~0.2 (hip one-joint muscles) and 11 (soleus) (see table 1.4).

Generally, the $L_{T,Slack} / L_o$ ratios of proximally located MTUs are lower than those of distal MTUs. This allows for a larger portion of muscle mass to be located closer to proximal joints, thus reducing moments of inertia of the extremities (see section **1.5.2**). Relatively long tendons of the distal leg muscles are also advantageous for preventing excessive length changes of the relatively short muscle fibers, so they can operate at their optimal length ranges and low shortening velocities and thus maintain improved force production capabilities. The long tendons of the distal muscles of the human arm reduce force fluctuations due to MTU length changes, a phenomenon that may provide more reliable force feedback to the central nervous system during grip force control. However, length changes in MTUs with long tendons involve large tendon length changes, and this prevents the in-parallel muscle spindles from accurately detecting MTU length changes, which may in turn thwart reliable position control.

■ ■ ■ FROM THE LITERATURE ■ ■ ■

Control of Forces and Position at the Human Thumb

Source: Rack, P.M., and H.F. Ross. 1984. The tendon of flexor pollicis longus: Its effects on the muscular control of force and position at the human thumb. *J Physiol* 351:99-110

Two experiments were performed. In the first, fresh human flexor pollicis longus tendons obtained at autopsy were subjected to repeated sinusoidal stretching movements with amplitude less than 1% of tendon length. The recorded associated force changes were almost in phase with the imposed tendon movement (the difference was <4°). Changes in frequency of imposed tendon movement between 2 and 16 Hz did not change this effect, indicating low viscosity influence of the tendon. In the second experiment, the responses to sinusoidal movements of the thumb interphalangeal joints of normal subjects were examined by measuring exerted thumb forces and calculating tendon and muscle length changes based on the tendon material properties. During exertion of a steady force by the thumb, only a small portion of imposed movement reached the muscle fibers. It was suggested that the compliant tendons of thumb and finger muscles do not permit reliable position control (changes in muscle fiber length and in strain of the spindles do not correspond to angular joint displacement). However, the compliant tendons simplify the gripping force control by reducing external force fluctuations.

5.2 FORCE TRANSMISSION VIA SOFT TISSUE SKELETON (FASCIA)

Representative publications: Wood-Jones 1944; Garfin et al. 1981; Maas et al. 2005; Benjamin 2009

The term *soft-tissue skeleton* typically refers to fascia. The term *fascia* originates from Latin and means "band or bandage." Fascia is a soft connective tissue that penetrates the body and surrounds muscles, bones, nerves, blood vessels, and internal organs. Fascia has multiple functions in the body—it provides mechanical support, protection, and stability for internal tissues and organs; it serves as a sensory organ through innervated fascial proprioceptors; and it plays important roles in hemodynamics and postinjury infection defense. This section deals mostly with the mechanical functions of fascia relevant to body motion and force transmission between muscles and bones and among muscles within and between fascial compartments.

5.2.1 Structure of Fascia

Representative publications: *Terminologia Anatomica* 1998; Kjaer 2004; Stecco et al. 2008a

Many anatomists distinguish superficial fascia, which is located immediately under the skin, and deep fascia, which, as follows from its name, is located deeper in the body and surrounds the muscles, nerves, blood vessels, and bones. The superficial fascia represents a layer of interlacing, loosely organized collagenous and elastin fibers that also contains fat tissue and many blood vessels. The deep fascia is connected to the superficial fascia but is composed of dense fibrous connective tissue that is much tougher than the superficial fascia. The deep fascia of the limbs consists of two or three parallel layers of collagen bundles that form a sheath approximately 1 mm thick. The layers are separated from each other by loose connective tissue that allows the layers to slide over one another. This structure allows the deep fascia to be strongly resistant to traction in different directions.

The deep fascia gives rise and connects to sheets of connective tissues that envelope muscle groups (e.g., crural fascia, the deep fascia of the leg), individual muscles (epimysium), muscle fascicles (perimysium), individual muscle fibers (endomysium), and bones (periosteum). All these sheets of connective tissues are interconnected and form a continuous network that holds together muscles, muscle compartments, and individual muscle fibers (figure 5.5; see also figure 1.15). Tendons of many skeletal muscles attach to bones not only directly but also through fascia.

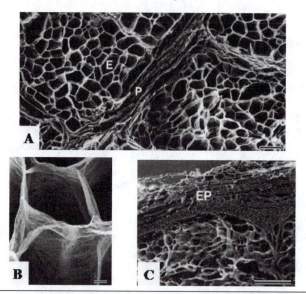

Figure 5.5 Structure of intramuscular connective tissue. Skeletal muscle (bovine semitendinosus muscle) extracellular network; electron micrographs after removal of muscle protein. *(a)* Endomysium (E) is a cylindrical sheath housing individual muscle fibers, and perimysium (P) is composed of several layers of collagen sheets, ×150. Horizontal bar = 100 μm. *(b)* Endomysial, membranous sheath. ×1,100. Horizontal bar = 10 μm. *(c)* Epimysium (EP) composed of two layers. ×70. Horizontal bar = 300 μm.

From T. Nishimura, A. Hattori, and K. Takahashi, 1994, "Ultrastructure of the intramuscular connective tissue in bovine skeletal muscle: A demonstration using the cell-maceration/scanning electron microscope method," *Acta Anatomica* 151:250-257. Reprinted with permission of Karger AG, Basel.

The structure of connective tissue networks suggests that forces generated by individual muscle fibers, fascicles, and single muscles are distributed within and between neighboring fibers, fascicles, and muscles (the lateral, or myofascial, force transmission, see sections **1.4** and **3.1.5**). Although significance of the contribution of the lateral force transmission in intact musculoskeletal systems is still debated, the transmission is substantial during excessive muscle stretching as well as in some clinical situations, such as a tendon transfer.

5.2.2 Muscle–Tendon–Fascia Attachments

Representative publications: Stecco et al. 2008b; Benjamin 2009

It is not uncommon that muscles and tendons in the vicinity of entheses attach to fascial expansions that link them to neighboring structures including bones and other muscles. Fascial expansions reduce stress concentration by providing an increased area of force transmission. Another functional consequence of muscle and tendon attachments to fascia is that tendon forces at both ends of the muscle could be unequal (because a fraction of the generated muscle force is transmitted laterally through fascia). A further consequence of muscle and tendon attachments to fascial expansions is that position of the line of muscle force action near the joint is dependent on the magnitude of force developed by the muscle (see section **5.3**).

Examples of such fascial extensions are numerous. The bicipital aponeurosis originating from the tendon of the short head of biceps brachii surrounds the flexor muscles of forearm and merges with the antebrachial deep fascia. The quadriceps tendon gives origin to a fascial expansion connected to the upper pole of the patella, and its fascial sheet passes anterior to the patella and contributes fibers to the patella tendon. The Achilles tendon is attached not only to the calcaneus but also, via fascial extensions, to the plantar aponeurosis and fibrous septa of the heel pad.

▪ ▪ ▪ *FROM THE LITERATURE* ▪ ▪ ▪

Pectoral Girdle Muscles Are Connected to the Brachial Fascia Through Fascial Expansions

Source: Stecco, C., A. Porzionato, V. Macchi, A, Stecco, E. Vigato, A. Parenti, V. Delmas, R. Aldegheri, and R. De Caro. 2008. The expansions of the pectoral girdle muscles onto the brachial fascia: morphological aspects and spatial disposition. *Cells Tissues Organs* 188:320-329

(continued)

From the Literature   *(continued)*

The authors investigated fascial expansions of the pectoral girdle muscles (i.e., pectoralis major, latissimus dorsi, and deltoid) and their possible connections to the brachial fascia. Thirty shoulder specimens from 15 unembalmed adult cadavers were studied using dissections. In addition, in vivo radiological studies were performed in 20 subjects using magnetic resonance imaging. It was found that the clavicular part of the pectoralis major muscle sent a fibrous expansion onto the anterior portion of the brachial fascia, its costal part onto the medial portion and medial intermuscular septum. The latissimus dorsi muscle showed a triangular fibrous expansion onto the posterior portion of the brachial fascia. The posterior part of the deltoid muscle inserted muscular fibers directly onto the posterior portion of the brachial fascia, its lateral part onto the lateral portion and the lateral intermuscular septum.

Retinacula are regional specializations of the deep fascia that envelopes muscles and tendons close to joints. Retinacula prevent tendons from bowstringing and permit low friction sliding of tendons. Examples of retinacula are the flexor and extensor retinacula of the hand and the flexor retinaculum of foot.

5.2.3 Fascia as Soft Tissue Skeleton (Ectoskeleton)

Representative publications: Wood-Jones 1944; Huijing et al. 2007

Widespread connections of deep fascia to muscles and tendons led British anatomist Frederic Wood-Jones (1879-1954) to coin the term *ectoskeleton* (soft tissue skeleton). He was especially intrigued by the fact that such powerful muscles as gluteus maximus and tensor fascia latae attach predominantly to deep fascia rather than bone. The concept of the ectoskeleton is supported by experimental studies demonstrating substantial changes in myofascial force transmission when certain fascial connections to individual muscles or fascia itself are compromised.

■ ■ ■ *FROM THE LITERATURE* ■ ■ ■

Fasciotomy of the Anterolateral Compartment Reduces Exerted Muscle Force by 10% to 16% and Muscle Pressure by 50%

Source: Garfin, S.R., C.M. Tipton, S.J. Mubarak, S.L. Woo, A.R. Hargens, and W.H. Akeson. 1981. Role of fascia in maintenance of muscle tension and pressure. *J Appl Physiol* 51(2):317-320

The goal of the study was to determine whether fasciotomy of the anterolateral compartments in an animal model affects muscle tension (measured by a force transducer attached to the tendon) and interstitial fluid pressure (measured by catheters in the muscle belly). Muscle tension and pressure were recorded in the tibialis cranialis muscle after low- and high-frequency stimulation of the peroneal nerve to produce twitch- and tetanic-type contractions. Fasciotomy decreased muscle force during the low-frequency stimulation by 16% (35.3 ± 4.9 to 28.4 ± 3.9 N) and during the high-frequency stimulation by 10% (60.8 ±4.9 to 54.8 ± 3.9 N). Muscle pressure decreased 50% after fasciotomy under both conditions, 15 ± 2 to 6 ± 1 mmHg and 84 ± 17 to 41 ± 8 mmHg, respectively. The authors concluded that that fascia is important in the development of muscle tension and changes in interstitial pressure.

Fascia plays important functional roles during postural support and movement. Let us consider several examples.

5.2.3.1 Plantar Fascia and the Windlass Mechanism

Representative publications: Ker et al. 1987; Benjamin 2009

The plantar fascia spans the foot from the calcaneal tuberosity to the distal aspects of the metatarsophalangeal joints, where it divides into five strips connecting to each toe. The fascia consists of mostly collagen fibers oriented from the heel to the metatarsal heads. It supports the arch of the foot during standing, landing, and the stance phase of gait.

During the support period in walking, before and immediately after midstance, the fascia is stretched, reaching around 10% of elongation at the beginning of toe-off. This stretching results from both Achilles tendon force and foot flattening. During this period, the fascia behaves as an elastic band assisting in energy conservation. However, its function changes during the late stance, when it plays a major role in the so-called *windlass mechanism*. The term comes from the analogy between the function of the plantar fascia and a cable being wound around the drum of a windlass. Windlasses were used for centuries to lift heavy weights, for instance, to raise a bucket of water from a well. In the foot, the role of the cable is played by a plantar fascia that is wound around the heads of the metatarsals.

As it was mentioned previously, the fascia is attached anterior to the proximal phalanges and posterior to the calcaneal tuberosity. The toes pull the fascia forward when they are dorsiflexed, as in late stance. This shortens the distance between the calcaneus and the metatarsal heads and forces the longitudinal foot arch to rise (figure 5.6). Because the toe extension can be due to body weight (e.g., during toe standing), the arch rise may not require direct action of muscle forces. The forces in the plantar fascia wound around the metatarsal heads may act not only along the fascia but also on the heads, toward the center of the curvature. The fascia is essentially a wraparound tendon. Mechanics of such tendons are analyzed in chapter 2, section **2.1.3.4.**

There is evidence that the plantar fascia experiences stretching and loading during early stance of walking. This early preloading of plantar fascia could improve conditions for the windlass mechanism.

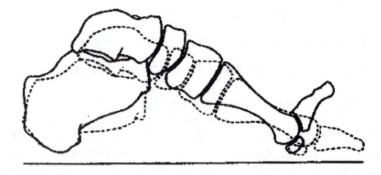

Figure 5.6   The windlass mechanism. Because of the action of the plantar fascia, the toes' dorsiflexion (extension) forces the foot arch to rise.

■ ■ ■ **FROM THE LITERATURE** ■ ■ ■

Contribution of the Windlass Mechanism to Gait Mechanics

Source: Caravaggi, P., T. Pataky, J.Y. Goulermas, R. Savage, and R. Crompton. 2009. A dynamic model of the windlass mechanism of the foot: Evidence for early stance phase preloading of the plantar aponeurosis. *J Exp Biol* 212(Pt 15):2491-2499

The authors estimated the temporal elongation of the plantar aponeurosis during normal walking using a subject-specific multisegment rigid-body model of the foot. Reflective markers were attached to bony landmarks to track the kinematics of the calcaneus, metatarsus, and toes during barefoot walking. Ultrasonography measurements were performed on three subjects to determine (a) the location of the origin of the plantar aponeurosis on the plantar aspect of the calcaneus and (b) the radii of the metatarsal heads. The results show that the plantar aponeurosis experienced tension significantly above rest during early stance phase in all subjects ($p < .01$), thus providing support for a hypothesis that the plantar aponeurosis is actively preloaded prior to forefoot contact. The mean maximum tension exerted by the plantar aponeurosis was 1.5 time body weight over the three subjects.

Plantar fascia assists in storing and release of strain energy during landings and subsequent takeoffs. For instance, total energy storage by the foot for a 79 kg man running at 4.5 m/s was estimated at 17 J. Removal of plantar fascia reduced the energy recoil by approximately 15%.

5.2.3.2 Fascia Lata and Iliotibial Tract

Representative publications: Evans 1979; Benjamin 2009

The fascia lata is a deep fascia of the thigh that forms an ectoskeleton for thigh muscles. It is thicker on the lateral aspect of the leg. This thickening is called the *iliotibial tract*. The iliotibial tract is also a tendon of tensor fascia latae muscle and gluteus maximus (figure 5.7).

The iliotibial tract is stretched in a standing person. It is shortest when the knee is fully flexed and the hip maximally abducted. Because of its location with respect to the hip and knee joints and its tensile strength, the iliotibial tract contributes to a passive abduction moment with respect to the hip and knee during activities like walking and standing. The contribution of this passive moment to the total hip abduction moment during standing is estimated at about 40%, and the contribution is much higher when the pelvis is tilted in the frontal plane, thereby increasing iliotibial tract stretch. At the knee the iliotibial tract maintains stability of the lateral compartment of the joint and contributes to an extension moment at the knee.

Examine the supporting function of fascia lata and the iliotibial tract on yourself. Stand and shift your weight to one leg such that the pelvis drops on the non–weight-bearing side. Try to relax muscles on the weight-bearing leg and palpate the lateral aspect of your thigh by the ipsilateral hand. This tension is caused by fascia stretch in this leg position, and if you are able to completely relax your leg muscles, the body support in the frontal plane will be provided mostly by the fascia.

Figure 5.7 The iliotibial tract.

5.3 MUSCLE MOMENT ARMS

This section introduces the notion of a *muscle moment arm*, the geometric characteristic of a muscle that determines the transformation from a muscle force to a muscle moment of force. Essentially, the moment arm is a coefficient in the following equation: *Moment of Force = Force × Moment Arm*. The basic mechanics of moments (described in section **1.1.1.2** of *Kinetics of Human Motion*) is valid but measurement of muscle moment arms is nontrivial, requiring sophisticated techniques.

We start by defining muscle moment arm vectors (section **5.3.1**) and then discuss methods used to determine muscle moment arms (section **5.3.2**). Finally, we consider factors that affect the magnitude of muscle moment arms in human joints (section **5.3.3**).

5.3.1 Muscle Moment Arm Vectors and Their Components

Representative publications: Dostal and Andrews 1981; Wood et al. 1989a, 1989b; Pandy 1999

5.3.1.1 Moment Arms as Vectors

As defined in section **1.1.1.2** of *Kinetics of Human Motion* (equation 1.8), the moment of force about a point O is a vector $\mathbf{M}_O$ obtained as a cross-product of vectors $\mathbf{r}$ and $\mathbf{F}$, where $\mathbf{r}$ is the position vector from O to any point along the line of force $\mathbf{F}$:

$$\mathbf{M}_O = \mathbf{r} \times \mathbf{F}. \quad [5.1]$$

The magnitude of moment is $M_O = F(r\sin\theta) = Fd$, where θ is the angle between vectors $\mathbf{r}$ and $\mathbf{F}$ and d is the moment arm, the shortest distance from the O to the line of action of $\mathbf{F}$. Equation 5.1 can be represented in a slightly different form, as a product of the muscle force magnitude and *the moment arm vector* $\mathbf{d}$:

$$\mathbf{M}_O = \mathbf{r} \times \mathbf{F} = \mathbf{r} \times F\mathbf{u} = F(\mathbf{r} \times \mathbf{u}) = F\mathbf{d}, \quad [5.2]$$

where $\mathbf{u}$ is a unit vector along the line of force action (figure 5.8). The *moment arm vector* $\mathbf{d}$ equals the moment per unit of force. The moment arm vector is also called the *normalized moment vector*. Note that in equation 5.2, muscle force $\mathbf{F}$ acts on a bone forming the joint.

When defined for a muscle, the cross-product

$$\mathbf{d} = (\mathbf{r} \times \mathbf{u}) \quad [5.3]$$

is the muscle moment arm vector. The magnitude d of the muscle moment arm vector $\mathbf{d}$ is equal to the perpendicular distance from the joint center (point O) to the line of muscle force action, or the shortest distance between the muscle force line of action and point O (figure 5.8), that is, $r\sin\theta$, where r is the magnitude of vector $\mathbf{r}$ and θ is the angle between vectors $\mathbf{r}$ and $\mathbf{u}$.

A concept of the moment arm vector is not generally used in classical mechanics, where moment arms are commonly treated not as vectors but as line segments. The concept was introduced in biomechanics because denoting a moment of muscle force as a product of the force magnitude F and a vector $\mathbf{d}$ is a convenient notation that represents the essence of muscle moment production. The moment arm vector $\mathbf{d}$ depends on human anatomy and for a given joint position cannot be voluntarily changed.

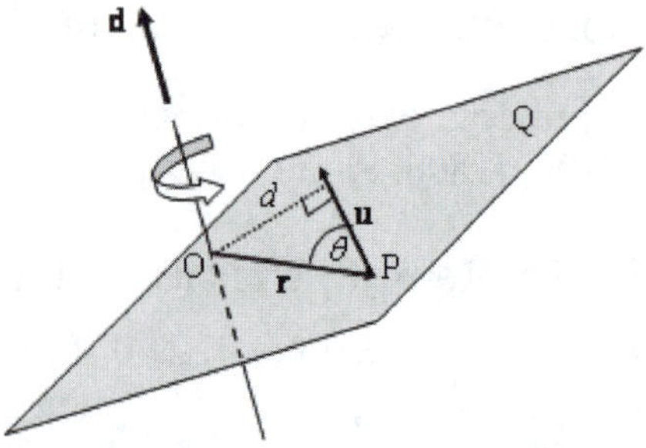

The direction of the muscle moment arm vector is the same as the direction of the moment of force $\mathbf{M}_O$ and can be defined by the right-hand rule—when the right-hand fingers are curling from vector $\mathbf{r}$ to vector $\mathbf{F}$, the extended thumb shows the direction of the moment of force vector and the muscle moment arm vector. The vector is along the moment axis; if $\mathbf{F}$ were the only force acting on the body, the body would rotate about this axis. Note (this is important!) that by definition the muscle moment arm vector $\mathbf{d}$ is not along the moment arm d shown in figure 5.8. As follows from equation 5.2, if a moment arm vector $\mathbf{d}$ is known, the moment of force can be found by multiplying $\mathbf{d}$ by the muscle force magnitude F. As already mentioned, the magnitude of the moment arm vector is numerically equal to the moment of force produced by a force of unit magnitude.

As an example, consider the humerus and scapula, which form a two-link kinematic chain con-nected at the glenohumeral joint

Figure 5.8 The definition of the muscle moment arm vector $\mathbf{d}$ (the moment per unit of muscle force). Muscle force $\mathbf{F}$ (not shown in the figure) is applied at point P and acts along the line prescribed by unit vector $\mathbf{u}$. The moment arm vector is cal-culated as $\mathbf{d} = \mathbf{r} \times \mathbf{u}$, where $\mathbf{r}$ is a vector from the joint center O to any point along the line of muscle force action. The direc-tion of $\mathbf{d}$ is orthogonal to the plane Q in which vectors $\mathbf{r}$ and $\mathbf{u}$ are located. d is the muscle moment arm, that is, the perpen-dicular distance from the joint center O to the line of muscle force action. d is a line segment (not a vector). Its length equals the magnitude of $\mathbf{d}$. Vector $\mathbf{d}$ and line seg-ment d are orthogonal. The vector $\mathbf{d}$ is along the moment axis, the same axis as the vector of moment of force $\mathbf{F}$. The axis is not specified in advance; its orientation is the consequence of the $\mathbf{r}$ and $\mathbf{u}$ spatial arrangements. Compare this figure with figure 1.3 in *Kinetics of Human Motion*.

and to which several muscles attach including teres major (figure 5.9). We shall assume that a muscle path can be represented by a straight line (such models are described in section **1.6.1**), and we shall use the classic methods of vector analysis.

In the reference system *XYZ* specified by the unit vectors $\mathbf{i}$, $\mathbf{j}$, and $\mathbf{k}$, the coordinates of the points *A* and *B* are (A_x, A_y, A_z) and (B_x, B_y, B_z), respectively. To avoid clutter, point coordinates and unit vectors are not depicted in the figure. The vectors $\mathbf{r}_A$ and $\mathbf{r}_B$ are then equal:

$$\mathbf{r}_A = A_x\mathbf{i} + A_y\mathbf{j} + A_z\mathbf{k}$$

$$\mathbf{r}_B = B_x\mathbf{i} + B_y\mathbf{j} + B_z\mathbf{k}. \qquad [5.4]$$

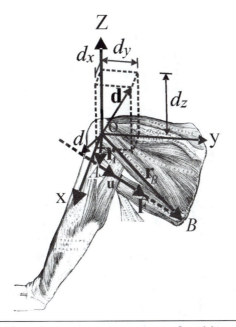

Figure 5.9 Definition of the moment arm vector **d** and its scalar components d_x, d_y, and d_z. The glenohumeral joint is formed by the humerus and scapula and shown with several muscles crossing the joint including the teres major muscle (TM). The origin of the orthogonal coordinate system O is located at the joint center, and the coordinate axes X, Y, and Z correspond to anatomical axes of rotation at the joint. The long dashed line passing through the midpoints of origin and insertion of the muscle is a linear approximation of the line of action of the TM muscle. The direction of the line is defined by the unit vector **u** (equation 5.2). The TM force vector is **F** = F**u**. The position vectors $\mathbf{r}_A$ and $\mathbf{r}_B$ are directed from the joint center O to the midpoint of TM insertion area (the point of TM force application) and to the midpoint of TM origin, respectively. **d** is the muscle moment arm vector; d_x, d_y, and d_z are its scalar components that specify the relative turning effects of a unit TM force on the humerus with respect to the corresponding axes of rotation at the joint. Note that d (the perpendicular distance from the line of muscle force action to joint center O) is a scalar.

The muscle force vector **F** equals F**u,** where F is the magnitude of **F** and **u** is the unit vector. Vector **u** specifying direction from A to B can be found from the difference between the vectors $\mathbf{r}_B$ and $\mathbf{r}_A$:

$$\mathbf{u} = \frac{\mathbf{r}_B - \mathbf{r}_A}{\left|\mathbf{r}_B - \mathbf{r}_A\right|},$$
[5.5]

where the vertical lines in the denominator signify vector magnitude. If B and A are the points of muscle origin and insertion and the straight line model of muscle path is accepted, vector **u** specifies the line of force action of the muscle of interest. Because r is known (r corresponds to r_A in figure 5.9), equation 5.3 is used to determine d.

Assuming that the force and the speed of muscle shortening do not change, increasing the muscle moment arm (e.g., because of a tendon transfer) will increase the joint moment and decrease the joint angular velocity. The proof of this statement is left to the reader.

5.3.1.2 Muscle Moment Arms About Rotation Axes

Representative publications: Rugg et al. 1990; Nakajima et al. 1999; Ackland et al. 2008

Equations 5.1 and 5.2 define the moment of muscle force about a point, for example, a joint center (if such a center exists; see chapter 4 on joint geometry and joint kinematics in *Kinematics of Human Motion*). The moments represent the tendency to cause rotation, and rotations are performed about axes. Therefore, in practice the topic of interest is the moment of muscle force about a predefined rotation axis. This axis (axes) can be the same as the axis of the muscle moment or can be different. In the former case, (a) the task is planar, (b) the bone movement is in the plane of muscle line of action, (c) the axis of rotation is orthogonal to the muscle line of action, and (d) the muscle moment arm vector is along the joint axis of rotation. The moment magnitude equals *Fd*, where the symbols are explained previously.

In three dimensions a muscle could generate a moment about an axis that is different from the axis (axes) of interest (e.g., the flexion–extension axis or an instantaneous rotation axis). The moment of force M_{oo} about an axis $O\text{-}O$ equals a component of moment M_O along this axis (the mechanics are explained in section **1.1.1.2.2** in *Kinetics of Human Motion*). Note that M_{oo} is a scalar whereas M_O is a vector; it has both magnitude and direction. The same is valid for the muscle moment arm vector d and the moment arm about a predefined axis, for example, moment arm d_x about axis X. Because the moment M_O and the vector d are along the same axis, the results of the vector's decomposition will be similar.

The muscle moment vector can be defined for the instantaneous joint axis of rotation. According to Chasles' theorem (see section **1.2.5.3** in *Kinematics of Human Motion*), any rigid-body motion (e.g., a motion of bone A with respect to bone B at a joint) can be obtained as the rotation around an axis (e.g., instantaneous axis of bone rotation), called the *screw* or *helical axis*, and a translation parallel to this axis. To determine the turning effect of the muscle force applied to the bone with respect to the instantaneous axis of bone rotation, the translation of the bone along that axis typically is not considered. The turning effect of muscle force on the bone with respect to the axis around which the bone instantly rotates is characterized by the moment of force about the axis:

$$M_{oo} = \mathbf{u}_{oo} \cdot Fd = \mathbf{u}_{oo} \cdot (\mathbf{r} \times \mathbf{F}) = \mathbf{u}_{oo} \cdot (\mathbf{r} \times F\mathbf{u}), \qquad [5.6]$$

where M_{oo} is the moment of muscle force about the instantaneous axis of bone rotation specified by the unit vector $\mathbf{u}_{oo}$; $\mathbf{F}$ is the vector of muscle force; F is the magnitude of the vector of muscle force applied to the bone whose direction is defined by the unit vector $\mathbf{u}$; and $\mathbf{r}$ is the position vector from any point on the axis of rotation to any point on the line of muscle force action. Equation 5.6 is a mixed triple product of the three vectors: $\mathbf{u}_{oo}$, $\mathbf{r}$, and $\mathbf{F}$ (this is essentially equation 1.23 from *Kinetics of Human Motion*).

The projection d_{oo} of muscle moment arm vector on axis *O-O* can be defined as the moment per unit of muscle force that causes rotation of the bone about the instantaneous axis of rotation:

$$d_{oo} = \frac{M_{oo}}{F} = \mathbf{u}_{oo} \cdot (\mathbf{r} \times \mathbf{u}) = \mathbf{u}_{oo} \cdot \mathbf{d}. \qquad [5.7]$$

As can be seen from equation 5.7, the magnitude of the muscle moment arm about axis *O-O* is equal to the projection of the vector $\mathbf{d} = \mathbf{r} \times \mathbf{u}$ onto the instantaneous axis of rotation $\mathbf{u}_{oo}$. The magnitude of the moment arm also equals the perpendicular, or shortest, distance d between the axis of rotation, specified by $\mathbf{u}_{oo}$, and the line of muscle force action, specified by $\mathbf{u}$, multiplied by the sine of the angle ϕ between the two lines:

$$d \sin \phi = \left| \mathbf{u}_{oo} \cdot (\mathbf{r} \times \mathbf{u}) \right|. \qquad [5.8]$$

When the line of muscle force action is along the axis of rotation, $\sin\phi = 0$ and the force does not produce any moment about this axis. When $\mathbf{u}_{oo}$ and $\mathbf{u}$ are at a right angle to each other, $\sin\phi = 1$ and the moment is maximal.

At the joints, three situations can be discerned: (a) the orientation of the instantaneous axis is changed during a joint motion but the axis still intersects the same center, a fixed joint center; bone B rotates but does not translate about bone A; (b) bone B moves—rotates and translates—in a plane, the instant center of rotation migrates, but the orientation of the instant axis of rotation does not change; and (c) both rotation and translation occur in three dimensions.

■ ■ ■ *FROM THE* L*ITERATURE* ■ ■ ■

Moment Arms of the Achilles Tendon and Tibialis Anterior Are Approximately Similar for the Fixed and Instant Centers of Rotation

Source: Rugg, S.G., R.J. Gregor, B.R. Mandelbaum, and L. Chiu. 1990. *In vivo* moment arm calculations at the ankle using magnetic resonance imaging (MRI). *J Biomech* 23(5):495-501

In this study the authors measured in vivo moment arms of the Achilles tendon and tibialis anterior muscles in 10 adult male subjects using a series of sagittal plane magnetic resonance images. The moment arm was defined as the shortest distance between the joint center of rotation and the center of the muscle's tendon. The position of the center of rotation for a given joint angle was determined using a modification of the graphic method of Reuleaux. The assumption of the fixed center of rotation at the ankle increases the measured moment arm of the Achilles tendon by 3.1% and the tibialis anterior by 2.5%. For both fixed and instantaneous centers of rotation, moment arms increased by approximately 20% for the Achilles tendon and decreased by approximately 30% for the tibialis anterior when the ankle moved from maximum dorsiflexion to maximum plantar flexion. The authors concluded that the averaged moment arm lengths for the Achilles tendon and the tibialis anterior were relatively unaffected by the use of a fixed versus moving center of rotation.

5.3.1.3 Muscle Moment Arms About Anatomical Axes: Muscle Functions at a Joint

Representative publications: Youm et al. 1976; Dostal et al. 1986

Determining vector **d** components along the anatomical axes (i.e., the flexion–extension, abduction–adduction, and internal–external rotation axes) is done using the same equations as for the helical axes, equations 5.4 to 5.6; this allows us to classify the muscles according to their functions, for example, as flexors or extensors. Strictly speaking, any muscle generates a moment of force at a joint about only one axis. So, it can be said that a muscle has only one function. The muscle moment axis, however, does not coincide typically with anatomical axes, the axes that are selected by anatomists to describe joint movements. Hence, the muscle functions are the consequence of the reference axis selection. If the anatomical axes were defined differently, the functions would change. In what follows, however, we neglect this consideration and speak about the muscle functions defined with respect to the joint anatomical axes.

Muscle functions are routinely described in anatomy textbooks. For instance, the functions of the biceps brachii muscle include elbow flexion, forearm supination, shoulder flexion assistance, abduction, and joint stabilization. The functions vary with joint posture; for example, the biceps works as a supinator only when the elbow is in some measure flexed. Although qualitative description of the muscle functions is relatively simple, the quantitative explication is far from trivial.

A basic mechanics procedure to quantify the functional role of a muscle at a joint involves two steps: (1) determining the moment arm vector of a muscle about a joint center and (2) decomposing the vector into three components parallel to the joint axes. For a joint with orthogonal axes of rotation intersecting at one center, the muscle moment arm vector **d** can be represented as the sum of its three vector components:

$$\mathbf{d} = d_x\mathbf{i} + d_y\mathbf{j} + d_z\mathbf{k}, \qquad [5.9]$$

where d_x, d_y, and d_z are the scalar components of the moment arm vector. The muscle moment arm vector components $d_x\mathbf{i}$, $d_y\mathbf{j}$, and $d_z\mathbf{k}$ specify the relative turning effects of the muscle force applied to the bone with respect to the corresponding axes of rotation *X*, *Y*, and *Z* intersecting at point *O* (the entire technique is described in section **1.1.1.2.2** in *Kinetics of Human Motion*). This procedure should be performed for particular joint configurations.

If a high degree of accuracy is required, the practical realization of this approach is not easy. Determining location of the joint center, the moment arm vector, and position of the joint axes is not a trivial task and is the subject of intense investigation. Note that (a) the assumption about the existence of the joint center is not always valid: the axes of joint rotation may not intersect at one point (see chapters 4 and 5 in *Kinematics of Human Motion*), and (b) the axes—for instance, the axes determined following the joint rotation convention (see section **2.1.5** in *Kinematics of Human Motion*)—may not be orthogonal and hence represent an oblique coordinate system.

A convenient way to visualize the moment arm components is to plot them in projections on a plane, as is done in figure 5.10. Such plots allow for easy understanding of the muscle functions at the joints. The values of the moment arm components that can be inferred from such plots are only accurate when the lines of muscle force action are orthogonal to the plane of projection.

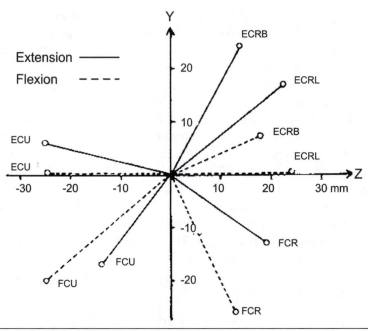

Figure 5.10 Moment arms of some muscles serving the wrist joint. Projections are on the transversal plane running perpendicular to the long axis of the forearm through the joint center. The center of rotation of the wrist joint was determined to be at the head of the capitate. The data were obtained on a cadaver at the maximum extension (solid lines) and maximum flexion (dashed lines) of the hand. A representative example: five muscles of the 24 wrist movers are shown. ECRL, extensor carpi radialis longus; ECRB, extensor carpi radialis brevis; ECU, extensor carpi ulnaris; FCU, flexor carpi ulnaris; and FCR, flexor carpi radialis. Coordinate axes: Y, extension–flexion; Z, abduction–adduction.

Adapted from Y. Youm, D.C.R. Ireland, B.L. Sprague, and A.E. Flatt, 1976, Moment arm analysis of the prime wrist movers. In *Biomechanics V-A: International Series on Biomechanics,* edited by P. Komi (Baltimore: University Park Press), 355-365.

The figure illustrates the moment arm components for the main muscles acting at the wrist joint. The muscles exert moments of force in the same angular direction as their moment arms. Note that there is not a single muscle that is a pure wrist flexor or extensor; all the muscles contribute to the abduction–adduction moments (e.g., the flexor carpi ulnaris flexes and adducts the hand, whereas the flexor carpi radialis flexes and abducts the hand). Also note that moment arms of all the muscles are different at flexed and extended postures. Similar findings are typical for other joints with 2 and 3 degrees of freedom.

Depending on the relative magnitude of each moment arm component, the dominant, intermediate, and less significant muscle functions at the joint can be identified. For example, in anatomical position of the femur and pelvis (hip flexion angle is 0°), the longest moment arm component of the anterior portion of the gluteus medius lies along the abduction–adduction axis (about 7 cm), the second largest component lies in the direction of internal–external rotation (about 2 cm), and the smallest component is directed along the flexion–extension axis (1 cm, figure 5.11). Thus, in this particular joint position the primary function of the anterior portion of the gluteus medius is abduction of the femur, its secondary function is internal rotation, and its minor function is extension. If hip flexion angle increases to 90°, the muscle no longer contributes significantly to femur rotation with respect to the abduction–adduction axis; its primary function becomes internal rotation (7 cm moment arm component) and its secondary function is femur flexion (2 cm; figure 5.11). This example demonstrates that muscle function can change when joint position changes. More formally, the magnitude of the muscle moment arm vector and its components can depend on the orientation of the bones to which the muscle is attached and, thus, on the joint angles.

Agonists and antagonists. The agonist (from Latin *agoniste—* "competitor") is a muscle responsible for producing a specific joint motion. The antagonist muscle opposes an agonist. Agonist and antagonist muscles are found in pairs, called *antagonistic pairs*; for example, biceps brachii and triceps brachii are antagonists for the elbow flexion–extension. In the research literature the terms *agonist* and *antagonist* are also commonly used with a broader meaning, as modifiers indicating whether the muscle either assists (agonist) or resists (antagonist) something (such as another muscle, joint torque production, or task performance).

In anatomy textbooks, agonist muscles are commonly defined in accordance with the basic meaning of the term: as the muscles that lie on the same side of a bone, act on the same joint, and move it in the same direction. The antagonist muscles act on the same joint but move it in opposite directions. According to this approach, the terms describe the functions of the muscles that are completely defined by their anatomy (e.g., muscle A is always an anatomical agonist–antagonist to muscle B). In joints with several degrees of freedom, muscles commonly generate moments of force with respect to more than one anatomical axis; see as an example figure 5.10. In such cases, some pairs of muscles can act as agonists in one direction and as antagonists in another direction. For instance, the flexor carpi ulnaris and flexor carpi radialis are agonists in wrist flexion and are antagonists in abduction–adduction; anterior and

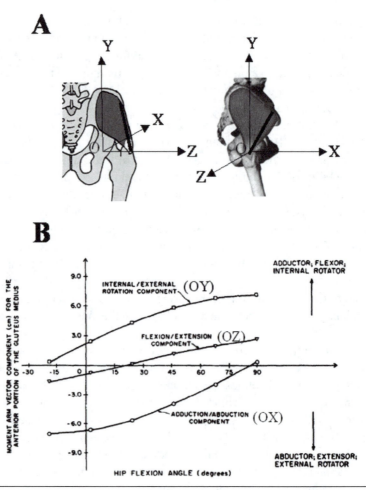

Figure 5.11 Dependence of the anterior portion of the gluteus medius moment arm vector and its components on hip flexion angle. *(a)* Diagram illustrating the anterior portion of the gluteus medius approximated by a straight line connecting the midpoints of insertion and origin of this portion of the muscle on the greater trochanter and ilium (thick black line). Left, rear view; right, lateral view. The origin of the coordinate system OXYZ is at the femoral head center. The axes *X*, *Y*, and Z correspond to femur rotations in adduction–abduction, internal–external, and flexion–extension directions. *(b)* Components of the moment arm vector for the anterior portion of the gluteus medius (in centimeters) as a function of the hip flexion angle (in degrees).

Adapted from *Journal of Biomechanics* 14(11), W.F. Dostal and J.G. Andrews, "A three-dimensional biomechanical model of hip musculature," 803-812, copyright 1981, with permission from Elsevier.

posterior portions of the deltoid are agonists in shoulder abduction and act as antagonists in flexion–extension.

In biomechanics, the muscle agonist and antagonist actions at a joint are often defined with respect to the resultant joint moment (joint torque). (For the definition of *joint moment*, a.k.a. *joint torque*, see section **2.1** in *Kinetics of Human Motion*.) A *joint torque agonist* (or simply *joint agonist*) generates the moment at the same direction as the resultant joint moment, whereas a *joint torque antagonist* (*joint antagonist*) generates the moment opposite to the resultant joint moment. Because the joint resultant moment can be in different directions, a given muscle can be a joint agonist in one situation and joint antagonist in another.

There is also another meaning of the terms *agonist* and *antagonist* in the literature. Muscles are classified as *task agonists* or *task antagonists* (*functional agonists* or *functional antagonists*) depending on whether their action assists or resists a given task. A muscle can be at the same time a functional agonist with respect to one task and a functional antagonist with respect to another task. For instance, the biceps brachii is a powerful elbow flexor and supinator. During a forceful elbow curl with arm pronation, the biceps is a task agonist with respect to the elbow flexion and a task antagonist with respect to the pronation efforts.

In complicated cases, it is advisable to specify the meaning of the terms *agonist* and *antagonist* (e.g., muscle A is agonist to muscle B, acts as antagonist to the joint moment, or is a task agonist—aiding task performance). Further discussion on agonist and antagonist muscle functions is provided in chapter 6.

■ ■ ■ *FROM THE LITERATURE* ■ ■ ■

Nineteen Muscle Segments of Three Shoulder Muscles Are Organized as Agonists or Antagonists

Source: Brown, J.M., J.B. Wickham, D.J. McAndrew, and X.F. Huang. 2007. Muscles within muscles: Coordination of 19 muscle segments within three shoulder muscles during isometric motor tasks. *J Electromyogr Kinesiol* 17(1):57-73

Using surface electromyography, the authors investigated the coordination of 19 intramuscular segments (such muscles within muscles were called *actons* in section **1.8**) of three shoulder muscles (pectoralis major,

(continued)

From the Literature   *(continued)*

deltoid, latissimus dorsi) of 20 male subjects during the production of rapid isometric force impulses (i.e., approximate time to peak 400 ms) in four different movement directions of the shoulder joint (flexion, extension, abduction, and adduction). It was shown that the muscle acton's moment arm and its mechanical line of action in relation to the intended direction of shoulder movement (e.g., flexion, extension, abduction, or adduction) influence the timing and intensity of the muscle acton's activation. For any particular motor task, individual muscle actons can be functionally classified as prime mover, synergist, or antagonist, classifications that are flexible from one movement to another.

When the topic of interest is the functioning of several (many) muscles at several joints, a powerful tool for analysis of muscle actions is muscle Jacobians; the elements of the muscle Jacobians correspond to the moment arms of the muscles about the preselected joint axes (muscle Jacobians are described in section **5.3.4** here and in section **3.4.2** of *Kinetics of Human Motion*).

5.3.1.4 Moment Arms of Muscles With Curved Paths: Quadriceps Moment Arm

Representative publications: Jensen and Davy 1975; Pandy 1999

In the previous analysis, the line of muscle force action was assumed to be a straight line connecting midpoints of muscle insertion and origin. Although such an assumption greatly simplifies the determination of a muscle moment arm, only a few muscles or their parts (actons) can be modeled this way (see section **1.6**). As a result, representing muscle centroids, or lines of force action, by straight lines from the origin to insertion might lead to substantial errors in determining muscle moment arm vector.

▪ ▪ ▪ *FROM THE LITERATURE* ▪ ▪ ▪

Curved Representation of Neck Muscle Paths Gives More Accurate Estimates of Muscle Moment Arm Than Straight Line Representations

Source: Vasavada, A.N., R.A. Lashera, T.E. Meyera, and D.C. Lin. 2008. Defining and evaluating wrapping surfaces for MRI-derived spinal muscle paths. *J Biomech* 41(7):1450-1457

The authors obtained MRI cross-sections of 18 neck muscles from the base of the skull to the second thoracic vertebra in one typical male subject. The images were taken in seven postures of the subject's neck and head. For each posture the centroid on each muscle slice was calculated, and centroids of the consecutive slices were connected to obtain the muscle centroid path. In most neck muscles, centroid paths deviated significantly from straight lines. For example, in semispinalis capitis muscle, a head extensor, the difference in moment arm magnitudes between the centroid and straight-line representations averaged across seven head postures was 35%. This was determined using the anteroposterior distance from vertebral center to the muscle path (figure 5.12).

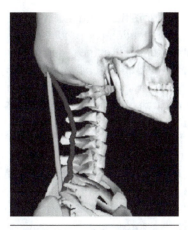

Figure 5.12 Straight-line and centroid path representations of spinalis capitis in the neutral head posture. Note the large differences in the distance from vertebrae centers (moment arm magnitudes) for the two models.

Adapted from *Journal of Biomechanics* 41(7), A.N. Vasavada, R.A. Lasher, T.E. Meyer, and D.C. Lin, "Defining and evaluating wrapping surfaces for MRI-derived spinal muscle paths," 1450-1457, copyright 2008, with permission from Elsevier.

The muscles with curved path can be represented by several line segments connected at via points (see section **1.6.3** and figure 1.19). One of the muscle line segments, or a muscle tendon, has its ends connected to different bones forming a joint of interest. The points of the terminal contact of the muscle or tendon with the bones at the joint are called the *effective origin* and *effective insertion*, respectively. Their coordinates determine the local muscle path at the joint which is a straight line. When the muscle path at a joint is known, the muscle moment arm can be determined using the methods described previously; see equation 5.7 in section **5.3.1.2**, where the unit vector **u** of muscle force should be along the line from the effective origin to the effective insertion (see figure 5.17*b*).

Muscle line segments that do not cross the joint do not immediately generate a moment of force about the joint. Hence, the distances from the joint rotation axis to these segments are irrelevant for the joint torque production. If the muscle–tendon path at the joint is perpendicular to the axis of rotation, the moment arm is equal to the shortest distance d between the muscle–tendon path and the axis of rotation. If the angle between the plane of muscle path at the joint and the rotation axis is $\phi \neq \pi/2$, then the shortest distance d should be multiplied by $\sin\phi$ (see equation 5.8).

The force magnitude in each muscle segment can be the same (uniform force transmission) or the forces can be different. However, because all muscle line segments are connected in series, all segment forces affect each other. Consider as an example torque production during knee extension.

The force of the quadriceps on the patella is not equal to the force of the patellar tendon on the tibia. This happens because the patella acts as a lever of the first class (the levers are discussed in *Kinetics of Human Motion*, section **1.1.6**) with the fulcrum at the point of patellofemoral contact and the two forces at the opposite sides of the fulcrum, the quadriceps force F_q and the patella force F_p. If the patella is at rotational equilibrium, the moments of force exerted by F_q and F_p are opposite and equal. Hence, the force transmitted to the patella is $F_p = F_q \times$ (*Moment Arm of F_q / Moment Arm of F_p*), where the moment arms are with respect to the instant center of rotation of the patella (the fulcrum). So, the knee extension moment is

$$M = F_p d = F_q d \frac{\text{moment arm of } F_q}{\text{moment arm of } F_p}, \qquad [5.10]$$

■ ■ ■ *FROM THE LITERATURE* ■ ■ ■

Forces at Quadriceps and Patella Are Different

Source: Huberti, H.H., W.C. Hayes, J.L Stone, and Shybut G.T. 1984. Force ratios in the quadriceps tendon and ligamentum patellae. *J Orthop Res* 2(1):49-54

The authors measured forces in the quadriceps tendon and ligamentum patellae in six human cadaver knee joints loaded through a range of flexion angles from 30° to 120°. The force ratio (patellar force/quadriceps force) reached a maximum value of 1.27 at 30° and minimum of 0.7 at 90° and 120° knee flexion. The ratio is determined by the changing location of the patellofemoral contact area relative to the insertions of the tendon and ligament. The patella is not a simple pulley that serves only to change the direction of equal forces in the quadriceps tendon and ligamentum patellae.

where d is the shortest distance between the patella and the knee axis of rotation.

The kinematics of human knee (tibiofemoral) motion is still a matter of debate in the biomechanics literature (see section **5.4.1** in *Kinematics of Human Motion*). If the axis of rotation and the patellar tendon are assumed to be perpendicular to each other, the distance d is the moment arm of the patellar-tendon force. However, near full knee extension, the conjunct axial rotation of the tibia with respect to the femur occurs (the so-called *screw-home mechanism*) and the lines of the patellar tendon and the axis of rotation do not stay perpendicular to each other. Thus, to find the moment arm, the product $d_{oo} = d\sin\phi$ should be used.

5.3.1.5 Moment Arms of Multijoint Muscles: Paradoxical Muscle Action

Representative publications: Kamper et al. 2002; Nimbarte et al. 2008

As an introductory example, consider the tendon of flexor digitorum profundus (FDP) muscle (figure 5.13). The tendon attaches to the distal phalange. The FDP is a multijoint muscle serving several joints; anatomy textbooks describe it as a flexor of the wrist, midcarpal, metacarpophalangeal (MCP), and interphalangeal joints. As shown in the figure, the distances from the joint centers to the central line of the tendon—which were defined above as the muscle moment arms—increase from the distal to the proximal joints. The distance at the MCP joint, which is the joint of interest, is 1.0 cm. The question is this: can this distance be considered a moment arm of the FDP muscle if we take into account that the FDP has no attachments with the proximal phalange of the finger and the corresponding metacarpal bone? In other words, will the FDP force generate a flexion moment with respect

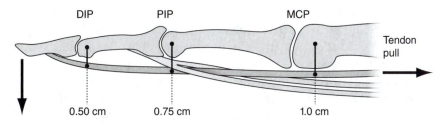

Figure 5.13 An index finger with approximate distances from the joint centers to the FDP tendon. Suppose the hand is fixed, the fingertip is free to move, and the tendon is pulled. What kind of movement at the MCP joint do you expect, flexion or extension? If you see the extension, will you label the FDP the MCP extensor in this case?

to the MCP joint center that equals the product *Muscle Force × Distance* from joint center to tendon centroid? For static tasks, the answer is positive. For the tasks involving or allowing fingertip movements, the answer is less evident and requires a more detailed biomechanical analysis. For instance, for an extended finger and idealized finger model (no passive resistance at the joints), pulling the FDP tendon causes a slight MCP extension, not flexion. Hence, the extension torque is generated. Following the basic equation *Moment of Force = Force × Moment Arm,* one has to conclude that the moment arm in this case is in extension and is not represented by the distance from the joint center to the tendon centroid shown in the figure, which suggests that the torque should be generated in flexion.

The paradox is explained by the fact that the FDP has no direct connections with the bones forming the joint and hence, strictly speaking, the definition of moment of force given previously (equation 5.1) is not valid. The torque in this case is an equivalent joint torque (explained in section **2.1.2** of *Kinetics of Human Motion*); it results from joint forces acting on the link ends and depends on the position and resistance at other joints of the kinematic chain. In the preceding example, introducing passive resistance at the joints results in MCP flexion.

■ ■ ■ *FROM THE LITERATURE* ■ ■ ■

In Multilink Chains Flexor Shortening May Result in Joint Extension

Source: Kamper, D.G., G.T. Hornby, and W.Z. Rymer. 2002. Extrinsic flexor muscles generate concurrent flexion of all three finger joints. *J Biomech* 35(12):1581-1589

A computer simulation of the index finger was created. The model consisted of a planar open-link chain composed of three revolute joints and four links, driven by the change in length of the flexor muscles. Passive joint characteristics were obtained from system identification experiments involving the application of angular perturbations to the joint of interest. Simulation results reveal that in the absence of passive joint torque, shortening of the extrinsic flexors results in proximal interphalangeal flexion (80°) but distal interphalangeal (8°) and MCP (7°) joint extension. However, the inclusion of normal physiological levels of passive joint torque results in simultaneous flexion of all three joints (63° for distal interphalangeal, 75° for proximal interphalangeal, and 43° for MCP).

5.3.2 Methods for Determination of Muscle Moment Arms

Representative publications: Dostal and Andrews 1981; An et al. 1984; Maganaris 2004

Although equations 5.1 through 5.3 are straightforward, their practical application is bound with difficulties: the vectors **r** and **F** should be precisely known. Determining the moment arms for individual muscles and joints is a field of extensive research.

The methods for determining muscle moment arms are classified as (1) geometric and (2) functional. The geometric methods are directly based on the definition of the moment arm vector magnitude as the shortest distance from the rotation center or axis to the line of force action in the plane perpendicular to the axis of rotation. If the coordinates of the center or axis and the line of muscle force action are known, the moment arm vector can be easily determined (the relevant mechanics theory was explained in sections **5.3.1** of this book and **1.1.1.2** in *Kinetics of Human Motion*). The methods for determining the joint axes of rotation are explained in section **4.2** in *Kinematics of Human Motion*. The second challenge is to determine the line of muscle force action. Depending on how this line is determined, the methods are classified as either (1a) anatomical or (1b) imaging methods. Functional methods, explained in detail in section **5.3.2.2**, are based on experimental determination of the moment arms.

5.3.2.1 Geometric Methods

Representative publications: Rugg et al. 1990; Blemker and Delp 2005

For muscles with long tendons, the line of muscle force action is usually assumed to be along the tendon and passing at the center of the tendon attachment area with a bone. For muscles with broad attachment areas (see section **1.8**), the moment arms can be determined for separate parts of the muscle (actons).

A common assumption is that the line of muscle force is along the muscle centroid line (explained in section **1.6.2**). This assumption is only approximately valid (unfortunately, a better approach does not exist). First, centroid models, as was explained in **1.6.2**, neglect muscle pennation. Second, equating the line of muscle force action with the muscle centroid implicitly assumes that (a) the muscle specific force (i.e., the force per unit of the muscle physiological cross-sectional area) is the homogeneous characteristic of the muscle; that is, it is the same at various regions of the muscle cross section; and (b) the angle of pennation of all muscle fibers is the same (otherwise the same fiber force would contribute differently to the muscle force). If these ambiguities are neglected (as is typically done), the moment arm can be determined geometrically by using anatomical or imaging methods.

5.3.2.1.1 Anatomical Geometric Methods

The methods of this group are based on geometric modeling of joints and muscles. In the simplest two-dimensional geometric models, muscle paths are considered to be straight lines except when the muscles wrap around bone structures or joints, as modeled by two-dimensional geometric shapes. In three-dimensional models, more complex geometric surfaces are typically used to guide the muscle centroid path.

5.3.2.1.1.1 Planar Geometric Models

Representative publications: Landsmeer 1961; Goslow et al. 1973; Pedotti 1977; Armstrong and Chaffin 1978

In planar geometric models, several assumptions are typically made: bones forming the joints move in one plane, location of the joint rotation center is constant with respect to the joint, muscle attachments on the bone can be represented by midpoints of attachment areas, and the muscle path is represented by a straight line or a set of straight lines with or without curved segments wrapping around defined geometric shapes. In models in which muscles or tendons wrap around joints, additional assumptions are necessary, for example, whether the muscle–tendon is free sliding versus attached to the wrapping surface.

To illustrate how muscle moment arms can be determined using this approach, let us consider three two-dimensional models of finger flexor tendon developed by Landsmeer (1961). In these models, spatial relations between the tendon and joint allow for determination of the tendon moment arm.

In the first model, it is assumed that the tendon is held in secure contact with the curved articular surface of the proximal bone of the joint and that this surface can be represented as a trochlea; that is, the joint surface is grooved (figure 5.14a). The model neglects the distance from the bone surface to the tendon center. Then the moment arm of the tendon with respect to the joint center equals the radius (d_1) of the curvature of the bone at the joint. If necessary, the tendon displacement (Δl) can

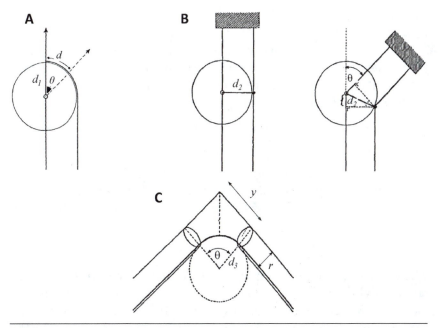

Figure 5.14 Landsmeer's three geometric models of the finger flexor tendon. *(a)* Model I. The tendon is held in secure contact with the curved articular surface of the proximal bone of the joint. The tendon moment arm d_1 equals the radius of the joint. *(b)* Model II. The tendon path is held in a position constrained by the tendon sheath along the bisection of the joint angle. The tendon moment arm d_2 equals the distance from the joint center to the geometric tendon constraint in both the extended position (left) and a flexed joint (right). *(c)* Model III. The tendon is bowstringing and curved smoothly around a joint. For small joint rotation angles θ, the tendon moment arm d_3 is approximately equal to the radius of the joint; at greater joint angles, the moment arm exceeds the joint radius and is determined by equation 5.14.

From J.M. Landsmeer, 1961, "Studies in the anatomy of articulation: I. The equilibrium of the 'intercalated' bone," *Acta Morphologica Neerlando-Scandinavica* 3: 287-303. Adapted by permission of Taylor & Francis Ltd., www.tandfonline.com.

be calculated from the moment arm as a function of the joint rotation angle θ:

$$\Delta l = d_1 \theta. \qquad [5.11]$$

Try to derive this equation. We will consider this relation again in section **5.3.2.2.1**.

The second model describes the situation when the tendon is not securely held against the articular surface but can be moved away from the joint (bowstringing) during joint flexion until it reaches a position constrained by the tendon sheath (a pulley) at the bisection of the joint angle (figure 5.14b). In this model the tendon moment arm (d_2) is the distance from the joint center to the geometric tendon constraint (a point), and the relation between the tendon displacement along the tendon length (Δl) and joint rotation angle (θ) is

$$\Delta l = 2d_2 \sin(\theta / 2). \qquad [5.12]$$

The third model was developed for the case when the tendon is bowstringing and runs through tendon sheath so that the tendon path is curved smoothly around a joint (figure 5.14c). The relation between the tendon and joint angular displacement is described by the following equation:

$$\Delta l = 2[y - \theta / 2 \{r - y / \tan(\theta / 2)\}], \qquad [5.13]$$

where r is the distance between the tendon and the long axis of the distal bone and y is the distance along the distal bone from the joint center to the point at which the tendon begins to curve. If the rotation angle is small ($\theta < 20°$), equation 5.13 simplifies to equation 5.11: $\Delta l = r\theta$, in which the distance r equals the tendon moment arm d_3. For larger joint rotations, the moment arm d_3 is described by equation 5.14 and is a function of the joint rotation angle:

$$d_3 = y\left[\frac{1 - \cos(\theta / 2)}{\sin(\theta / 2)}\right] + r. \qquad [5.14]$$

Note that $d_3 > r$.

Similar planar geometric models have been developed for muscles of the lower (figure 5.15) and upper extremities and for muscles of the trunk. The advantages of such models are their simplicity and their ability to depict an important feature of musculoskeletal geometry—the dependence of muscle moment arms on joint angles. Once the model parameters are determined (on cadavers or on living subjects with various imaging methods) and the joint rotations are measured using motion capture systems, the muscle moment arms can be calculated using equations analogous to 5.14. The main limitations of planar geometric models arise from the fact that the muscle path does not typically run in one anatomical plane and often cannot be represented as a set of straight lines and simple two-dimensional geometric shapes.

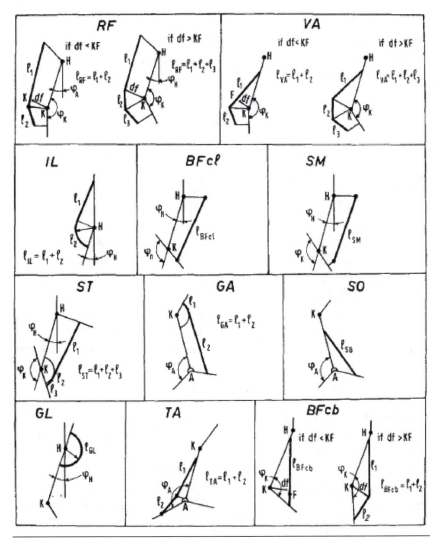

Figure 5.15 Diagram illustrating geometric models of muscle paths with respect to the joint centers for major leg muscles operating in the sagittal plane. Letters A, K, and H designate the ankle, knee, and hip joints; l is length of muscle or its part; and ψ is a joint angle. RF, rectus femoris; VA, vastii; IL, iliacus; BFcl, biceps femoris long head; SM, semimembranosus; ST, semitendinosus; GA, gastrocnemii; SO, soleus; GL, glutei; TA, tibialis anterior; and BFcb, biceps femoris short head.

With kind permission from Springer Science+Business Media: *Biological Cybernetics*, "A study of motor coordination and neuromuscular activities in human locomotion," 26(1), 53-62, A. Pedotti, fig. 3. © Springer-Verlag 1977.

5.3.2.1.1.2 Three-Dimensional Geometric Models

Representative publications: Garner and Pandy 2000; Charlton and Johnson 2001; Gao et al. 2002; Blemker and Delp 2005; Marsden and Swailes 2008; Ramsay et al. 2009

Developing three-dimensional muscle models requires an accurate description of the muscle path between the identified muscle attachment points. This is typically achieved by selecting appropriate points (via points) or geometric shapes around which the muscle wraps and introducing rules for selection of a specific muscle path out of multiple possibilities. Additionally, the muscle path can be adjusted by fine-tuning the shape of geometric objects and improving the smoothness of the path.

Calculating three-dimensional muscle paths is computationally expensive. Fortunately, several commercial and open-source software packages for musculoskeletal modeling are available to compute three-dimensional muscle paths and moment arms (e.g., SIMM and OpenSim, MusculoGraphics, AnyBody Technology, AnimatLab).

The process of modeling the muscle path in 3-D starts with connecting muscle attachment points by lines that have the property of elastic strings under tension. This property and additional assumptions—for example, that no friction exists between the string and wrapping surfaces, the string does not penetrate the surface, the string follows a geodesic line while in touch with the surface—ensure that the string assumes the shortest path possible, given the attachment points, via points, and wrapping surfaces. The most popular shapes are cylinders and spheres because they provide unique geodesic paths of wrapping, respectively, a *helix* (a smooth curve in 3-D for which the tangent line at any point makes a constant angle with an axis) and an arc of a great circle. For a number of muscles, these shapes are insufficient to correctly represent the muscle path and other shapes are used (e.g., tapered cylinders, ellipsoids, or nonanalytical surfaces). Positions of via points and wrapping surfaces along the muscle path (so-called *obstacle sets*) are selected to minimize the mismatch between the modeled path and the muscle centroid path determined experimentally for a given limb position (see figure 5.16). In some algorithms, the muscle path and the corresponding moment arms found for one body configuration can be computed rather accurately for other configurations, thus making it unnecessary to measure the moment arms for all possible joint angles. The moment arms can be calculated using equations 5.7 and 5.8.

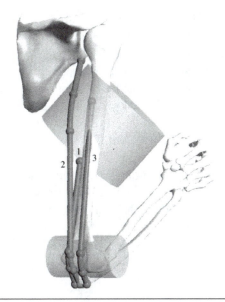

Figure 5.16   An obstacle-set model of the paths of the three heads of triceps brachii. The medial head (1) and lateral head (3) are each modeled as a single-cylinder obstacle set (obstacles not shown). The long head (2) is modeled using a double-cylinder obstacle set as shown. The locations of the attachment sites of the muscle and the locations and orientations of the obstacles were chosen to reproduce the centroid paths of each head. The geometry of the bones and centroid paths of the muscles were obtained from three-dimensional reconstructions of the anatomical structures.

Reprinted, by permission, from B.A. Garner and M.G. Pandy, 2000, "The obstacle-set method for representing muscle paths in musculoskeletal models," *Computer Methods in Biomechanics and Biomedical Engineering* 3(1): 1-30.

5.3.2.1.2 Imaging Geometric Methods

Representative publications: Wilson et al. 1999; Maganaris 2004

As follows from the name of this group of methods, measurements of muscle moment arms are conducted using an imaging technique such as X-ray, MRI, or computed tomography. These methods are especially useful when the joint of interest is planar and the axis of rotation in the joint is perpendicular to the plane of joint movement. If these conditions are met, images of the bones forming the joint at different joint angles are made, the instantaneous positions of the joint center of rotation are determined (described in section **4.2** of *Kinematics of Human Motion*), and the moment arm of the muscle crossing the joint is measured on the images as the perpendicular (shortest) distance from the line of muscle action (a tendon or muscle centroid) to the joint center of rotation.

If the muscle centroid path runs in the plane of joint operation and the position of the center of rotation is known, one image of the muscle and the joint in this plane allows one to determine the muscle moment arm at this particular joint angle. Images taken at different joint angles can be used to derive a moment arm–joint angle relation (figure 5.17). As evident from figure 5.17c, neglecting to accurately determine the instantaneous center of rotation at the joint can lead to substantial errors.

The preceding example illustrates a muscle (tendon) with a centroid path that runs largely in one plane in the vicinity of the enthesis; this allows for

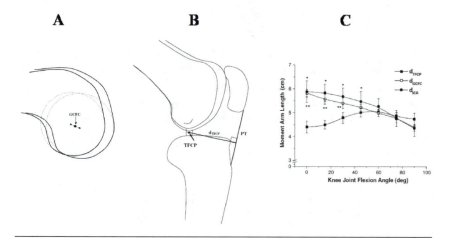

Figure 5.17 Determination of the patellar tendon moment arm with respect to the three joint center locations. The moment arm of the patellar tendon was determined from MRI with respect to the three points at the knee: (1) the geometric center of the femur posterior condyles (GCFC), shown in *a*; (2) the tibiofemoral contact point (TFCP), shown in *b*; and (3) the instant center of rotation (ICR), not shown. *(c)* Patella moment arm determined with respect to the three points at the knee as a function of knee joint flexion angle (0°, full knee extension). The moment arm estimates are (1) d_{GCFC}, (2) d_{TFCP}, and d_{ICR}, respectively.

Reprinted with kind permission from Springer Science+Business Media: *European Journal of Applied Physiology*, "A comparison of different two-dimensional approaches for the determination of the patellar tendon moment arm length," vol. 105, pp. 811-812, D.E. Tsaopoulos, V. Baltzopoulos, P.J. Richards, and C.N. Maganaris, fig. 2, fig. 3, fig. 4. © Springer-Verlag 2009.

accurate determination of the moment arm from a single two-dimensional image of the muscle and the joint. If muscles have more complex paths (see, e.g., figure 1.17), the muscle moment arms can still be determined but from multiple two-dimensional computer tomography or MRI images. First, the muscle path is reconstructed by connecting centroids of individual slices made in the proximity to the joint. Then the moment arm can be determined in a given plane as the shortest distance between the path and the center of joint rotation (assuming that the instantaneous axis of rotation is perpendicular to the line of muscle force action; otherwise equation 5.8 should be used).

5.3.2.2 Functional Methods

Representative publications: Brand et al. 1975; An et al. 1984; Young et al. 1993

Muscle moment arms can be determined without knowing the exact positions of the joint center and muscle line of action. These methods are called functional methods and are subclassified as kinematic and static methods. A moment arm determined with functional methods is the effective moment arm. Rather than representing an actual distance from point A to point B, an effective moment arm simply fits the equation Moment of Force = Force × Moment Arm. Existing functional methods are mainly designed for planar tasks in which the direction of tendon pull is in the plane orthogonal to the axis of rotation. If applied alone, their application to three-dimensional tasks can lead to erroneous results. Combining the methods with imaging techniques greatly improves their potential.

Functional methods are especially useful when the location of the muscle with respect to the joint, muscle architecture, and muscle broad attachment sites is so complex that the geometric and imaging methods might not provide accurate moment arm estimations. The supinator muscle of the forearm (also called supinator brevis) is an example of such a muscle. It originates from a broad area of two different bones (the humerus and the ulna) and soft tissues (radial collateral ligament of the elbow and the annular ligament) and wraps around the radius so that the midplane of the muscle is very close to bone surface and the muscle moment arm is difficult to determine with imaging techniques. Functional methods are also advantageous when large myofascial force transmission is expected. These methods give rather accurate results and are relatively easy to implement in most cases.

5.3.2.2.1 Tendon Excursion Method (Kinematic Method)

Representative publications: Storace and Wolf 1979; An et al. 1983; Spoor et al. 1990; Delp et al. 1999; Pandy, 1999; Langenderfer et al. 2006

The idea of the method can be explained using a simple analysis of two pulley wheels with different radii (d_1 and $d_2 = 2d_1$) fixed to common axle. Initially two ropes that are attached and wrapped around the wheels support loads at the same height (figure 5.18). When both wheels rotate by the same angle θ, load 1 will move by the distance $\Delta l_1 = d_1\theta$ and load 2 by $\Delta l_2 = d_2\theta = 2\,\Delta l_1$. This is because, by definition of an angle in radians, $\theta = s\,/\,d$ where $s = \Delta l$ is the length of the enclosed arc and d is the circle radius. The load displacement in both pulleys is proportional to the radius of the wheel, which is the moment arm of weight of the load (or rope tension) with respect to the center of wheel rotation. Therefore, the moment arm can be determined from the rope displacement at a given angle change: $d = \Delta l\,/\,\theta$. Note that this result corresponds to the first Landsmeer's model (see figure 5.14*a* and equation 5.11), which was developed for a tendon wrapping around a joint whose cross-section can be approximated by a circle.

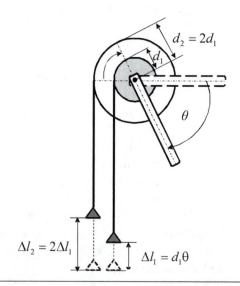

Figure 5.18 The relations among wheel rotation, wheel radius, and rope displacement. Two pulley wheels are rotated by the same angle θ. As a result the rope wrapped around the wheel with a smaller radius d_1 moves the load (depicted as triangle) by a distance of $\Delta l_1 = d_1\theta$, whereas the rope wrapped around the wheel with a longer radius $d_2 = 2d_1$ moves the load through a distance which is two times larger, $\Delta l_2 = d_2\theta = 2\,\Delta l_1$. The rope displacement for a given wheel rotation is proportional to the wheel radius, or the moment arm of weight of the load.

What if the tendon path around the joint is not an arc of a circle and the axis of rotation is moving as joint angle changes? It turns out that the instantaneous muscle moment arm is still a function of the tendon displacement and joint angle. This can be shown using *the principle of virtual work* according to which the sum of works of all forces and moments done during virtual displacements in a system with workless constraints is zero (the principle was discussed in section **2.2.2** of *Kinetics of Human Motion*). A typical joint can be considered a system with workless constraints because friction in the joint is very low and joint ligaments with the capsule can be regarded as inextensible in normal in vivo conditions. Then for a muscle force and the corresponding moment of force at the joint, the following equality can be written:

$$\delta l F + \delta \theta M = 0, \qquad [5.15]$$

where F is force exerted by the muscle (tendon), M is the corresponding moment of muscle force acting at the joint, and δl and $\delta \theta$ are the virtual (infinitesimal) tendon linear and joint angular displacements, respectively, that do not violate joint constraints (i.e., joint integrity). It follows from equation 5.15 that

$$M = -F \frac{\delta l}{\delta \theta} = -Fd, \qquad [5.16]$$

where

$$d = \frac{\delta l}{\delta \theta}. \qquad [5.17]$$

The ratio $\delta l / \delta \theta$ is the instantaneous value of the muscle moment arm that transforms muscle force into the moment of force (see equation 5.1). Thus, the muscle moment arm can be defined as the derivative of muscle length with respect to the joint angle. Such interpretation of the muscle moment arm allows for its determination by measuring tendon displacement as a function of joint angle. The slope of this relation at a given angle corresponds to the moment arm at this particular joint posture. In joints that can be described by Landsmeer's model 1 (see equation 5.11 and figure 5.14*a*), the tendon excursion–joint angle relation is linear with constant slope (i.e., constant muscle moment arm). For the majority of muscles, however, this relation is nonlinear, implying that the muscle moment arm changes with joint angle.

In a typical experiment for determining the muscle moment arm using the tendon excursion method, one bone of a cadaveric joint is attached to a rigid frame and one end of the muscle of interest is detached from the bone and connected to a displacement measuring device that also applies small force through a system of pulleys to prevent muscle bowstringing (figure 5.19*a*). By changing joint angle in one plane and measuring the corresponding tendon travel distance, the muscle length change–joint angle relationship is obtained (figure 5.19, *b* and *c*) and fitted by an analytical function (e.g., a polynomial) that can be differentiated to obtain the muscle moment arm as a function of joint angle (see equation 5.17). The tendon excursion method is also used to determine the dependence of muscle length on the joint angle. In this case the topic of interest is the muscle length–joint angle relation itself, not its derivative.

When the tendon excursion method is applied to revolute joints with 1 degree of freedom in which the line of muscle force action and the plane

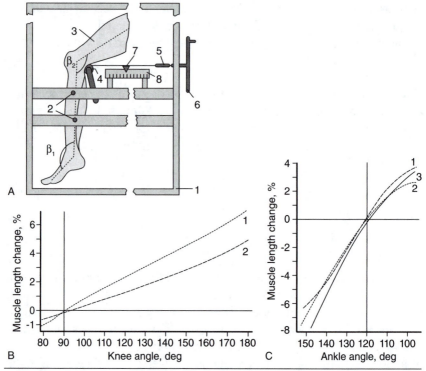

Figure 5.19 Experimental determination of tendon excursion of soleus, medial, and lateral gastrocnemius muscles as functions of knee and ankle joint angles. *(a)* Schematic of the measuring device. 1, metal frame; 2, pins fixing the limb to the frame; 3, cadaveric lower extremity; 4, pulley system; 5 and 6, load cell and lever regulating pulling force; 7 and 8, tendon displacement measurement device; β_1 and β_2, ankle and knee joint angles. *(b)* Muscle length change (normalized to shank length) as a function of knee joint angle; zero length change is set at 90°. 1 and 2, medial and lateral gastrocnemius muscles, respectively. *(c)* Muscle length change (normalized to shank length) as a function of ankle joint angle; zero length change is set at 120°. 1 and 2, medial and lateral gastrocnemius muscles, respectively; 3, soleus. The muscle moment arm of each muscle at a given joint angle can be determined as a tangent of the presented curve at the joint angle (see also equation 5.17).

Adapted from A.S. Aruin and B.I. Prilutsky, 1986, "Dependence of lengthening of the triceps surae muscle on knee and ankle joint angles," *Human Physiology* (in Russian, *Fiziologia Cheloveka*) 12: 244-248.

of joint movement are orthogonal to the joint axis of rotation, the method immediately yields the values of the muscle moment arm d, that is, the shortest distance from the axis of rotation to the line of muscle force action. Equation 5.17 can be applied. For joints with 2 or 3 rotational degrees of freedom, two experimental techniques are available: (a) the joint angle is systematically changed in the plane perpendicular to the rotation axis of interest, tendon elongation is measured, and then the obtained tendon elongation–joint angle relation is differentiated with respect to the angle; or (b) the joint angle is changed in a plane that is not perpendicular to the line of muscle length change. In such a case, equation 5.8 should be used: the derivative of the muscle length–joint angle relation is equal to the product of the shortest distance between the axis of rotation and the line of muscle force action multiplied by the sine of the angle between these two lines.

The method of tendon excursion has been typically used to determine moment arms in cadavers, which is a disadvantage compared with geometric and imaging methods used in vivo. In cadavers, the dependence of the muscle moment arm on the exerted muscle force due to changes in muscle shape and deformation of tissues around the muscle is difficult to assess. Another limitation of the tendon excursion method is the implicit assumption that the imposed joint angle change does not stretch the tendon and thus the resulting tendon displacement is associated only with the corresponding joint angle change. In reality, some tendon stretch is likely to occur because of tendon's high compliance at small loads (see figure 2.4 in section **2.1.1.1.1**) or because in some joint positions a slack in the tendon may occur. In these situations, the tendon excursion method underestimates the moment arm.

The use of the method in conjunction with ultrasound or MRI allows one to determine muscle moment arms in vivo and to explore moment arm dependence on muscle force.

■ ■ ■ *FROM THE LITERATURE* ■ ■ ■

Combining the Tendon Excursion Method With Imaging Techniques Opens New Horizons

1. Muscle Moment Arms Measured In Vivo With the Tendon Excursion Method

Source: Lee, S.S., and S.J. Piazza. 2008. Inversion-eversion moment arms of gastrocnemius and tibialis anterior measured in vivo. *J Biomech* 41:3366-3370

(continued)

The authors used ultrasonography to determine the frontal plane moment arms of tibialis anterior (TA) and the lateral and medial heads of gastrocnemius (LG and MG) using the tendon excursion method in 10 subjects at different ankle angles in the sagittal plane and different muscle activity levels. The subject was sitting upright with the right knee flexed at 90° with shank and thigh immobilized. The foot was strapped to a platform that could be rotated with respect to anterior–posterior axis when it was fixed at three different ankle plantar flexion positions: 0°, 15°, and 30°. Muscle moment arm measurements were conducted at three levels of muscle activity of each muscle: maximum voluntary contraction (MVC, >85% MVC), light contraction (10%-30% MVC) and rest (<10%) MVC. Tendon excursion was measured using an ultrasonographic transducer placed on the muscle parallel to the tibia. The insertions of muscle fascicles into the central aponeurosis were used for tracking tendon displacement in each of nine experimental conditions: three frontal joint angles (15° eversion, 0°, and 20° inversion), three plantar flexion joint angles (0°, 15°, and 30°), and three muscle activity levels (rest, light, MVC). The moment arms of each muscle depended on frontal plane angle. When the foot was everted, TA and LG had eversion moment arms, but MG had a slight inversion moment arm. The inversion moment arms of each muscle increased with ankle inversion angles. In neutral position, the inverter moment arm of MG exceeded that of LG. No significant effect of plantar flexion ankle position on the frontal moment arms was found. The increase in gastrocnemius activity at 15° eversion decreased the everter moment arms, whereas at 0° and 20° of inversion, the inverter moment arms of gastrocnemius muscles became longer with muscle activity. The authors concluded that the tendon excursion method can be used to assess muscle action in vivo.

2. Comparison of Muscle Moment Arms Determined by Tendon Excursion and Imaging (Two-Dimensional and Three-Dimensional) Methods

Source: Wilson, D.L., Q. Zhu, J.L. Duerk, J.M. Mansour, K. Kilgore, and P.E. Crago. 1999. Estimation of tendon moment arms from three-dimensional magnetic resonance images. *Ann Biomed Eng* 27:247-256

The authors developed two-dimensional and three-dimensional MRI methods for tendon moment arm measurement and compared these two methods with the results of the tendon excursion method in which tendon displacements were recorded using three-dimensional MRI. The three methods were applied to the flexor digitorum profundus at the third metacarpophalangeal joint of four able-bodied subjects. Initially, using serial three-dimensional static MRI slices, the investigators reconstructed the tendon path for each of several hand postures. To determine the tendon moment arm using the tendon excursion method, the investigators determined changes in tendon path length between successive joint rotation angles along with the corresponding joint angles.

In the three-dimensional imaging method for determining the tendon moment arm, the same images were used to find (1) orientation of the instantaneous axis of rotation at the third metacarpophalangeal joint, (2) the shortest distance between this line and the tendon path, and (3) the tendon line of action. The tendon moment arm was determined as the minimal distance between the tendon path and the instantaneous line of joint rotation multiplied by the sine of the angle between the tendon line of action and the line of joint rotation (see equation 5.8).

The tendon moment arm was also determined using a two-dimensional MRI slice in the tendon path plane reconstructed from three-dimensional image data sets. First, the center of joint rotation was calculated using a technique similar to the graphical method of Reuleaux (see for details book *Kinematics of Human Motion,* section **4.2.1.3**). Then the moment arm was measured on the two-dimensional MRI slice as the perpendicular distance from the joint center of rotation to a line drawn through the center of the tendon.

It was found that the three-dimensional tendon excursion method was more reproducible (6% variation) than the three-dimensional (12%) and two-dimensional (27%) imaging methods.

5.3.2.2.2 Load Application Method (Static Method)

Representative publications: Zatsiorsky et al. 1983; Buford et al. 2005; Lee et al. 2008

In this method, the tendon force and the moment of force are directly measured and then the moment arm is computed. The method can be used for muscles with very complex design and multiple attachment points. The load application method determines an effective muscle moment arm based on a static equilibrium between the moment of force developed by a muscle crossing the joint and attached to the bone and the resulting external moment (equation 5.18):

$$\sum_j F_{ej} d_{ej} = F_m d_m$$

[5.18]

$$d_m = \frac{\sum_j F_{ej} d_{ej} F_m}{F_m},$$

where F_m is the force exerted by the muscle, d_m is the effective muscle moment arm with respect to the axis of rotation of the segment, and F_{ej} and d_{ej} are externally measured forces and the corresponding moment arms with respect to the joint center. Typical external forces are a contact reaction force and force of gravity applied to the segment (figure 5.20).

Because this method is invasive, it is performed on cadavers. The muscle tendon is detached from one bone forming the joint and attached to a force application device such that the muscle position and line of action with respect to the joint are minimally disturbed. A known force is applied to

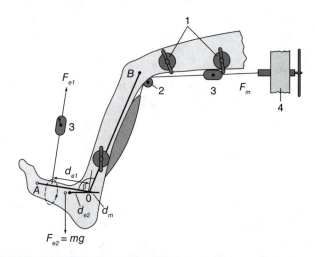

Figure 5.20 Load application for muscle moment arm determination. The method is based on equilibrium between the moment of muscle force *(F_m)* applied by a force application device (4) and recorded by a force transducer (3), on one hand, and the resultant moment of external forces applied to the segment (force F_{e1} and weight of the segment F_{e2}), on the other hand. The external forces and their moment arms (d_{e1} and d_{e2}) are measured or estimated, and the muscle moment arm d_m is calculated from equation 5.18.

Adapted from V.M. Zatsiorsky, A.S. Aruin, B.I. Prilutsky, and A.I. Shakhnazarov, 1985, "Determination of moment arms of ankle extensors by 'biomechanical' method," *Human Physiology* (in Russian, *Fiziologia Cheloveka*) 11: 612-622.

the muscle and the resulting external force is recorded at the body segment, to which muscle is attached, at a measured distance from the joint center. Gravitational force acting on the segment can be calculated from the known or estimated mass and center of mass position (see for details chapter 4 in *Kinetics of Human Motion*).

The use of the load application method has limitations. Normally it cannot be used to measure muscle moment arms in living subjects. Measurements conducted on cadavers are difficult to translate to specific subjects because of their different anatomy and dimensions. In addition, as seen from equation 5.18, a precise position of center of joint rotation (distance d_e) is needed to obtain accurate external moments. One key advantage of the method is that the moment arms can be determined for different loads and hence effects of the deformation of soft tissue on the moment arms values can be accounted for.

5.3.3 Factors Affecting Muscle Moment Arm

As demonstrated in several examples in the previous sections, the muscle moment arm does not typically have a constant value but depends on several factors, for example, joint angle, exerted muscle force, and bone dimensions (see figures 5.11*b* and 5.17*c*).

5.3.3.1 Moment Arm as a Function of Joint Angles

Representative publications: Nemeth and Ohlsen 1985; Spoor et al. 1990; Murray et al. 1995; Delp et al. 1999; Buford et al. 2001; Bremer et al. 2006; Ackland et al. 2008

Why the muscle moment arm often depends on a joint angle becomes clear after considering a simplified example—the gastrocnemius muscle (GA) represented by a straight line in the sagittal plane running from the origin on the femur to insertion on the calcaneus (figure 5.21*a*, see also figure 5.15, GA panel). The distances from ankle and knee joint centers to muscle attachment points (a and a_1, respectively), the shank length, b, and ankle and knee joint angles, θ_1 and θ_2, are known (figure 5.21*a*). Given this geometric model and considering only the case for which the shortest distance from the knee center to the muscle is longer than the radius of the knee joint r (i.e., the muscle does not wrap around the knee joint), the GA moment arm d with respect to the ankle joint center can be expressed as a function of ankle and knee joint angles:

$$d = \frac{ab_1 \sin(\theta_3)}{l}, \qquad [5.19]$$

where b_1 and l are functions of the ankle and knee angles:

$$b_1 = \sqrt{a_l^2 + b^2 - 2a_1 b \cos(\theta_2)},$$ [5.20]

$$\theta_3 = \pi - \theta_1 - \cos^{-1}\left(\frac{b^2 + b_l^2 - a_l^2}{2bb_l}\right),$$ [5.21]

and

$$l = \sqrt{a^2 + b_l^2 - 2ab_l \cos(\theta_3)}$$ [5.22]

If we plot the calculated moment arm d as a function of ankle joint angle (θ_2) for several values of knee angle ($\theta_2 = 40°$, $\theta_2 = 100°$, $\theta_2 = 160°$), we see that the moment arm reaches its peak values in the midrange of ankle angles and decreases when angles approach extreme values (figure 5.21b). We also see that knee angle affects the magnitude of the moment arm of this two-joint muscle at the ankle, albeit to a much lesser degree than ankle angle.

Despite its simplicity, this model generally explains experimental data obtained for muscles that do not wrap around joints near their attachments. In such muscles, moment arm peak values often occur in the middle of the joint range of motion. For example, the elbow flexion moment arms of biceps brachii and brachialis muscles achieve maximum at 90° of elbow

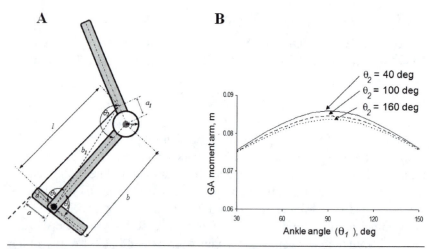

Figure 5.21   Illustration of the dependence of the gastrocnemius (GA) moment arm at the ankle on the ankle and knee joint angles. (*a*) Geometric representation of GA muscle as a straight line connecting points of origin and insertion. Model parameters: $a = 0.07$ m, $a_1 = 0.03$ m, $b = 0.4$ m. (*b*) GA moment arm with respect to the ankle as a function of ankle joint angle. Three lines correspond to different knee joint angles.

flexion (4-5 cm and 2-3 cm, respectively), whereas at 20° the moment arms are approximately only 1 to 2 cm for both muscles and at 120° of flexion decrease to about 3 to 4 cm and 1.5 to 2.5 cm, respectively (figure 5.22). Another elbow flexor, brachioradialis, has the maximum moment arm (6-9 cm) closer to the upper limit of elbow flexion range, which can be explained by its origin in the vicinity of the elbow (figure 5.22*a*).

For muscles that wrap around the joint (e.g., as in the first Landsmeer's model, see figure 5.14*a*), the moment arm does not change with joint angle.

Another important consequence of the geometric model of the gastrocnemius muscle (figure 5.21) is that adjacent joint angle changes in the sagittal plane (or about another degree of freedom at the same joint) may affect the moment arm of a multifunctional muscle (figure 5.21*b*). Such influences

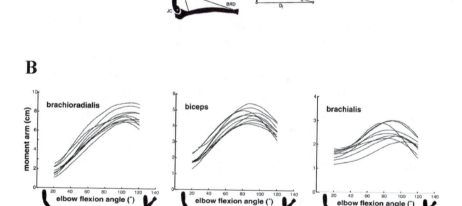

Figure 5.22 The moment arms of brachioradialis (BRD), biceps, and brachialis (BRA) as a function of elbow flexion angle. *(a)* Geometric representation of brachioradialis and brachialis muscles and their moment arm (ma), shorter (Ds), and longer (Dl) distances between the joint center (JC) and a muscle's attachment sites on the bone. *(b)* Elbow flexion moment arms versus elbow flexion angle measured by the tendon excursion method in 10 upper extremity specimens. 0° flexion is full extension.

Reprinted from *Journal of Biomechanics* 35(1), W.M. Murray, T.S. Buchanan, and S.L. Delp, "Scaling of peak moment arms of elbow muscles with upper extremity bone dimensions," 19-26, copyright 2002, with permission from Elsevier.

are small in joints and muscles for which considerable displacements of the muscle path are restricted by fasciae, tendon sheaths, and neighboring muscles. In other situations, substantial changes in the moment arm in the sagittal plane may occur as a result of joint angle changes in other joints or planes.

■ ■ ■ FROM THE LITERATURE ■ ■ ■

1. Moment Arms of Elbow Flexors and Extensors Depend on Elbow Angle and Supination/Pronation Position of the Forearm

Source: Murray, W.M., S.L. Delp, and T.S. Buchanan. 1995. Variation of muscle moment arms with elbow and forearm position. *J Biomech* 28:513-525

The moment arms of muscles crossing the elbow at different elbow flexion–extension and forearm supination–pronation angles have been determined. The measurements were conducted on two cadavers (one male, one female) using the tendon excursion method and a computer model. It was shown that the flexion–extension moment arms varied by 25% (for triceps) to 86% (for brachioradialis) while the elbow flexion angle changed from 25° to 120°. The flexion moment arm of biceps increased by 5 mm and 7 mm (~20%) in two cadavers when the forearm was fully supinated compared with neutral forearm position. Similarly, the peak supination moment arm of biceps decreased by 34% (0.38 cm, female specimen) and 49% (0.66 cm, male specimen) when elbow flexion angle changed from 85° to 45°.

2. Joint Angles at Finger Joints Substantially Affect the Joint Moments and Endpoint Index Finger Force

Source: Kamper, D.G., H.C. Fischer, and E.G. Cruz. 2006. Impact of finger posture on mapping from muscle activation to joint torque. *Clin Biomech (Bristol, Avon)* 21(4):361-369

The authors examined in vivo the mapping from activation of multijoint muscles of the index finger to joint torques and the endpoint force as a function of finger posture. Seven muscles of the index finger of five healthy subjects were individually stimulated using intramuscular electrodes. Stimulations were performed using 12 different combinations of flexion angles at the three finger joints while fingertip forces and moments were recorded. It was found that the joint moment magnitudes (and thus the magnitude and direction of the endpoint finger force) resulting from the stimulations were significantly dependent on finger posture ($p < .05$), typically changing by more than 60%. Such big changes could not be explained by length-dependent changes in the muscle force (estimated to be 10%-20%). It was concluded that the changes in finger joint moments and endpoint force were caused by substantial dependence of the muscle moment arms on finger joint angles. The authors recommended that this postural dependence be considered when constructing biomechanical models of the hand or planning tendon transfers for the fingers.

5.3.3.2 Moment Arm as a Function of Exerted Muscle Force

Representative publications: Agee et al. 1998; Imran et al. 2000; Graichen et al. 2001; Maganaris 2004; Lee and Piazza 2008

As discussed in sections **5.1** and **5.2**, force exerted by the muscle is transmitted to the bone through elastic entheses and in many cases also through soft tissue skeleton (ectoskeleton; i.e., tendon sheath). One would expect that the application of muscle force to these elastic tissues could deform them and change the muscle path orientation, that is, the line of muscle action, with respect to the joint, and thus the muscle moment arm. Another expectation is that moment arm changes would depend on the magnitude of muscle force—the higher the force, the greater the change in the moment arm.

Several experimental studies performed in vivo have confirmed that for some muscles there is dependence between the muscle moment arm and the level of muscle activity. For example, the moment arm of the Achilles tendon measured using two-dimensional MRI imaging increases by 22% to 27% at the maximum voluntary contraction (MVC) compared with rest (figure 5.23*a*). An even greater increase of 44% occurs in the moment arm of tibialis anterior at MVC (figure 5.23*b*). Although a small shift in the instantaneous center of rotation at the joint during MVC can contribute to these results, the major factor explaining the increase in the muscle moment arms with an increasing level of muscle activity appears to be deformation of activated muscles and passive soft tissue around the joint (figure 5.23, *a* and *b*). When the triceps surae and tibialis anterior are strongly activated, the distance between the superficial and deep aponeuroses (thickness of the muscle) increases, as confirmed by ultrasound measurements. In addition, surrounding fasciae or retinacular bands deform while muscles develop high forces. These changes lead to a shift in the line of muscle force action and thus the moment arm.

In some situations, a force development by one muscle can affect the moment arm of neighboring muscles because of muscle attachments to a soft skeleton.

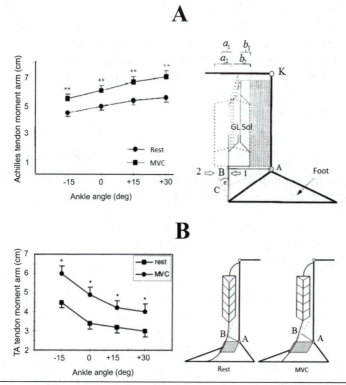

Figure 5.23 The moment arm of Achilles tendon *(a)* and tibialis anterior muscle *(b)* at rest and maximum voluntary contraction (MVC) determined using sagittal plane MRI in six subjects at different ankle angles. Negative angles, dorsiflexion; positive angles, plantar flexion; 0°, neutral ankle position. *(a)* (left): Achilles tendon moment arm determined at rest (diamonds) and MVC (squares); right: Model of triceps surae explaining the dependence of Achilles tendon moment arm on muscle activity (see text). *(b)* (left): Tibialis anterior moment arm determined at rest (squares) and MVC (circles); right: Model of tibialis anterior explaining the moment arm dependence on muscle activity (see text).

(a) From C.N. Maganaris, V. Baltzopoulos, and A.J. Sargeant, 1998, "Changes in the Achilles tendon moment arm from rest to maximum isometric plantarflexion: In vivo observations in man," *The Journal of Physiology* 510(3): 977-985. Reprinted by permission of John Wiley & Sons Ltd.

(b) Reprinted from *Clinical Biomechanics* 14(9), C.N. Maganaris, V. Baltzopoulos, and A.J. Sargeant, "Changes in the tibialis anterior tendon moment arm from rest to maximum isometric dorsiflexion: In vivo observations in man," 661-666, copyright 1999, with permission from Elsevier.

■ ■ ■ *FROM THE LITERATURE* ■ ■ ■

Differential Loading Affects Moment Arms of Neighboring Tendons

Source: Agee, J.M., T.R. Maher, and M.S. Thompson. 1998. Moment arms of the digital flexor tendons at the wrist: Role of differential loading in stability of carpal tunnel tendons. *J Hand Surg Am* 23:998-1003

Using 6 cadaver specimens and the tendon excursion method, the authors determined the moment arms of the flexor pollicis longus and the four flexor digitorum superficialis tendons with respect the flexion–extension axis at the wrist while each of the 9 carpal tunnel tendons was loaded with a baseline tension of 85 g. Applying a higher 540 g load to individual flexor digitorum superficialis tendons and the flexor pollicis longus while loading the remaining tendons with the baseline 85 g tension significantly changed the moment arms from those measured under baseline load (up to 50%). The results were explained by the mechanism according to which the increased tension of a given muscle–tendon unit may cause the corresponding tendon to translate around the lower tensioned tendons until it stabilizes closer to the more rigid wall of the carpal tunnel.

5.3.3.3 Scaling of Moment Arms

Representative publications: Lew and Lewis 1977; Jorgensen et al. 2001; Murray et al. 2002; Matias et al. 2009

The muscle moment arm magnitude depends linearly on dimensions of the bones and joints to which the muscle is attached or around which it is wrapped. It seems self-evident that muscle moment arms of an average-built person should be smaller than those of a 2.20 m tall basketball player. Theoretical analysis and experimental results support this intuitive view. Equations 5.19 through 5.22 show that for a simple geometric musculoskeletal model, the muscle moment arm depends on the segment length and distances from the joint center to the muscle attachment points. An especially close relation is expected between the moment arm and the shorter attachment distance for muscles that do not wrap around the joint (see figure 5.22a). Indeed as evident from figure 5.22a, the moment arm of such a muscle cannot exceed the shorter attachment distance D_s and close to this distance at midrange of joint angles. If a muscle wraps around a joint, the moment arm magnitude is expected to be related to the joint diameter (see equation 5.11 and figure 5.14*a*).

The preceding theoretical considerations have been supported by experimental results. For example, the moment arm of a number of upper extremity muscles is highly correlated with parameter D_s; the latter explains 97% of the variation in peak moment arm across muscles (figure 5.24). The anterior–posterior dimension of the joint explains up to 60% of the peak moment variation for muscles that wrap around the joint, for example, for the elbow joint, triceps brachialis, extensor carpi radialis longus, and pronator teres. Length of the segment along which the muscle path runs normally is not as good a predictor of the peak moment arm as are D_s and a joint dimension. Therefore, the moment arms measured in different subjects and specimens should be normalized to the shorter attachment distance D_s (for muscles that do not wrap around a joint and have the centroid path well approximated by a straight line near enthesis) or to a joint linear dimension for muscles that wrap around the joint.

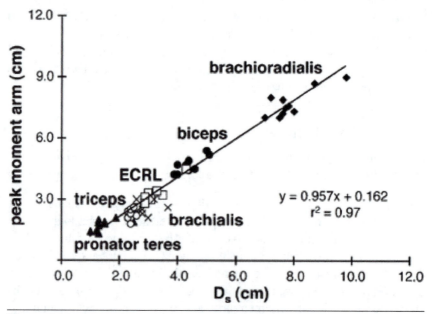

Figure 5.24   Relation between the peak moment arm and the shorter distance (D_s) between the axis of rotation and a muscle's attachment sites (see figure 5.22*a*) for the 58 muscles studied. Peak moment arm is significantly correlated with D_s, which accounts for 97% of the variation in peak moment arm across muscles.

Reprinted from *Journal of Biomechanics* 35(1), W.M. Murray, T.S. Buchanan, and S.L. Delp, "Scaling of peak moment arms of elbow muscles with upper extremity bone dimensions," 19-26, copyright 2002, with permission from Elsevier.

Most two-dimensional and three-dimensional geometric models used for moment arm estimations require knowledge of model parameters such as muscle attachment coordinates and curvature (diameter) of wrapping surfaces. These model parameters are normally measured on a group of subjects or specimens and are then scaled to a specific subject in the study. Thus, scaling of bone landmarks is an important procedure for obtaining subject-specific moment arm data. Several scaling techniques have been developed for this purpose (see, e.g., chapter 4 in *Kinetics of Human Motion*). In one such technique, bones are scaled homogeneously; that is, they are stretched uniformly by different amounts in three mutually orthogonal directions. The three scalar transformation parameters specifying the amount of stretching are derived from spatial coordinates of four palpable bony landmarks between a reference bone and the subjects' bone. Several similar scaling procedures have been developed either to permit less restrictive or more realistic assumptions on deformation properties of anatomical structures or to reduce the number of bony landmark coordinates that must be measured.

■ ■ ■ *FROM THE LITERATURE* ■ ■ ■

Scaling Method for Estimation of Muscle Attachments Based on Three Bony Landmarks

Source: Matias, R., C. Andrade, and A.P. Veloso. 2009. A transformation method to estimate muscle attachments based on three bony landmarks. *J Biomech* 42:331-335

To obtain subject-specific muscle attachment locations based on homologous points measured in a specimen, the authors developed a scaling method that allows for an algebraic transformation using three bony landmarks per segment. Bony landmarks and muscle attachment information were taken from the published data set of 17 muscles from seven cadaveric scapulae and humeri. This data set was also used to assess quality of the scaling method. The developed method demonstrated an overall root mean square error for the scapula and humerus muscles of 7.6 mm and 11.1 mm, respectively. The estimated moment arm of teres minor at 30° and 60° of glenohumeral elevation in the scapula plane with 0° axial rotation was shown to be within published ranges. The method, however, does not account for inter- and intrasubject variability, which is a problem common to all scaling methods.

Another factor affecting the magnitude of muscle moment arms and related to scaling is the subject's gender. Reported moment arm values for the homologous muscles are typically greater in males than females. This difference is partly explained by the fact that on average men are taller, and body dimensions affect moment arm values (see figure 5.24). Another possible contributing factor is greater muscle strength in men, which is related to the muscle physiological cross-sectional area, and thus muscle thickness, but not to muscle length (can the reader explain why?). As discussed earlier in this chapter (see, e.g., figure 5.23), larger muscle thickness may lead to an increase in the moment arm. Furthermore, greater muscle forces developed by males cause structural adaptations of entheses and bone attachment sites, which typically lead to an increase in their size and thus to larger muscle moment arms.

Interindividual differences in the muscle moment arms may be important for external force generation and speed production. Imagine two equally strong and fast muscles with the moment arms of different magnitude. If these muscles exert equal forces, the muscle with a larger moment arm will generate a larger moment of force. However, when the muscles shorten at the same speed, the muscle with the smaller moment arm will generate a larger angular velocity at the joint (explain why this happens).

▪ ▪ ▪ *FROM THE LITERATURE* ▪ ▪ ▪

Good Sprinters Have Shorter Achilles Tendon Moment Arms

Source: Lee, S.S., and S.J. Piazza. 2009. Built for speed: Musculoskeletal structure and sprinting ability. *J Exp Biol* 212(Pt 22):3700-7707

The authors measured the plantar flexion moment arms of the Achilles tendon, lateral gastrocnemius fascicle lengths and pennation angles, and anthropometric characteristics of the foot and lower leg in collegiate sprinters and height-matched nonsprinters. The Achilles tendon moment arms of the sprinters were 25% smaller on average in sprinters than in nonsprinters ($p < .001$), whereas the sprinters' fascicles were 11% longer on average ($p = .024$). The ratio of fascicle length to moment arm was 50% larger in sprinters ($p < .001$).

5.3.4 Transformation of Muscle Forces to Joint Moments: Muscle Jacobian

Representative publication: Tsuji 1997

Analysis of transformation from muscle force to joint moment for a single muscle is facilitated with the use of the muscle moment arm derived by vector analysis (see equations 5.3, 5.7, and 5.8) or the principle of virtual work (see equation 5.15). The muscle force–joint moment transformation for several muscles crossing multiple joints can be conveniently performed using muscle Jacobian matrices whose elements are moment arms of the individual muscles about joints' axes of rotation.

To derive this matrix, let us apply the principle of virtual work (see equation 5.15) to a multijoint muscle system. For such a system, the total work done by the muscles in the system (or at each joint) over infinitesimal muscle length changes is equal to total work of the resultant joint moments done over infinitesimal joint angle displacements:

$$\delta l_1 F_1 + \delta l_2 F_2 + ... + \delta l_m F_m + \delta \alpha_1 M_1 + \delta \alpha_2 M_2 + ... + \delta \alpha_n M_n = 0, \quad [5.23]$$

where δl_i and $\delta \alpha_j$ are infinitesimal changes in length of the ith muscle and in jth joint angle, respectively. F_i and M_j are the ith muscle force and the jth joint moment, and m and n are the number of muscles and rotational degrees of freedom in the system, respectively. For each muscle in the model, the muscle length is a function of system joint configuration (generalized joint coordinates), that is, $l_i = l_i(\delta \alpha_1, \delta \alpha_2, \ldots, \delta \alpha_n)$, and this implies that the virtual muscle length changes can be expressed as functions of the virtual joint angle changes and partial derivatives of the muscle length with respect to the joint angles:

$$\delta l_1 = \frac{\partial l_1}{\partial \alpha_1} \delta \alpha_1 + \frac{\partial l_1}{\partial \alpha_2} \delta \alpha_2 + ... + \frac{\partial l_1}{\partial \alpha_n} \delta \alpha_n$$

$$\delta l_2 = \frac{\partial l_2}{\partial \alpha_1} \delta \alpha_1 + \frac{\partial l_2}{\partial \alpha_2} \delta \alpha_2 + ... + \frac{\partial l_2}{\partial \alpha_n} \delta \alpha_n \qquad [5.24]$$

$$... \qquad ... \qquad ... \qquad ... \qquad ...$$

$$\delta l_m = \frac{\partial l_m}{\partial \alpha_1} \delta \alpha_1 + \frac{\partial l_m}{\partial \alpha_2} \delta \alpha_2 + ... + \frac{\partial l_m}{\partial \alpha_n} \delta \alpha.$$

If a muscle does not cross a particular joint, its length partial derivative with respect to the angle change at that joint is zero and thus the angle changes at this joint do not affect the muscle length. Substituting the virtual muscle length changes in equation 5.23 with the expressions 5.24, one obtains equation 5.25:

$$M_1 = \frac{\partial l_1}{\partial \alpha_1} F_1 + \frac{\partial l_2}{\partial \alpha_1} F_2 + \dots + \frac{\partial l_m}{\partial \alpha_1} F_m$$

$$M_2 = \frac{\partial l_1}{\partial \alpha_2} F_1 + \frac{\partial l_2}{\partial \alpha_2} F_2 + \dots + \frac{\partial l_m}{\partial \alpha_2} F_m \qquad [5.25]$$

$$\dots \qquad \dots \qquad \dots \qquad \dots \qquad \dots$$

$$M_n = \frac{\partial l_1}{\partial \alpha_n} F_1 + \frac{\partial l_2}{\partial \alpha_n} F_2 + \dots + \frac{\partial l_m}{\partial \alpha_n} F_m$$

or in matrix form

$$\mathbf{M} = \mathbf{G}^T\mathbf{F}, \qquad [5.26]$$

where $\mathbf{M}$ and $\mathbf{F}$ are the vectors of joint moments and muscle forces, respectively. Matrix $\mathbf{G}$ is

$$\mathbf{G} = \begin{bmatrix} \dfrac{\partial l_1}{\partial \alpha_1} & \dfrac{\partial l_1}{\partial \alpha_2} & \cdots & \dfrac{\partial l_1}{\partial \alpha_n} \\[2ex] \dfrac{\partial l_2}{\partial \alpha_1} & \dfrac{\partial l_2}{\partial \alpha_2} & \cdots & \dfrac{\partial l_2}{\partial \alpha_n} \\[2ex] \cdots & \cdots & \cdots & \cdots \\[2ex] \dfrac{\partial l_m}{\partial \alpha_1} & \dfrac{\partial l_m}{\partial \alpha_2} & \cdots & \dfrac{\partial l_m}{\partial \alpha_n} \end{bmatrix}. \qquad [5.27]$$

See also equation 3.33 in *Kinetics of Human Motion*.

Matrix $\mathbf{G}$ is called *muscle Jacobian*. Each element of this matrix represents the muscle moment arm with respect to a particular joint (see equation 5.17). The muscle Jacobian is a convenient tool for performing the transformations from muscle forces to joint moments (equation 5.26) or from small joint angle to small muscle length changes:

$$d\mathbf{L} = \mathbf{G}d\boldsymbol{\alpha}. \qquad [5.28]$$

The muscle Jacobian is widely used to study the role of muscle moment arms in regulation of apparent limb, joint, and muscle stiffness (for details, see section **3.4** in *Kinetics of Human Motion*).

5.4 SUMMARY

Muscle force is transmitted to the bones through pathways that include (1) elastic tissues in series with the muscle (aponeurosis and tendon), (2) the soft tissue skeleton (ectoskeleton), and (3) tendon–bone junctions (enthesis).

The connection between a muscle and its tendon is called a muscle–tendon junction (MTJ). At sites where muscle fibers meet collagen fibers, the force is mainly transmitted through shear stresses exerted on neighboring fibers. The tendon–bone junction, the site of insertion of tendon or fascia into bone, is called enthesis.

Force patterns recorded at the ends of an isolated muscle–tendon unit (MTU) are different from force patterns generated by muscle fibers alone due to the series elastic component of MTU. When muscles are activated in a fixed joint position (isometric action), muscle fascicles shorten and the in-series tendinous tissues are stretched. This stretch distorts the classic force–length relation—the force at the ascending limb decreases while the force on the descending limb increases; thus, the peak isometric force occurs at a longer length. The extent to which the MTU force–length relation is modified by tendon elasticity depends on the ratio $L_{T,Slack} / L_o$ (described in sections **1.5.2** and **4.4.2**), where $L_{T,Slack}$ and L_o are tendon slack length and optimal fiber length, respectively. In general, the longer the tendon compared with muscle fiber, the more muscle fiber shortening and tendon stretch occur during isometric muscle action.

The term *soft-tissue skeleton* typically refers to fascia. Tendons of many skeletal muscles attach to bones not only directly but also through fascia. Forces generated by individual muscle fibers, fascicles, and single muscles are distributed within and between neighboring fibers, fascicles, and muscles (lateral, or myofascial, force transmission, see sections **1.3** and **3.1.5**).

The moment of force about a point O is a vector $\mathbf{M}_O$ and is obtained as the cross-product of vectors $\mathbf{r}$ and $\mathbf{F}$, where $\mathbf{r}$ is the position vector from O to any point along the line of force $\mathbf{F}$:

$$\mathbf{M}_O = \mathbf{r} \times \mathbf{F}. \qquad [5.1]$$

The magnitude of moment is $M_O = F\,(r \sin \theta) = Fd$, where θ is the angle between vectors $\mathbf{r}$ and $\mathbf{F}$ and d is the shortest distance from the O to the line of action of $\mathbf{F}$, the moment arm. Equation 5.1 can also be represented as a product of the muscle force magnitude and the moment arm vector $\mathbf{d}$:

$$\mathbf{M}_O = \mathbf{r} \times \mathbf{F} = \mathbf{r} \times F\mathbf{u} = F(\mathbf{r} \times \mathbf{u}) = F\mathbf{d}, \qquad [5.2]$$

where **u** is a unit vector along the line of force action. The moment arm vector **d** equals the moment per unit force. The moment arm vector is also called the normalized moment vector.

When defined for a muscle, the cross-product

$$\mathbf{d} = (\mathbf{r} \times \mathbf{u}) \qquad [5.3]$$

is the muscle moment arm vector. The magnitude d of the muscle moment arm vector **d** is equal to the perpendicular distance from the joint center (point O) to the line of muscle force action, or to the shortest distance between the muscle force line of action and point O. The direction of the muscle moment arm vector coincides with the direction of the moment of force $\mathbf{M}_O$ and can be defined by the right-hand rule—when the right-hand fingers are curling from vector **r** to vector **F**, the extended thumb shows the direction of the moment of force vector and the muscle moment arm vector.

A concept of the moment arm vector is not generally used in classical mechanics where moment arms are commonly treated not as vectors but as the line segments. The concept was introduced in biomechanics, where denoting a moment of muscle force as a product of the force magnitude F and a vector **d** is convenient and represents the essence of the muscle moment production. The moment arm vector **d** depends on human anatomy and for a given joint position cannot be changed, whereas the force magnitude F is a control variable that can be changed by the controller.

In three dimensions, a muscle can generate a moment about an axis that is different from the axis (axes) of interest (e.g., the flexion–extension axis or an instantaneous rotation axis). The moment of force M_{oo} about an axis O-O equals a component of moment $\mathbf{M}_O$ along this axis (the mechanics are explained in section **1.1.1.2.2** in *Kinetics of Human Motion*):

$$M_{oo} = \mathbf{u}_{oo} \cdot F\mathbf{d} = \mathbf{u}_{oo} \cdot (\mathbf{r} \times \mathbf{F}) = \mathbf{u}_{oo} \cdot (\mathbf{r} \times F\mathbf{u}), \qquad [5.6]$$

where the axis is specified by unit vector $\mathbf{u}_{oo}$; **F** is the vector of muscle force; F is the magnitude of the vector of muscle force applied to the bone with the direction defined by the unit vector **u**; and **r** is the position vector from any point on the axis of rotation to any point on the line of muscle force action. Equation 5.6 is a mixed triple product of the three above vectors: $\mathbf{u}_{oo}$, **r**, and **F**.

The projection d_{oo} of muscle moment arm vector on axis O-O can be defined as the moment per unit of muscle force that causes rotation of the bone about the axis:

$$d_{oo} = \frac{M_{oo}}{F} = \mathbf{u}_{oo} \cdot (\mathbf{r} \times \mathbf{u}) = \mathbf{u}_{oo} \cdot \mathbf{d} \qquad [5.7]$$

The magnitude of the moment arm about an axis is equal to the perpendicular, or shortest, distance d between the axis of rotation and the line of muscle force action multiplied by the sine of the angle ϕ between the two lines:

$$d \sin \phi = |\mathbf{u}_{oo} \cdot (\mathbf{r} \times \mathbf{u})|. \qquad [5.8]$$

When the line of muscle force action is along the axis of rotation, $\sin\phi = 0$ and the force does not produce any moment about this axis. When $\mathbf{u}_{oo}$ and $\mathbf{u}$ are at a right angle to each other, $\sin\phi = 1$ and the moment is maximal. The muscle moment arm vector components $d_x\mathbf{i}$, $d_y\mathbf{j}$, and $d_z\mathbf{k}$ specify the relative turning effects of the muscle force applied to the bone with respect to the corresponding axes of rotation X, Y, and Z intersecting at point O, where d_x, d_y, and d_z are the scalar components of the moment arm vector.

In anatomy, the agonistic muscles are commonly defined as the muscles that lie on the same side of a bone, act on the same joint, and move it in the same direction. Correspondingly, the antagonists act on the same joint and move it in the opposite direction. Joint torque agonists (or simply joint agonists) generate the moment at the same direction as the resultant joint moment, whereas joint torque antagonists (joint antagonists) generate the moment opposite to the resultant joint moment. Muscles can also be classified as task agonists or task antagonists (functional agonists or functional antagonists) depending on whether their action assists or resists a given task.

The methods for determining muscle moment arms are classified as (1) geometric and (2) functional. Geometric methods are directly based on the definition of the moment arm vector magnitude as the shortest distance from the rotation center or axis to the line of force action in the plane perpendicular to the axis of rotation. They are further classified as (1a) anatomical methods and (1b) imaging methods. In simplest two-dimensional geometric models, muscle paths are considered straight lines except when the muscles wrap around bone structures or joints modeled by two-dimensional geometric shapes.

Muscle moment arms can be determined without knowledge of the exact positions of the joint center and muscle line of action. These methods are called functional methods and are subclassified as kinematic and static methods. A moment arm determined with such a method is termed the effective moment arm, because it might not represent an actual physical distance; rather an effective moment arm simply fits the equation *Moment of Force = Force × Moment Arm.*

The muscle moment arm equals the derivative of muscle length with respect to the joint angle:

$$d = \frac{\delta l}{\delta \theta}, \qquad [5.17]$$

where δl and $\delta \theta$ are the virtual (infinitesimal) tendon linear and joint angular displacements, respectively. The relation can be derived from the principle of virtual work according to which the sum of works of all forces and moments done during virtual displacements in a system with workless constraints is zero. Such interpretation of the muscle moment arm allows it to be determined experimentally by measuring tendon displacement as a function of joint angle. The slope of this relation at a given angle corresponds to the moment arm at this particular joint posture (the tendon excursion method).

The static, or load application, method determines an effective muscle moment arm based on a static equilibrium between the moment of force developed by a muscle crossing the joint and attached to the bone and the resulting external moment.

The moment arms of muscles that do not wrap around bones and joints typically change with joint angle and reach their maximum in the midrange of joint motion. Moment arms of multijoint muscles or muscles with several functions at a single joint often depend on joint angle changes in other joints or planes. The moment arms of many muscles also depend on exerted force because of deformation of ectoskeleton and elastic entheses and increasing thickness of the muscle, which cause the muscle path to shift with respect to the joint.

The magnitude of muscle moment arms is different in people of different body size. In particular, the moment arm magnitudes depend on linear dimensions of the bones and joints to which the muscle is attached or around which it is wrapped. It seems that the best predictors of the moment arm magnitude are (a) the shorter attachment distance from the joint center, for the muscles that do not wrap around a joint, and (b) a joint linear dimension, for muscles that wrap around the joint.

The transformation from muscle forces to joint moments is afforded by a matrix of moment arms—the transpose muscle Jacobian—which can be derived evoking the principle of virtual work.

5.5 QUESTIONS FOR REVIEW

1. List morphological structures that participate in force transmission from a muscle to bone.

2. Define a muscle–tendon junction. Describe its morphological features, how they differ between slow-twitch and fast-twitch muscles, and how they contribute to injury risk.

3. Define the term *enthesis*. What role do entheses play in force transmission? Describe their morphology, mechanical properties, and potential injury mechanisms.

4. How and why does the ratio $L_{T,Slack} / L_o$ (tendon slack length divided by optimal fiber length) affect the isometric force–length relation of the muscle?

5. Describe the structure of fascia and its major functional roles. Give several examples of fascia in the human body.

6. What is the ectoskeleton? What role does it play in muscle force transmission?

7. Give definitions of moment arm vector, its magnitude, and its scalar components. Use a straight-line representation of the muscle path.

8. Define the muscle moment arm for a muscle with a curved path.

9. What is the paradoxical action of multijoint muscles? How can the paradox be explained in terms of anatomical connections?

10. Define the functional role of muscles using muscle moment arm components. At one joint what is the maximum number of functions a muscle could have? What about multijoint muscles?

11. Define and discuss agonist and antagonist muscles. Provide examples.

12. Define a muscle moment arm with respect to the instantaneous center or axis of rotation. How can the magnitude of the muscle moment arm vector be found if the instantaneous line of rotation and the line of muscle forces action are not orthogonal?

13. Describe how planar and three-dimensional anatomical geometric methods can be used to determine muscle moment arm. Give examples for muscles with linear and curved muscle paths.

14. How does the obstacle set method work for determining muscle moment arm?

15. List imaging methods that can be used for determining muscle moment arms. Describe the use of these methods for a single plane muscle action around the joint and when the muscle path crosses several planes.

16. What are advantages and disadvantages of the functional methods for determining muscle moment arms?

17. Explain the tendon excursion method for determining muscle moment arms. Derive the equation

$$d = \frac{\delta l}{\delta \theta}.$$

18. Explain the load application method for determining muscle moment arms. Derive the equation

$$d_m = \frac{\sum_j F_{ej} d_{ej}}{F_m}.$$

19. List factors that affect the magnitude of the muscle moment arm.

20. How does the muscle moment arm depend on bone and joint size and muscle attachment locations?

21. Define and derive muscle Jacobian.

22. As a result of a surgery, a muscle moment arm was increased. Assuming that after the surgery the muscle exerts the same force and contracts at the same speed as before the surgery, explain how the surgery affected (a) the end-point force, (b) the joint angular velocity, and (c) the joint power.

5.6 LITERATURE LIST

Ackland, D.C., P. Pak, M. Richardson, and M.G. Pandy. 2008. Moment arms of the muscles crossing the anatomical shoulder. *J Anat* 213(4):383-390.

Agee, J.M., T.R. Maher, and M.S. Thompson. 1998. Moment arms of the digital flexor tendons at the wrist: Role of differential loading in stability of carpal tunnel tendons. *J Hand Surg Am* 23(6):998-1003.

Alexander, R.M. 2002. Tendon elasticity and muscle function. *Comp Biochem Physiol A Mol Integr Physiol* 133(4):1001-1011.

An, K.N., K. Takahashi, T.P. Harrigan, and E.Y. Chao. 1984. Determination of muscle orientations and moment arms. *J Biomech Eng* 106(3):280-282.

An, K.N., Y. Ueba, E.Y. Chao, W.P. Cooney, and R.L. Linscheid. 1983. Tendon excursion and moment arm of index finger muscles. *J Biomech* 16(6):419-425.

Arampatzis, A., S. Stafilidis, G. DeMonte, K. Karamanidis, G. Morey-Klapsing, and G.P. Bruggemann. 2005. Strain and elongation of the human gastrocnemius tendon and aponeurosis during maximal plantarflexion effort. *J Biomech* 38(4):833-841.

Armstrong, T.J., and D.B. Chaffin. 1978. An investigation of the relationship between displacements of the finger and wrist joints and the extrinsic finger flexor tendons. *J Biomech* 11(3):119-128.

Arruda, E.M., S. Calve, R.G. Dennis, K. Mundy, and K. Baar. 2006. Regional variation of tibialis anterior tendon mechanics is lost following denervation. *J Appl Physiol* 101(4):1113-1117.

Aruin, A.S., and B.I. Prilutsky. 1986. Dependence of lengthening of the triceps surae muscle on knee and ankle joint angles. *Fiziologia Cheloveka* 12:244-248.

Benjamin, M. 2009. The fascia of the limbs and back—a review. *J Anat* 214(1):1-18.

Benjamin, M., H. Toumi, J.R. Ralphs, G. Bydder, T.M. Best, and S. Milz. 2006. Where tendons and ligaments meet bone: attachment sites ("entheses") in relation to exercise and/or mechanical load. *J Anat* 208(4):471-490.

Blemker, S.S., and S.L. Delp. 2005. Three-dimensional representation of complex muscle architectures and geometries. *Ann Biomed Eng* 33(5):661-673.

Brand, P.W., K.C. Cranor, and J.C. Ellis. 1975. Tendon and pulleys at the metacarpophalangeal joint of a finger. *J Bone Joint Surg Am* 57(6):779-784.

Bremer, A.K., G.R. Sennwald, P. Favre, and H.A. Jacob. 2006. Moment arms of forearm rotators. *Clin Biomech (Bristol, Avon)* 21(7):683-691.

Brown, J.M., J.B. Wickham, D.J. McAndrew, and X.F. Huang. 2007. Muscles within muscles: Coordination of 19 muscle segments within three shoulder muscles during isometric motor tasks. *J Electromyogr Kinesiol* 17(1):57-73.

Buford, W.L., Jr., F.M. Ivey, Jr., T. Nakamura, R.M. Patterson, and D.K. Nguyen. 2001. Internal/external rotation moment arms of muscles at the knee: Moment arms for the normal knee and the ACL-deficient knee. *Knee* 8(4):293-303.

Buford, W.L., Jr., S. Koh, C.R. Andersen, and S.F. Viegas. 2005. Analysis of intrinsic-extrinsic muscle function through interactive 3-dimensional kinematic simulation and cadaver studies. *J Hand Surg Am* 30(6):1267-1275.

Caravaggi, P., T. Pataky, J.Y. Goulermas, R. Savage, and R. Crompton. 2009. A dynamic model of the windlass mechanism of the foot: Evidence for early stance phase preloading of the plantar aponeurosis. *J Exp Biol* 212(Pt 15):2491-2499.

Charlton, I.W., and G.R. Johnson. 2001. Application of spherical and cylindrical wrapping algorithms in a musculoskeletal model of the upper limb. *J Biomech* 34(9):1209-1216.

De Smet, A.A., and T.M. Best. 2000. MR imaging of the distribution and location of acute hamstring injuries in athletes. *AJR Am J Roentgenol* 174(2):393-399.

Delp, S.L., W.E. Hess, D.S. Hungerford, and L.C. Jones. 1999. Variation of rotation moment arms with hip flexion. *J Biomech* 32(5):493-501.

Delp, S.L., and F.E. Zajac. 1992. Force- and moment-generating capacity of lower-extremity muscles before and after tendon lengthening. *Clin Orthop Relat Res*(284):247-259.

Doschak, M.R., and R.F. Zernicke. 2005. Structure, function and adaptation of bone–tendon and bone-ligament complexes. *J Musculoskelet Neuronal Interact* 5(1):35-40.

Dostal, W.F., and J.G. Andrews. 1981. A three-dimensional biomechanical model of hip musculature. *J Biomech* 14(11):803-812.

Dostal, W.F., G.L. Soderberg, and J.G. Andrews. 1986. Actions of hip muscles. *Phys Ther* 66(3):351-361.

Evans, P. 1979. The postural function of the iliotibial tract. *Ann R Coll Surg Engl* 61(4):271-280.

Finni, T., J.A. Hodgson, A.M. Lai, V.R. Edgerton, and S. Sinha. 2003. Nonuniform strain of human soleus aponeurosis-tendon complex during submaximal voluntary contractions in vivo. *J Appl Physiol* 95(2):829-837.

Gao, F., M. Damsgaard, J. Rasmussen, and S.T. Christensen. 2002. Computational method for muscle-path representation in musculoskeletal models. *Biol Cybern* 87(3):199-210.

Garfin, S.R., C.M. Tipton, S.J. Mubarak, S L. Woo, A R. Hargens, and W.H. Akeson. 1981. Role of fascia in maintenance of muscle tension and pressure. *J Appl Physiol* 51(2):317-320.

Garner, B.A., and M.G. Pandy. 2000. The obstacle-set method for representing muscle paths in musculoskeletal models. *Comput Methods Biomech Biomed Eng* 3(1):1-30.

Gordon, A.M., A.F. Huxley, and F.J. Julian. 1966. The variation in isometric tension with sarcomere length in vertebrate muscle fibres. *J Physiol* 184(1):170-192.

Goslow, G.E., Jr., R.M. Reinking, and D.G. Stuart. 1973. The cat step cycle: Hind limb joint angles and muscle lengths during unrestrained locomotion. *J Morphol* 141(1):1-41.

Graichen, H., K.H. Englmeier, M. Reiser, and F. Eckstein. 2001. An in vivo technique for determining 3D muscular moment arms in different joint positions and during muscular activation—application to the supraspinatus. *Clin Biomech (Bristol, Avon)* 16(5):389-394.

Huberti, H.H., W.C. Hayes, J.L Stone, and Shybut G.T. 1984. Force ratios in the quadriceps tendon and ligamentum patellae. *J Orthop Res* 2(1):49-54.

Huijing, P.A., R.W. van de Langenberg, J.J. Meesters, and G.C. Baan. 2007. Extramuscular myofascial force transmission also occurs between synergistic muscles and antagonistic muscles. *J Electromyogr Kinesiol* 17(6):680-689.

Imran, A., R.A. Huss, H. Holstein, and J.J. O'Connor. 2000. The variation in the orientations and moment arms of the knee extensor and flexor muscle tendons with increasing muscle force: A mathematical analysis. *Proc Inst Mech Eng H* 214(3):277-286.

Jensen, R.H., and D.T. Davy. 1975. An investigation of muscle lines of action about the hip: A centroid line approach vs the straight line approach. *J Biomech* 8(2):103-110.

Jorgensen, M.J., W.S. Marras, K.P. Granata, and J.W. Wiand. 2001. MRI-derived moment-arms of the female and male spine loading muscles. *Clin Biomech (Bristol, Avon)* 16(3):182-193.

Kamper, D.G., H.C. Fischer, and E.G. Cruz. 2006. Impact of finger posture on mapping from muscle activation to joint torque. *Clin Biomech (Bristol, Avon)* 21(4):361-369.

Kamper, D.G., T. George Hornby, and W.Z. Rymer. 2002. Extrinsic flexor muscles generate concurrent flexion of all three finger joints. *J Biomech* 35(12):1581-1589.

Ker, R.F., M.B. Bennett, S.R. Bibby, R.C. Kester, and R.M. Alexander. 1987. The spring in the arch of the human foot. *Nature* 325(7000):147-149.

Kjaer, M. 2004. Role of extracellular matrix in adaptation of tendon and skeletal muscle to mechanical loading. *Physiol Rev* 84(2):649-698.

Landsmeer, J.M. 1961. Studies in the anatomy of articulation: I. The equilibrium of the "intercalated" bone. *Acta Morphol Neerl Scand* 3:287-303.

Langenderfer, J.E., C. Patthanacharoenphon, J.E. Carpenter, and R.E. Hughes. 2006. Variation in external rotation moment arms among subregions of supraspinatus, infraspinatus, and teres minor muscles. *J Orthop Res* 24(8):1737-1744.

Lee, S.S., and S.J. Piazza. 2008. Inversion-eversion moment arms of gastrocnemius and tibialis anterior measured in vivo. *J Biomech* 41(16):3366-3370.

Lee, S.S., and S.J. Piazza. 2009. Built for speed: Musculoskeletal structure and sprinting ability. *J Exp Biol* 212(Pt 22):3700-7707.

Lee, S.W., H. Chen, J.D. Towles, and D.G. Kamper. 2008. Estimation of the effective static moment arms of the tendons in the index finger extensor mechanism. *J Biomech* 41(7):1567-1573.

Lew, W.D., and J.L. Lewis. 1977. An anthropometric scaling method with application to the knee joint. *J Biomech* 10(3):171-181.

Lieber, R.L., and C.G. Brown. 1992. Sarcomere length-joint angle relationships of seven frog hindlimb muscles. *Acta Anat (Basel)* 145(4):289-295.

Loren, G.J., and R.L. Lieber. 1995. Tendon biomechanical properties enhance human wrist muscle specialization. *J Biomech* 28(7):791-799.

Maas, H., H.J. Meijer, and P.A. Huijing. 2005. Intermuscular interaction between synergists in rat originates from both intermuscular and extramuscular myofascial force transmission. *Cells Tissues Organs* 181(1):38-50.

Maganaris, C.N. 2004. Imaging-based estimates of moment arm length in intact human muscle–tendons. *Eur J Appl Physiol* 91(2-3):130-139.

Maganaris, C.N., V. Baltzopoulos, and A.J. Sargeant. 1998. Changes in Achilles tendon moment arm from rest to maximum isometric plantarflexion: In vivo observations in man. *J Physiol* 510(Pt 3):977-985.

Maganaris, C.N., V. Baltzopoulos, and A.J. Sargeant. 1999. Changes in the tibialis anterior tendon moment arm from rest to maximum isometric dorsiflexion: In vivo observations in man. *Clin Biomech (Bristol, Avon)* 14(9):661-666.

Maganaris, C.N., M.V. Narici, L.C. Almekinders, and N. Maffulli. 2004. Biomechanics and pathophysiology of overuse tendon injuries: Ideas on insertional tendinopathy. *Sports Med* 34(14):1005-1017.

Marsden, S.P., and D.C. Swailes. 2008. A novel approach to the prediction of musculotendon paths. *Proc Inst Mech Eng H* 222(1):51-61.

Matias, R., C. Andrade, and A. Veloso. 2009. A transformation method to estimate muscle attachments based on three bony landmarks. *J Biomech* 42(3):331-335.

Monti, R.J., R.R. Roy, H. Zhong, and V.R. Edgerton. 2003. Mechanical properties of rat soleus aponeurosis and tendon during variable recruitment in situ. *J Exp Biol* 206(Pt 19):3437-3445.

Murray, W.M., T.S. Buchanan, and S.L. Delp. 2002. Scaling of peak moment arms of elbow muscles with upper extremity bone dimensions. *J Biomech* 35(1):19-26.

Murray, W.M., S.L. Delp, and T.S. Buchanan. 1995. Variation of muscle moment arms with elbow and forearm position. *J Biomech* 28(5):513-525.

Nakajima, T., J. Liu, R.E. Hughes, S. O'Driscoll, and K.N. An. 1999. Abduction moment arm of transposed subscapularis tendon. *Clin Biomech (Bristol, Avon)* 14(4):265-270.

Nemeth, G., and H. Ohlsen. 1985. In vivo moment arm lengths for hip extensor muscles at different angles of hip flexion. *J Biomech* 18(2):129-140.

Nimbarte, A.D., R. Kaz, and Z.M. Li. 2008. Finger joint motion generated by individual extrinsic muscles: A cadaveric study. *J Orthop Surg Res* 3:27.

Nishimura, T., A. Hattori, and K. Takahashi. 1994. Ultrastructure of the intramuscular connective tissue in bovine skeletal muscle: A demonstration using the cell-maceration/scanning electron microscope method. *Acta Anat (Basel)* 151(4):250-257.

Pandy, M.G. 1999. Moment arm of a muscle force. *Exerc Sport Sci Rev* 27:79-118.

Pedotti, A. 1977. A study of motor coordination and neuromuscular activities in human locomotion. *Biol Cybern* 26(1):53-62.

Rack, P.M., and H.F. Ross. 1984. The tendon of flexor pollicis longus: Its effects on the muscular control of force and position at the human thumb. *J Physiol* 351:99-110.

Ramsay, J.W., B.V. Hunter, and R.V. Gonzalez. 2009. Muscle moment arm and normalized moment contributions as reference data for musculoskeletal elbow and wrist joint models. *J Biomech* 42(4):463-473.

Rugg, S.G., R.J. Gregor, B.R. Mandelbaum, and L. Chiu. 1990. In vivo moment arm calculations at the ankle using magnetic resonance imaging (MRI). *J Biomech* 23(5):495-501.

Spoor, C.W., J.L. van Leeuwen, C.G. Meskers, A.F. Titulaer, and A. Huson. 1990. Estimation of instantaneous moment arms of lower-leg muscles. *J Biomech* 23(12):1247-1259.

Stecco, C., A. Porzionato, L. Lancerotto, A. Stecco, V. Macchi, J.A. Day, and R. De Caro. 2008a. Histological study of the deep fasciae of the limbs. *J Bodyw Mov Ther* 12(3):225-230.

Stecco, C., A. Porzionato, V. Macchi, A. Stecco, E. Vigato, A. Parenti, V. Delmas, R. Aldegheri, and R. De Caro. 2008b. The expansions of the pectoral girdle muscles onto the brachial fascia: Morphological aspects and spatial disposition. *Cells Tissues Organs* 188(3):320-329.

Storace, A., and B. Wolf. 1979. Functional analysis of the role of the finger tendons. *J Biomech* 12(8):575-578.

Terminology, F.C.o.A. 1998. *Terminologia Anatomica: International Anatomical Terminology.* Stuttgart, Germany: Thieme.

Tidball, J.G. 1991. Force transmission across muscle cell membranes. *J Biomech* 24(Suppl 1):43-52.

Tidball, J.G., G. Salem, and R. Zernicke. 1993. Site and mechanical conditions for failure of skeletal muscle in experimental strain injuries. *J Appl Physiol* 74(3):1280-1286.

Trotter, J.A. 2002. Structure-function considerations of muscle–tendon junctions. *Comp Biochem Physiol A Mol Integr Physiol* 133(4):1127-1133.

Tsaopoulos, D.E., V. Baltzopoulos, P.J. Richards, and C.N. Maganaris. 2009. A comparison of different two-dimensional approaches for the determination of the patellar tendon moment arm length. *Eur J Appl Physiol* 105(5):809-814.

Tsuji, T. 1997. Human arm impedance in multi-joint movements. In *Self-Organization, Computational Maps, and Motor Control*, edited by P. Morasso & V. Sanguineti. Amsterdam: Elsevier Science, pp. 357-381.

Vasavada, A.N., R.A. Lasher, T.E. Meyer, and D.C. Lin. 2008. Defining and evaluating wrapping surfaces for MRI-derived spinal muscle paths. *J Biomech* 41(7):1450-1457.

Wilson, D.L., Q. Zhu, J.L. Duerk, J.M. Mansour, K. Kilgore, and P.E. Crago. 1999. Estimation of tendon moment arms from three-dimensional magnetic resonance images. *Ann Biomed Eng* 27(2):247-256.

Wood-Jones, F. 1944. *Structure and Function as Seen in the Foot.* London: Baillière, Tindall and Cox.

Wood, J.E., S.G. Meek, and S.C. Jacobsen. 1989a. Quantitation of human shoulder anatomy for prosthetic arm control: I. Surface modelling. *J Biomech* 22(3):273-292.

Wood, J.E., S.G. Meek, and S.C. Jacobsen. 1989b. Quantitation of human shoulder anatomy for prosthetic arm control: II. Anatomy matrices. *J Biomech* 22(4):309-325.

Youm, Y., D.C.R. Ireland, B.I. Sprague, and A.E. Flatt. 1976. Moment arm analysis of the prime wrist movers. In *Biomechanics V-A*, edited by P.V. Komi. Baltimore, MD: University Park Press, pp. 355-365.

Young, R.P., S.H. Scott, and G.E. Loeb. 1993. The distal hindlimb musculature of the cat: Multiaxis moment arms at the ankle joint. *Exp Brain Res* 96(1):141-151.

Zatsiorsky, V.M., A.S. Aruin, B.I. Prilutsky, and A.I. Shakhnazarov. 1985. Determination of moment arms of ankle extensors by "biomechanical" method. *Hum Physiol* 11:616-622.

Zatsiorsky, V.M., L.M. Raitsin, V.N. Seluyanov, A.S. Aruin, and B.I. Prilutsky. 1983. Biomechanical characteristics of the human body. In *Biomechanics and Performance in Sport,* edited by W. Baumann. Cologne, Germany: Bundesinstitut für Sportwissenschaft, pp. 71-83.

CHAPTER

6

TWO-JOINT MUSCLES IN HUMAN MOTION

This chapter reviews anatomical and morphological features of two-joint muscles and several functional consequences of their design, such as creating moments of force at both joints (sections **6.1.1** and **6.1.2**). Functional roles of two-joint muscles during leg extensions are analyzed using methods of kinetics and kinematics in sections **6.2.1** and **6.2.2**. The mechanisms of mechanical energy transfer by two-joint muscles are explained in section **6.3.1**, whereas section **6.3.2** demonstrates how two-joint muscles can reduce mechanical energy expenditure in human locomotion.

6.1 TWO-JOINT MUSCLES: A SPECIAL CASE OF MULTIFUNCTIONAL MUSCLES

Representative publications: Bock 1968; Rozendal 1994; van Ingen Schenau and Bobbert 1993

As described in the previous chapter (see section **5.3.1.3**), a single-joint muscle can have several functions at a joint; that is, it can produce moments around several axes of joint rotation. For example, at a given hip joint position, the anterior portion of gluteus medius can contribute simultaneously to flexion, abduction, and internal rotation of the femur around the rotational axes at the hip (see figure 5.11). As discussed in chapter 5, most of the muscles in the human body serving a single joint have two or three functions. There are also muscles that have two or three times more functions than those crossing just one joint.

6.1.1 Functional Features of Two-Joint Muscles

Representative publications: Basmajian 1957; Fujiwara and Basmajian 1975; Markee et al. 1955; Wells and Evans 1987; van Ingen Schenau 1989; Jacobs and van Ingen Schenau 1992; de Looze et al. 1993; Bolhuis et al. 1998

Some muscles cross two, three, or more joints (*multijoint muscles* or *polyarticular muscles*) and have up to three functions at each spanned joint. In this chapter we focus only on *two-joint muscles* (or *biarticular muscles*), which cross just two joints, and on their actions in the sagittal plane. This is because two-joint muscles are common in the lower and upper extremities and the moment arms of these muscles are typically much longer with respect to flexion–extension rotation axes than other axes, and thus the effects of two-joint muscle actions in the sagittal plane are generally much greater.

Two-joint muscles of the lower extremity include gastrocnemius (ankle extensor, or plantar flexor, and knee flexor); rectus femoris (knee extensor and hip flexor); the two-joint hamstrings (hip extensors and knee flexors): long head of biceps femoris, semitendinosus, and semimembranosus; gracilis (knee and hip flexor), and sartorius (hip and knee flexor). Two-joint muscles of the upper extremity are less numerous: biceps brachii (elbow and shoulder flexors) and long head of triceps brachii (elbow and shoulder extensors). Many distal muscles in the human leg and arm cross more than one joint, but they are multijoint and span not only the knee or ankle and the elbow and wrist joints but also the metatarsophalangeal or metacarpophalangeal joints and interphalangeal joints through the long distal tendons and also ligaments. Examples of such muscles are plantaris and flexor hallucis longus in the leg and flexor carpi radialis and palmaris longus in the arm.

Most muscles in the human body have a number of functions at the joints they cross, and they contribute to joint moments with respect to different joint axes. The mechanical functions of a muscle can be determined using the scalar components of the muscle moment arm vector (see equation 5.9) and visualized by projecting the moment arm vector onto planes of joint axes (see, e.g., figure 5.10). A similar approach can be used to determine the contribution of a two-joint muscle to the rotational effect at each joint in the sagittal plane if we define the moment arm vector of a two-joint muscle as $\mathbf{r} = (r_1, r_2)$, where r_1 and r_2 are the moment arms of the muscle with respect to the flexion–extension axes of rotation at joints 1 and 2. Plotting such vectors in the plane formed by flexion–extension rotational axes of the two joints (figure 6.1) reveals the dominant and less significant functions of the muscle at the two joints. For instance, at given hip and knee joint angles (e.g., when the leg is close to full extension), the contribution of gracilis to the knee flexion is twice that of hip flexion, whereas tensor fasciae lata is mostly a hip flexor and to a much lesser extent is a knee extensor (figure 6.1). In contrast, sartorius has about equal contribution to the knee and hip flexion, rectus femoris is slightly more a knee extensor than a hip flexor (the corresponding moment arm values are approximately 6 cm and –4 cm), and the three hamstrings muscles contribute slightly more to hip extension than to knee flexion (6 cm and –4 cm).

Although two-joint muscles have intrigued scientists for many centuries, a systematic research of their anatomy and morphology, mechanical actions, and physiology started only in the 19th century. The interest in two-joint muscles arose partly because of difficulties in determining their

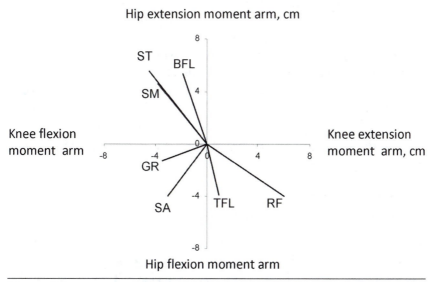

Figure 6.1 Moment arm vectors of two-joint leg muscles with respect to flexion–extension axes of the knee and hip joints. Positive direction of axes corresponds to extension, negative to flexion. BFL, biceps femoris long head; ST, semitendinosus; SM, semimembranosus; GR, gracilis; SA, sartorius; TFL, tensor fasciae latae muscle; RF, rectus femoris. Note that there is no two-joint muscle that contributes simultaneously to knee and hip extension. The moment arm data were estimated for a leg posture near full extension from the following studies: Dostal and Andrews 1981; Spoor and van Leeuwen 1992; Visser et al. 1990.

functional roles. For example, when one is rising from a chair, both (a) rectus femoris, which is a knee extensor and hip flexor, and (b) two-joint hamstrings, which are hip extensors and knee flexors (see figure 6.1), are active. (The sitting reader can quickly verify this fact by putting one hand on hamstrings and the other hand on rectus femoris in midthigh and standing up.) These muscles are considered anatomical antagonists, and their moment arm vectors are located in opposite quadrants in the knee–hip moment arm plane (see figure 6.1). Why, then, are rectus femoris and hamstrings coactivated during this task, when their actions oppose each other? Also, it seems that both muscles provide both a positive and a negative contribution to leg extension. Rectus femoris tends to extend the knee but opposes hip extension. The hamstrings contribute to hip extension and resist knee extension. This seemingly paradoxical behavior of rectus femoris and hamstrings is called Lombard's paradox after the American physiologist Warren Plimpton Lombard (1855-1939), who was one of the first to notice this effect and to propose a functional explanation for it (discussed later). Two-joint muscles can also act as stiff ropes, or ligaments, linking body segments that do not share a common joint. This action of two-joint muscles permits mechanical energy transfer between these segments and is called the *tendon action* of two-joint muscles. Additional interest in two-joint muscles has been stimulated by the investigations of their neural control, which is understandably complex, but this topic is outside the scope of this book.

6.1.2 Anatomical and Morphological Features of Two-Joint Muscles

Representative publications: Heron and Richmond 1993; Loeb and Richmond 1994; Yang et al. 1998; Ward et al. 2009

Two-joint muscles are long, spanning the full length of the segment that lies between the muscle attachments. Some two-joint muscles, such as gastrocnemius and distal multijoint muscles, achieve this by having a long tendon and relatively short muscle fibers with a large angle of pennation (see tables 1.2 and 1.4). This design permits a large physiological cross-sectional area and thus high force potential of the muscles (see sections **1.7** and **3.1.5.2.2.1**), which in the case of distal muscles helps the muscle to develop sufficient moments despite small mechanical advantage (relatively small moment arms at the distal joints). The long tendons of these two-joint and multijoint muscles allow for substantial tendon strain and storage and release of elastic energy during the stretch–shortening cycle, creating favorable conditions for high movement efficiency and power development (see section **7.6**). Furthermore, proximal muscle belly locations reduce distal segment inertia, thereby alleviating the proximal muscle burden to develop limb acceleration.

Most of the two-joint muscles in the proximal limb segments like semitendinosus, gracilis, and sartorius in the leg and biceps brachii in the arm have typically long muscle fibers, small pennation angles, and relatively short tendons (table 6.1, see also table 1.2). Even though certain muscle fibers span the whole muscle belly length and run from the proximal to the distal tendons, reaching lengths of 40 to 60 cm, as in sartorius, normally these muscles contain a larger proportion of shorter, nonspanning fibers tapering at one or both ends (see figure 1.12). Because proximal moment arms are large, and because movements at the distal joint can cause even further stretch, it is conceivable that the long lengths of two-joint muscle fibers are protective adaptations against overstretch. Long spanning fibers constitute a noticeable proportion of fibers in the proximal two-joint leg muscles.

Table 6.1   Architecture of Proximal Two-Joint Muscles of the Human Leg and Arm

Muscle and the number of studied specimens	Muscle length (ML), cm	Fiber length (FL), cm	Pennation angle, °	Physiological cross-sectional area, cm²
Gracilis, $n = 19$	28.69 ± 3.29	22.78 ± 4.38	8.2 ± 2.5	2.2 ± 0.8
Biceps femoris long head, $n = 18$	34.73 ± 3.65	9.76 ± 2.62	11.6 ± 5.5	11.3 ± 4.8
Semitendinosus, $n = 19$	29.67 ± 3.86	19.30 ± 4.12	12.9 ± 4.9	4.8 ± 2.0
Semimembranosus, $n = 19$	29.34 ± 3.42	6.90 ± 1.83	15.1 ± 3.4	18.4 ± 7.5
Biceps brachii short head, $n = 8$	23.4 ± 4.2	14.5 ± 3.2	0	2.1 ± 0.6
Biceps brachii long head, $n = 6$	21.6 ± 4.5	12.8 ± 3.2	0	2.5 ± 1.1
Triceps brachii long head, $n = 8$	29.1 ± 5.2	9.3 ± 2.8	8.0 ± 2.0	10.5 ± 5.8

Ward et al. (2009) studied 21 human lower extremities (height of the specimen 168.5 ± 9.3 cm, mass 82.7 ± 15.3 kg). Murray et al. (2000) studied 10 unembalmed upper extremities from nine cadavers (humerus, radius, and ulna lengths 32.0 ± 1.4 cm, 23.9 ± 1.3 cm, and 25.7 ± 1.5 cm, respectively). Values in the table are expressed as mean ± standard deviation.

Data from Ward et al. (2009) and Murray et al. (2000).

■ ■ ■ *FROM THE LITERATURE* ■ ■ ■

Approximately 30% of Sartorius Motor Units Include Fibers That Span the Full Muscle Length

Source: Harris, A.J., M.J. Duxson, J.E. Butler, P.W. Hodges, J.L. Taylor, and S.C. Gandevia. 2005. Muscle fiber and motor unit behavior in the longest human skeletal muscle. *J Neurosci* 25:8528-8533

The authors studied the location and timing of activation within motor units of sartorius muscle, the longest muscle in the human body (see table 1.2), in five male volunteers. They recorded propagation of action potentials through muscle fibers using intramuscular electrodes at multiple sites along sartorius during steady voluntary contractions. It was found that approximately 30% of sartorius motor units included fibers that crossed all recording sites and spanned essentially the full length of the muscle (i.e., >50 cm). Approximately 60% of these fibers had endplates located closer toward one or the other end of the fiber. Given that the speed of action potential propagation along the fiber was 3.9 ± 0.1 m/s, it would take sarcolemmal activation, started at the endplate, more than 100 ms to reach the other end of the fiber. This considerable time lag could conceivably cause asynchronous contractions along the fiber, that is, shortening already activated region and stretching as-yet passive regions. The results of this study suggest that within muscles with long fibers and the shorter nonspanning fibers, such as sartorius, lateral force transmission would be advantageous for faster force development and for preventing excessive fiber strain.

As a consequence of the large number of tapering nonspanning fibers in proximal two-joint muscles, lateral force transmission through the endomysial and perimysial connective tissue plays a dominant role in these muscles (see sections **1.4** and **3.1.4**). It promotes mechanical stability inside a muscle by redirecting forces exerted by the activated regions through parallel connective tissues and thus preventing the excessive shortening of activated muscle regions and overstretching of passive or less active regions. Lateral force transmission is also important for preventing possible instabilities and overstretches of very long spanning fibers caused by relatively slow action potential propagation. As a result of the slow activation transmission speed, shortening of one activated fiber region leads to stretching of another region.

6.2 FUNCTIONAL ROLES OF TWO-JOINT MUSCLES

Representative publications: Cleland 1867; Lombard 1903; Elftman 1940; Peters and Rick 1977; van Ingen Schenau et al. 1987; Prilutsky 2000a, 2000b

We limit the current discussion only to the functions of muscles at the joints that they cross; that is, we do not consider here the role of the muscles in creating the interaction forces and moments and their mechanical effects on nonspanned joints, although the muscles can contribute to acceleration of joints they do not cross (contribution of interaction torques and forces are discussed in section **5.4.3.2** in *Kinetics of Human Motion*). While reading this chapter keep in mind that muscle forces acting at a joint do not exclusively determine joint motion or joint acceleration at that joint. They only contribute to the resultant joint torque or joint moment, and then the joint torque contributes to the angular acceleration at the joint. Both the moment generated by a given muscle and the resultant moment of all the muscles crossing the joint can be in opposite direction to the joint motion or acceleration (see section **5.4** on joint torques and joint forces in human motion in *Kinetics of Human Motion*).

A number of approaches have been attempted over the years to decipher the mechanical functions and roles of two-joint muscles at the joints. The approaches include analysis of muscle force contribution to the joint torques (kinetic analysis) and analysis of length changes in the two-joint muscle (kinematic analysis). We start with the kinetic analysis.

6.2.1 Kinetic Analysis of Two-Joint Muscles: Lombard's Paradox

Representative publications: Molbec 1965; Carlsoo and Molbech 1966; Andrews and Hay 1983; Andrews 1987

Traditionally in biomechanics, the functional role of a muscle at a joint is classified using a standard kinetic method that involves the following steps:

1. Quantitatively describe a muscle and joint of interest in terms of the muscle line of action in a vicinity of the joint with respect to the joint center and joint axes *(X, Y, Z)*. With this information, the muscle moment arm components with respect to joint axes can be found (see section **5.3.1.3**).

2. Determine the moment of the muscle force (see equation 5.2) and its components acting on the distal segment with respect to the same joint axes at a given joint position (M_{OX}^m, M_{OY}^m, M_{OZ}^m).

3. Determine the resultant moment at the joint with respect to the same joint axes as step 2 (M_{OX}^R, M_{OY}^R, M_{OZ}^R) during motion of interest (by a means of inverse dynamics, see sections **2.1.1** and **2.1.2** in *Kinetics of Human Motion*).

4. Compare the components of the muscle joint moment $\mathbf{M}_O^m$ and the resultant joint moment $\mathbf{M}_j^R$ about each joint axis.

If the directions of muscle and resultant moment components about an axis are the same (i.e., their corresponding products are positive $M_j^m \cdot M_j^R > 0$, where $j = X, Y, Z$), the muscle is defined as a *joint agonist* with respect to a given axis during a given motion (this issue was briefly mentioned in **5.3.1.3**). For example, the resultant knee moment (in the sagittal plane) is an extension moment when one rises vertically from a chair (figure 6.2), and vastii, being knee extensors, actively produce the knee extensor moment during this task. Similar situations occur in many human movements, for example, jumping, landing, and the stance phase of running. During these tasks vastii are classified as joint agonists. If the muscle and resultant moments at the joint act in opposite directions ($M_j^m \cdot M_j^R < 0$), the muscle is called a *joint antagonist* for a given task. According to this definition of muscle function at a joint, a muscle can be an agonist with respect to one joint or joint axis and an antagonist with respect to another. For instance, iliacus is a hip extension antagonist but can also be a hip abduction and internal rotation agonist.

In the case of two-joint muscles, it is rather common that a muscle producing sagittal moments at the two joints has an agonistic action at one joint and an antagonistic action at the other. Several examples of these situations can be observed in figure 6.2. Rectus femoris has agonistic action at the knee (knee extension) and antagonistic action at the hip (hip flexion) when rising from a chair. Active two-joint hamstrings have the opposite functions: antagonistic at the knee (flexion) and agonistic at the hip (extension).

The kinetic method of classification of muscle function permits multiple roles of muscles. In some situations two-joint muscles, which act in one plane, can act against themselves by unavoidably producing an antagonistic action at one joint (e.g., quadrants I and III, figure 6.3). Two-joint muscles can also contribute to the resultant moments at both joints as

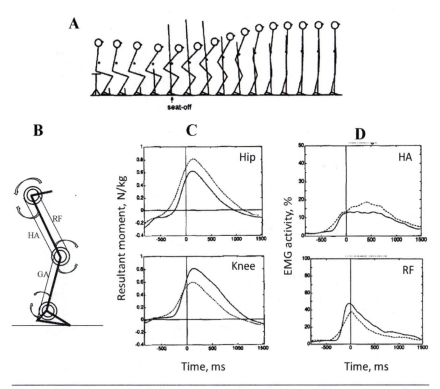

Figure 6.2   Resultant joint moments at the hip and knee and muscle activity of two-joint hamstrings (HA) and rectus femoris (RF) muscles during rising from a chair. *(a)* Stick figures of a typical example of rising from a chair. The small arrow corresponds to the instant of seat-off. The black dot anterior to the hip indicates the position of mass center of the body. The magnitude and direction of the vertical lines correspond to the magnitude and direction of the ground reaction force. *(b)* A diagram of the leg illustrating moment directions at the hip, knee, and ankle (arrows point to extension). Thin lines denote two-joint muscles: HA, RF, and gastrocnemius (GA). *(c)* Resultant hip and knee moments normalized to body mass during rising from a chair (positive values correspond to extension). Solid and dashed lines correspond to two techniques of performing task (natural and with increased trunk flexion). The vertical line corresponds to the seat-off instant. Mean data of 9 subjects. *(d)* Low-pass filtered electromyographic (EMG) activity of HA and RF normalized to a maximum voluntary contraction. The rest is the same as in part *c*.

(a, c, d) Reprinted from *Journal of Biomechanics* 27(11), C.A.M. Doorenbosch, J. Harlaar, M.E. Roebroeck, and G.J. Lankhorst, "Two strategies of transferring from sit-to-stand; the activation of monoarticular and biarticular muscles," 1299-1307, copyright 1994, with permission from Elsevier.

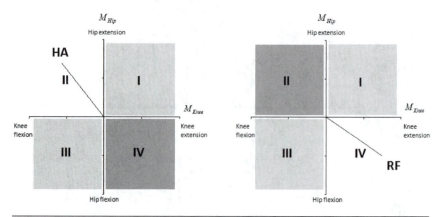

Figure 6.3 Functional roles of two-joint hamstrings (HA) and rectus femoris (RF) muscles determined by the kinetic classification method. The moment arm vectors of rectus femoris and hamstrings (specifying the muscle moment direction in the plane, see figure 6.1) are plotted in the plane of knee and hip resultant joint moments $(M_{Knee}$ and $M_{Hip})$. The positive direction is defined as extension and negative as flexion. Each muscle can act as an agonist in both joints (nonshaded quadrants), as an agonist in one joint and an antagonist in the other (lightly shaded quadrants), and as antagonists in both joints (heavily shaded quadrants).

agonists (figure 6.3*a,* quadrant II, HA; and figure 6.3*b,* quadrant IV, RF). This latter behavior intuitively seems to be more efficient.

Indeed, two-joint muscles are often highly active during the movement phases in which they act as agonists at both joints. In early swing of walking and running, for example, high RF electromyographic (EMG) activity coincides with hip flexion and knee extension moments; during late swing, high HA EMG activity contributes to the hip extension and knee flexion resultant moments (figure 6.4*a*). During load lifting with straight legs (so-called back-lift), the two-joint gastrocnemius and hamstrings operate in accordance with their agonistic functions at both joints (figure 6.4*b*).

In one quadrant of the joint moment plane (e.g., quadrant IV for HA and quadrant II for RF, figure 6.3), a two-joint muscle produces antagonistic actions at both joints. This situation seems the most unfavorable use of two-joint muscles. As expected, in many such situations (see, e.g., figure 6.4*b,* RF), two-joint muscles are not active and thus do not oppose the ongoing task.

The kinetic method of classifying the functions of two-joint muscles in the sagittal plane suggests muscle agonistic and antagonistic roles—a muscle is a complete agonist when it has agonistic functions at both joints, and it is a complete antagonist when it acts as an antagonist at both joints.

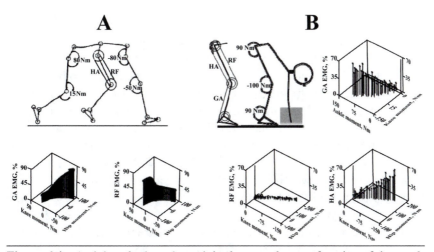

Figure 6.4 Activity of selected two-joint leg muscles as a function of the resultant moments at the joints they cross during swing phase of running *(a)* and load lifting *(b)*. *(a)* Top panel demonstrates leg positions and the values of the resultant hip and knee moments in early, mid-, and late swing during fast walking (2.7 m/s). The bottom panels show values of low-pass filtered electromyogram (EMG) of two-joint hamstrings (HA) and rectus femoris (RF) as functions of the hip and knee joint moments for each percentage of swing phase. Electromyogram was shifted in time to account for electromechanical delay between muscle excitation and muscle force development (see section **3.1.2.2**). Note that the moment direction convention in the panels is the same as in figure 6.3. In early swing, the resultant moments at the hip and knee are flexor and extensor, respectively, and RF, acting as an agonist at both joints, has the highest activity, whereas HA has antagonistic activity at both joints and the smallest EMG magnitude. In midswing the knee and hip moments are close to zero. In late swing the hip moment is extensor and knee moment flexor; HA in this phase is agonist at both joints and has the largest EMG magnitude, whereas RF, acting as an antagonist at both joints, has the lowest EMG. *(b)* The stick figure demonstrates the subject's leg position in the beginning of lifting a load (18 kg) and the corresponding resultant joint moments. The panels show EMG activity of two-joint gastrocnemius (GA), HA, and RF during load lifting as a function of joint moments. The joint moment convention is the same as in figure 6.3: extension moments are positive, flexion negative. Note that GA and HA act as agonists at both joints, demonstrating high EMG activity in early lift and diminishing activity with joint moments magnitudes as lifting progresses. Rectus femoris, in contrast, would have an antagonistic action at both joints if activated, but it is not active.

(a) Graphs reprinted with kind permission from Springer Science+Business Media: *Experimental Brain Research,* "Coordination of two-joint rectus femoris and hamstrings during the swing phase of human walking and running," 120(4), 1998, 479-486, B.I. Prilutsky, R.J. Gregor, and M.M. Ryan, fig. 2.

(b) Graphs reprinted from *Journal of Biomechanics* 31(11), B.I. Prilutsky, T. Isaka, A.M. Albrecht, and R.J. Gregor, "Is coordination of two-joint leg muscles during load lifting consistent with the strategy of minimum fatigue?," 1025-1034, copyright 1998, with permission from Elsevier.

In several movements, such as rising from a chair or jumping, the anatomical antagonists, two-joint hamstrings and rectus femoris, are active together (figure 6.2, *c* and *d*) and thus each produces not only antagonistic action at one of the joints but also antagonistic actions against each other. Given that a joint agonist is supposed to contribute positively to the joint moment and presumably to the ongoing task, which of the two muscles should be called agonist and which should be called antagonist? A precise answer to this question cannot be obtained if we use the standard kinetic method for muscle classification. As mentioned earlier, W.P. Lombard suggested an explanation for this seemingly paradoxical behavior of two-joint muscles: Lombard suggested that both muscles can be functional agonists if they satisfy certain conditions, which will be considered in the next sections. In short, *functional agonists* (or *task-agonists*) are muscles that assist the ongoing task even if they have an antagonistic action at a joint (see section **5.3.1.3**).

6.2.2 Kinematic Analysis of Two-Joint Muscles: Solution of Lombard's Paradox

Representative publications: Enklaar 1954; Landsmeer 1961; Andrews 1985; Kuo 2001

The kinematic procedures proposed to determine functional roles of two-joint muscles in multijoint systems are based on analysis of muscle length changes during motion. It is postulated that if an active two-joint muscle is shortening during propulsive motion, the muscle acts as a *functional* or *task agonist* and assists the ongoing motion. The rationale is that when a muscle shortens and produces force it is supplying mechanical energy to the system by generating positive power $P_m = F_m \cdot V_m$ (where P_m, F_m, and V_m are muscle power, force, and velocity; see introduction to chapter 4).

If the length of a two-joint muscle increases during propulsive motion, the muscle resists motion by absorbing mechanical energy of the moving body segments (the developed muscle power is negative). In this case, the two-joint muscle is called a *task antagonist*. Later in the chapter the reader will see that two-joint muscles can still contribute to the ongoing propulsive motion if their length is constant or increasing. If a goal of a motor task is to decelerate the body (e.g., during landing) or a segment (e.g., the shank during a terminal swing of gait, see figure 6.4*a*), muscles that elongate and absorb mechanical energy should be called task agonists and those generating energy, task antagonists.

Let us first demonstrate how length of two-joint muscles can change depending on joint kinematics and muscle moment arms. Consider a three-link chain with two joints served by 1 two-joint muscle producing a force (figure 6.5). Links 1 and 3 move through equal angular displacements in clockwise direction from the positions shown on the left to the positions on the right under the action of external moments (not shown in the figure). The points of muscle attachments on the bones in panels *(a)*, *(b)*, and *(c)* are different, which corresponds to different muscle moment arm values at the corresponding joints. The larger the distance between the muscle attachment point and the joint rotation axis, the larger the linear distance traveled by the attachment point. This is because the muscle displacement is related to the angle change and moment arm at the joint (see equations 5.11 and 5.17). Hence the changes in the muscle lengths in all three cases are different. The length is constant in *(a)*, decreases in *(b)*, and increases in case *(c)*.

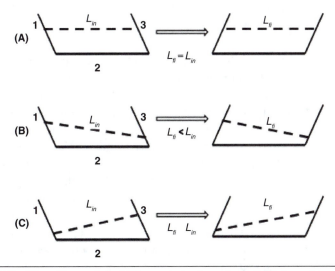

Figure 6.5 The effects of the differences in the points of origin and insertion of a two-joint muscle on its length at the same angular displacements at both joints. The solid lines are links of a 3-link kinematic chain and the dashed line is a two-joint muscle attached to links 1 and 3. The angular displacements are being caused by external moments (not shown), while the muscle is generating force. L_{in} and L_{fi} are initial and final muscle lengths. *(a)* The muscle length does not change (isometric action). *(b)* Muscle length decreases (concentric action). *(c)* Muscle length increases (eccentric action).

We recommend that as an exercise, the reader ponder what would happen to muscle lengths if a kinematic chain were simultaneously served by the two muscles shown in parts *(b)* and *(c)* and producing equal forces. One example of such architecture can be seen in figure 6.1; the biceps femoris long head (BFL) has a longer moment arm for hip extension and a smaller moment arm for knee flexion than its anatomical agonist semimembranosus (SM).

Under the same motion at the joints, the length of one such muscle decreases while the muscle does positive work (i.e., generates mechanical energy), whereas the length of the second muscle increases and does negative work (i.e., absorbs mechanical energy; see chapter 4). Hence, the shortening muscle stretches the second muscle while the second muscle resists this stretch. Note that according to a typical anatomical definition of the agonist and antagonist muscles (see section **5.3.1.3**), these muscles should be called agonists: they lie on the same side of the bone, act on the same joints, and move the joints in the same direction. It seems that this definition does not work properly here. This example illustrates the difficulties associated with precisely defining the agonist and antagonist roles for two-joint muscles in multijoint chains.

In the example shown in figure 6.5, the muscle moment arms (the shortest distances from the joint center to the line of muscle force action) change with the joint angle. Because these changes do not affect the following analysis, we shall replace figure 6.5 with another schematic where the muscle moment arms are fixed and 2 two-joint muscles are present (figure 6.6). This schematic is also more realistic because it represents an actual mechanism: a leg with a knee (K) and hip (H) and 2 two-joint muscles: (1) knee extensor and hip flexor (corresponds to rectus femoris) and (2) hip extensor and knee flexor (two-joint hamstrings).

In the model the muscle moment arms are constant and set to be, for example, $r_{K1} = 6$ cm, $r_{K2} = -4$ cm, $r_{H1} = -4$ cm, $r_{H2} = 6$ cm for muscles 1 (rectus femoris) and 2 (two-joint hamstrings) at the knee and hip joints, respectively (the values are from figure 6.1; positive values correspond to extension, negative to flexion). Note that $|r_{K1}| > |r_{H1}|$ and $|r_{K2}| < |r_{H2}|$; also $|r_{K1}| > |r_{K2}|$ and $|r_{H1}| < |r_{H2}|$. From $|r_{K1}| > |r_{H1}|$ it follows that equal changes of the H and K angles induce larger length changes of muscle 1 at K than at H. When both joints extend over the same angular range, the length of muscle 1 decreases.

For instance, if the joint angles change by 1 radian, the muscle length will decrease by 6 cm at the knee joint and increase by 4 cm at the hip joint. Hence, the decrease–increase difference is 2 cm. For muscle 2, length changes are similar. The length will decrease by 6 cm at the hip joint and increase by 4 cm at the knee joint.

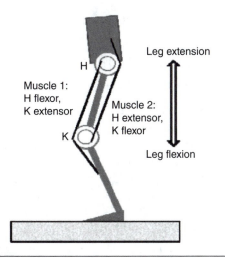

Figure 6.6 A multilink chain with 2 two-joint muscles. The muscles are anatomical antagonists at the K and H joints. Note that both joints extend during leg extension. Muscle 2 serves as a joint agonist at the hip and muscle 1 serves as a joint antagonist. At the knee the functions are reversed: muscle 1 assists joint movement while muscle 2 resists joint movement. We leave to the reader consider these muscles' functions during leg flexion.

Given the length of muscles 1 and 2 at the initial flexed leg position, under what conditions will the two muscles shorten during leg extension? When a joint angle changes by $\Delta\theta$ rad, the corresponding muscle length change is $\Delta L = r\Delta\theta$, and when the joint angular velocity is $\dot{\theta}$ the muscle velocity is $V = r\dot{\theta}$. In the following analysis, we define muscle-shortening velocity as positive (corresponding to the direction of muscle force that tends to shorten the muscle). Because length and velocity changes occur at both muscle ends, we can write

$$V_1 = r_{K1}\dot{\theta}_K + r_{H1}\dot{\theta}_H$$
$$V_2 = r_{K2}\dot{\theta}_K + r_{H2}\dot{\theta}_H,$$

[6.1a]

where all the symbols have been defined previously. Equations 6.1a can be written as scalar products of vectors, $V_1 = \mathbf{r}_1 \cdot \dot{\boldsymbol{\theta}}$ and $V_2 = \mathbf{r}_2 \cdot \dot{\boldsymbol{\theta}}$, where $\boldsymbol{\theta} = (\dot{\theta}_K, \dot{\theta}_H)^T$ is the vector of joint angular velocities and $\mathbf{r}_1 = (r_{K1}, r_{H1})$ and $\mathbf{r}_2 = (r_{K2}, r_{H2})$ are the vectors of muscle moment arms for muscles 1 and 2, respectively (see also figure 6.1, muscles RF, BFL, ST, and SM).

Assume that at some joint angular velocities, the length of a muscle stays constant, an isometric case (note that *isometric* refers here to the constant length of a two-joint muscle, not to a constant joint angle; figure 6.7). In the isometric case, the dot product of two vectors defining muscle velocity equals zero:

$$V_1 = \mathbf{r_1} \cdot \dot{\boldsymbol{\theta}} = r_{K1}\dot{\theta}_K + r_{H1}\dot{\theta}_H = 0$$
$$V_2 = \mathbf{r_2} \cdot \dot{\boldsymbol{\theta}} = r_{K2}\dot{\theta}_K + r_{H2}\dot{\theta}_H = 0.$$

[6.1b]

The dot products of vectors are zero when the vectors are orthogonal (see p. 12 in *Kinetics of Human Motion*). Hence, we may conclude that in isometric muscle conditions, the vector of muscle moment arms is orthogonal to the vector of joint angular velocity. By abusing the rules of mathematical drawings, these vectors are plotted in figure 6.7 by overlaying the plane of the joint angular velocities, $\dot{\theta}_K$ and $\dot{\theta}_H$, and the plane of the joint moment arms (r_K and r_H, figure 6.1). Compare the moment arm vector directions with the moment arm vectors of RF, BFL, ST and SM in figure 6.1. Equation 6.1b defines two kinematic constraints on joint angular velocities (two straight lines in the joint velocity plane). When joint velocity values are on the isometric constraint line of muscle 1, the length of muscle 1 stays constant. Deviation from the line to the right or left will cause muscle 1 to shorten or lengthen, respectively. The same is true for the isometric constraint of muscle 2 except that a shift from the line to the left leads to muscle shortening and shift to the right to lengthening. Note the differences in the slopes of the two isometric constraint lines. The slope reflects the ratio of the angle changes at the two joints that are required to maintain a constant length of the muscle during leg extension. The fact that the slope of the isometric constraint line of muscle 1 exceeds 45° means that to keep muscle 1 at constant length, hip joint velocity should be higher than knee joint velocity.

The joint velocity plane is divided by the isometric constraint lines into four sectors: (a) both muscles shorten, (b) both muscles lengthen, (c) muscle 1 shortens while muscle 2 lengthens, and (d) muscle 2 shortens and muscle 1 lengthens (figure 6.7*b*). Thus, given realistic moment arms of muscles 1 and 2 (see figure 6.1), both muscles can shorten during leg extension and both muscles can be called task agonists according to the kinematic definition given previously. We can see now that in some movements with simultaneous knee and hip extension (e.g., rising from a chair, figure 6.2), the hamstrings and rectus femoris, two-joint anatomical antagonists, can both shorten and therefore can contribute to the ongoing leg extension and be called task agonists.

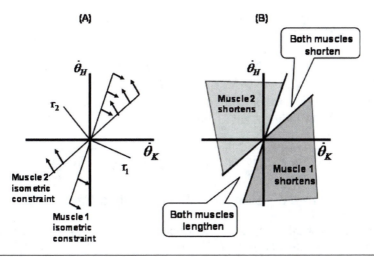

Figure 6.7 The relation between joint angular velocities and muscle length. Positive angular velocities correspond to joint extension. *(a)* The angular velocity pairs that are on the isometric constraints correspond to constant muscle length. This means that along these constraints, the joints move in such a way that length of the muscle does not change. When velocities deviate from the isometric constraint, the muscle length changes. The constraints divide the joint velocity plane into the areas where the muscle shortens (denoted by arrows) or lengthens. r_1 and r_2 are moment arm vectors of muscles 1 and 2. *(b)* Isometric constraints divide the joint velocity plane into four sectors: (a) both muscles shorten, (b) both muscles lengthen, (c) muscle 1 shortens while muscle 2 lengthens, and (d) muscle 2 shortens and muscle 1 lengthens.

Adapted, by permission, from A.D. Kuo, 2001, The action of two-joint muscles: The legacy of W.P. Lombard. In *Classics in movement science,* edited by M.L. Latash and V.M. Zatsiorsky (Champaign, IL: Human Kinetics), 294.

Let us consider another case when the length of one muscle (e.g., muscle 1) stays constant while the other muscle shortens ($V_2 > 0$). The corresponding equations are

$$V_1 = \mathbf{r_1} \cdot \dot{\mathbf{\theta}} = r_{K1}\dot{\theta}_K + r_{H1}\dot{\theta}_H = 0$$

$$V_2 = \mathbf{r_2} \cdot \dot{\mathbf{\theta}} = r_{K2}\dot{\theta}_K + r_{H2}\dot{\theta}_H > 0.$$

[6.2]

Recall that we have defined moment arms as negative for flexor moments and positive for extensor moments (see figure 6.1). From 6.2 the conditions can be derived in which muscle 1 is isometric, muscle 2 is shortening, and both joints extend ($\dot{\theta}_H > 0$, $\dot{\theta}_H > 0$):

$$r_{K2} - r_{H2}\frac{r_{K1}}{r_{H1}} > 0. \qquad [6.3]$$

Note that in this and the previous examples, joint K extends although muscle 2 exerts a flexor moment at the joint.

■ ■ ■ *FROM THE LITERATURE* ■ ■ ■

Lombard's Paradox: Is It Really a Paradox?

Source: Lombard, W.P. 1903. The action of two-joint muscles. *American Physical Education Review* 9:141-145

In this short but seminal treatise, W.P. Lombard discussed the possibility that a pair of anatomically antagonistic two-joint muscles may cease to be antagonists in certain conditions and, instead of opposing each other, may reinforce one another by becoming pseudo-antagonists. Also each of these muscles can produce extension at a joint for which it has a flexor action. This fact is known as Lombard's paradox. In Lombard's words:

"A muscle can cause the extension of a joint which it can flex. In order that this can occur the following conditions are essential:

- It must have a better leverage at the end by which it acts as extensor.
- There must be a two-joint muscle that flexes the joint which the muscle in question extends, and extends the joint which it flexes.
- It must have sufficient leverage and strength to make use of the passive tendon action of the other muscle." (pp. 143-144)

All these requirements are satisfied in the example presented in figure 6.7; equations 6.2 and 6.3 reveal the conditions under which "a muscle can cause the extension of a joint which it can flex." This muscle (muscle 2, hamstrings) can do this if equation 6.3 is true and muscle 1 satisfies an isometric constraint; that is, it acts as an inextensible tendon.

The kinematic analysis of two-joint muscles demonstrates that rectus femoris and two-joint hamstrings, which are anatomical antagonists, can either shorten or avoid lengthening during leg extension (figure 6.7*b*). Thus, these muscles can be task agonists during certain phases of leg extension, according to this kinematic classification of muscle function.

6.3 MECHANICAL ENERGY TRANSFER AND SAVING BY TWO-JOINT MUSCLES

One question about two-joint muscles that has intrigued many anatomists and physiologists is why these muscles exist. If they have evolved over millions of years, they are likely to have certain advantages over one-joint muscles with the same actions at the corresponding joints (flexion–extension, adduction–abduction, and internal-external rotation), which would have been able to develop the same joint moments and thus produce the same joint motion. One possible advantage of two-joint muscles could be efficiency with which mechanical energy is distributed along the limb segments.

6.3.1 Tendon Action of Two-Joint Muscles

Representative publications: Lombard 1903; Elftman 1939b; Morrison 1970; Gregoire et al. 1984; Bobbert and van Ingen Schenau 1988; Gregersen et al. 1998; Carroll et al. 2008

Tendon action is the term introduced to describe muscle behavior in which muscle acts as a tow rope and transmits tension and energy from the muscle attachment point on one segment to the attachment point on the other segment (see also section **6.2.2.3** in *Kinetics of Human Motion*). All muscles are capable of exhibiting this behavior, but tendon action of two-joint muscles is considered more common. Before describing the possible functional significance of tendon action, we present several intuitive definitions, discuss methods of its evaluation, and show several mechanical models illustrating its effects.

6.3.1.1 Illustrative Examples of Tendon Action of Two-Joint Muscles

Representative publications: van Ingen Schenau 1989; Bobbert et al. 1986a; Prilutsky and Zatsiorsky 1994; Prilutsky et al. 1996a

Consider again a three-link kinematic chain with two joints and a two-joint muscle attached to the distal links (figure 6.8). Let us assume that

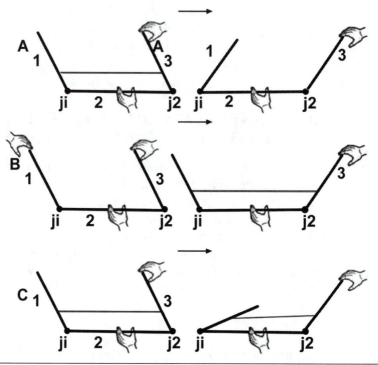

Figure 6.8 Illustrations of the tendon action of a two-joint muscle (thin line) crossing two joints (j1 and j2) of a three-link kinematic chain. External moment applied to link 3 causes motion of link 1. *(a)* The muscle acts as an inextensible string. Work done on the muscle due to displacement of its attachment point at joint 2 equals work done by the muscle over movement of its attachment point at joint 1. *(b)* The muscle acts as an elastic string, and link 1 movement is prevented by an external force. Work done at joint 2 by an external force is completely absorbed by the muscle, so no energy is transported from the attachment on link 3 to the attachment on link 1. *(c)* The muscle acts as a prestretched elastic string; its length in the initial chain configuration exceeds its rest length. Application of an external moment to link 3 and release of link 1 change the configuration to that shown in the right panel. Link 1 receives energy both transmitted from link 3 via tendon action and generated by the muscle because of elastic recoil.

the two-joint muscle is an inextensible string. If one grasps the middle link with one hand and rotates link 3 to the right as shown on the left panel of figure 6.8*a,* link 1 will also move to occupy the position shown in the right panel. This is because tension at the right attachment point, introduced by an externally applied force (hand), is the same as at the left attachment point. (Link 1 is attached to an external load, not shown, which must be overcome by the external force applied to link 3.)

In this example, the hand produces an external moment (with respect to joint 2) and moves attachment point on link 3 by distance $-\Delta L_2$ in the direction opposite to the muscle force direction, which is shortening. The work done by force F over muscle length change $-\Delta L_2$ at joint 2 is negative: $-W_2 = -\Delta L_2 \cdot F$ (subscripts indicate work and length changes at joint 2). The muscle length change due to angular displacement at joint 1 (ΔL_1) is in the direction of muscle force, so work done by muscle force at joint 1 is positive: $W_1 = \Delta L_1 \cdot F$. Because $|-\Delta L_2| = |\Delta L_1|$ (the total muscle length does not change), the total work is zero: $W = F(\Delta L_1 - \Delta L_2) = 0$, or $|-W_2| = |W_1|$. Hence, work done by an external force on joint 2, or energy introduced to the system at joint 2, equals the positive work done by the muscle at joint 1. In other words, mechanical energy is transferred through the muscle from the attachment point in link 3 to link 1, or from joint 2 to joint 1. The action of a two-joint muscle resulting from motion and energy transmission is called tendon action.

In the next example (figure 6.8*b*), the muscle is represented by an elastic string. In addition, link 1 is prevented from moving by an external force while link 3 is rotated by an external force. Work done on the muscle at joint 2 is the same as in the previous example: $-W_2 = -\Delta L_2 \cdot F$. Work done by muscle force at joint 1 is zero ($W_1 = 0$) because the muscle attachment point did not move. As a result, work done by the muscle is negative and equals $W = -W_2$. The energy introduced to the system by an external force at joint 2 is completely absorbed by the muscle, and no energy was used to move muscle attachment point on link 1 or to change the angle at joint 1.

In the last example (figure 6.8*c*), the muscle is represented as an elastic string whose initial length exceeds its rest length. Application of an external moment to link 3 and simultaneous release of link 1 change the configuration to that shown in the right panel. Work done on the muscle because of angular displacement at joint 2 is $-W_2 = -\Delta L_2 \cdot F$. Work done by muscle force to move attachment point on link 1 is $W_1 = |-\Delta L_2 F| + \Delta L_e F$. The product $|-\Delta L_2|F$ is the energy transported via tendon action of the muscle from muscle attachment on link 3 to link 1. The term $\Delta L_e F$ is positive work done, or energy generated, by the elastic string itself. The total work done by the muscle is $W = F(\Delta L_1 - \Delta L_2 + \Delta L_e) = F\Delta L_e > 0$. In this example, link 1 moved a greater angular distance and the displacement of muscle attachment on link 1 is greater because, in addition to the tendon action, the muscle shortened because of its elasticity and thus generated positive work.

Another, more realistic physical model illustrating tendon action of two-joint muscles of the leg is shown in figure 6.9. It consists of four links representing the main leg segments (foot, shank, thigh, and pelvis) that are interconnected by three joints: hip, knee and ankle. Two-joint gastrocnemius

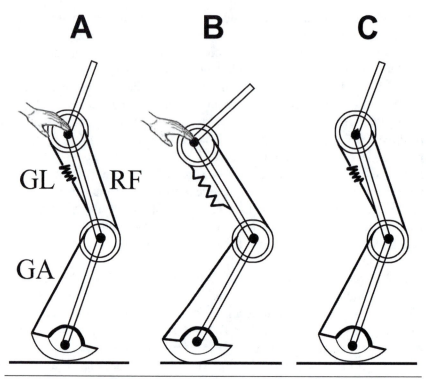

Figure 6.9 Illustration of tendon action of two-joint muscles during flexion (from position *a* to position *b*) and extension (from position *b* to position *c*). Two-joint muscles GA and RF are represented by inextensible strings and one-joint hip extensors by a spring. For other explanations see the text.

(GA, ankle extensor and knee flexor) and rectus femoris (RF, knee extensor and hip flexor) are represented by inextensible strings. The role of one-joint hip extensors (GL, gluteal muscles) is played by a spring. If one presses down on the hip axis of rotation of the model (figure 6.9*a*), the leg will flex in all three joints and occupy the position depicted in figure 6.9*b*. The stretched spring GL, which crosses the flexed hip, absorbs the work done by the external force applied to the hip (W_{tot}). Notice that the ankle and knee joints have also flexed under the external force. This occurs because pressing down on the leg causes the foot to rotate counterclockwise. This rotation displaces the distal attachment point of GA downward, so GA pulls the shank down and causes the knee to flex by applying tension at GA's proximal attachment point. The knee flexion in turn displaces the distal attachment of RF and it pulls the pelvis, causing the hip joint to flex. This example illustrates how tendon action of GA and RF can distribute the energy of the body along the leg. The example also provides a means for quantitative characterization of tendon action based on mechanical energy analysis.

Leg flexion occurs despite resistance of the inextensible strings GA and RF at the ankle and knee, respectively, whose tension (F_{st}) develops extension moments (M_j) at the joints $M_j = d_j F_{st}$, where d_j is the string moment arm at joint j. Because flexion occurs at the ankle, knee, and hip joints while the joint moments are in extension, work done at the joints is negative and equal (see chapter **6** in *Kinetics of Human Motion*):

$$W_a = \int_{t_o}^{t_1} M_a \cdot \dot{\theta}_a \, dt < 0, \qquad [6.4]$$

$$W_k = \int_{t_o}^{t_1} M_k \cdot \dot{\theta}_k \, dt < 0, \qquad [6.5]$$

$$W_h = \int_{t_o}^{t_1} M_h \cdot \dot{\theta}_h \, dt < 0, \qquad [6.6]$$

where W_a, W_k, and W_h are work done at the ankle, knee, and hip, respectively; $M_j \cdot \dot{\theta}_j$ is the product of joint moment and angular velocity at joint j, that is, the joint power P_j; and t_o and t_1 are the initial and final time instances of the motion: t_o corresponds to leg configuration in figure 6.9*a* and t_1 to figure 6.9*b*. Note that joint moments and joint velocities are functions of time.

If we recall the principle of virtual work (see sections **5.3.2.2.1** of this book and **2.2.2** of *Kinetics of Human Motion*) and that the muscle moment arm can be expressed as the ratio of small changes of muscle length and joint angle (see equation 5.17) or muscle and joint velocities, the work of moment at each joint can also be written in terms of tension and velocities of the (inelastic) strings and the (elastic) spring due to the corresponding joint angle changes:

$$W_a = \int_{t_o}^{t_1} F_{GA} \cdot \frac{V_{GA,a}}{\dot{\theta}_a} \cdot \dot{\theta}_a \, dt = \int_{t_o}^{t_1} F_{GA} \cdot V_{GA,a} \, dt < 0, \qquad [6.7]$$

$$W_k = \int_{t_o}^{t_1} (F_{GA} \cdot \frac{V_{GA,k}}{\dot{\theta}_k} \cdot \dot{\theta}_k - F_{RF} \cdot \frac{V_{RF,k}}{\dot{\theta}_k} \cdot \dot{\theta}_k) \, dt =$$

$$\int_{t_o}^{t_1} (F_{GA} \cdot V_{GA,k} - F_{RF} \cdot V_{RF,k}) \, dt < 0, \qquad [6.8]$$

$$W_h = \int_{t_o}^{t_1} (F_{RF} \cdot \frac{V_{RF,h}}{\dot{\theta}_h} \cdot \dot{\theta}_h - F_{GL} \cdot \frac{V_{GL,h}}{\dot{\theta}_h} \cdot \dot{\theta}_h) \, dt =$$

$$\int_{t_o}^{t_1} (F_{RF} \cdot V_{RF,h} - F_{GL} \cdot V_{GL,h}) \, dt < 0, \quad [6.9]$$

where terms $F_i \cdot V_{i,j}$ are the products of force i and the rate of length change in the string i due to the corresponding joint velocity at joint j. Note that the knee and hip joints are crossed by two muscles: GA and RF, and RF and GL, respectively.

Adding equations 6.7 through 6.9 and regrouping the terms yields an expression for the total work done by the leg muscles during flexion:

$$W_{tot} = W_a + W_k + W_h = \int_{t_o}^{t_1} \left(F_{GA} \cdot V_{GA,k} - F_{GA} \cdot V_{GA,a} \right) dt +$$

$$\int_{t_o}^{t_1} \left(F_{RF} \cdot V_{RF,H} - F_{RF} \cdot V_{RF,k} \right) dt + \quad [6.10]$$

$$\int_{t_o}^{t_1} \left(F_{GL} \cdot V_{GL,h} \right) dt < 0.$$

The integral terms in parentheses on the right-hand side of equation 6.10 are powers developed by the three "muscles" GA, RF, and GL during leg flexion. Accordingly, each integral represents the work done by the muscle. Note that power and work of GA and RF are zero because the length of the muscles does not change.

We can see now that the ankle moment absorbs mechanical energy; that is, its power and work are negative (equations 6.4 and 6.7). However, the muscles crossing the ankle (GA) do not do negative work; that is, they do not absorb energy—the GA does zero work (the first integral in equation

6.10). We can say that energy absorbed by the ankle moment $-\int_{t_o}^{t_1} F_{GA} \cdot V_{GA,a} \, dt$

is used to flex the knee by GA, which does positive work $\int_{t_o}^{t_1} F_{GA} \cdot V_{GA,k} \, dt$

at the knee. The positive work done on the knee is due to the energy transported from the distal attachment point of GA at the ankle to the proximal attachment at the knee, or simply from the ankle to the knee. The energy transported to the knee is absorbed by the knee moment; the knee moment does negative work (equations 6.5 and 6.8). Again, muscles crossing the knee (GA and RF) do not absorb this energy because they do zero work (first two integrals in 6.10). Instead, the energy absorbed at the knee $\int_{t_o}^{t_l} (F_{GA} \cdot V_{GA,k} - F_{RF} \cdot V_{RF,k})\, dt$ is used to flex the hip via RF, which does positive work at the hip: $\int_{t_o}^{t_l} (F_{RF} \cdot V_{RF,h})\, dt$. The GL spring absorbs this energy. In fact, because energy loss to heat in the system can be neglected in our example, the total work done on the system sums to negative work done by GL, energy absorption (see equation 6.10).

The preceding example demonstrates that the tendon action of GA and RF permits energy absorbed at distal joints to be transported in the distal-to-proximal direction and absorbed by the hip extensor.

If the leg model is released from the flexed position (figure 6.9*b*), it will extend all three joints and occupy an extended posture (figure 6.9*c*). Extending the hip joint by GL causes the pelvis to rotate counterclockwise and pull the proximal attachment of RF upward. Because the RF length does not change, this causes knee extension (tendon action of RF). Knee extension in turn pulls the proximal attachment point of GA upward, through which the GA tendon action (pulling the distal attachment point on the foot) causes ankle extension.

In this case, mechanical energy released by the spring GL contributes to positive work of the hip moment (equation 6.9 multiplied by –1) and to positive work done by the knee and ankle moments (equations 6.8 and 6.7 multiplied by –1). Thus, during leg extension, the tendon actions of GA and RF allow for energy transport in the proximal-to-distal direction.

▪ ▪ ▪ *FROM THE LITERATURE* ▪ ▪ ▪

Tendon Action of Two-Joint Muscles Permits Transfer of Mechanical Energy Between Body Segments

Source: Bobbert, M.F., P.A. Huijing, and G.J. van Ingen Schenau. 1986a. An estimation of power output and work done by the human triceps surae muscle-tendon complex in jumping. *J Biomech* 19:899-906

Using a musculoskeletal model of the human gastrocnemius muscle and recorded kinematics and ground reaction force during one-legged

(continued)

jumping of 10 subjects, the authors computed the amount of energy transported from the knee to the ankle during the last 90 ms before toe-off. (The method for calculation of energy transfer by the gastrocnemius is similar to that explained later in the chapter, see section **6.3.1.2.2.**) According to the estimates, the amount of energy transported from the knee to the ankle (i.e., generated by proximal muscles and used to extend the ankle) was 25% of the positive work done by the ankle joint moment. The authors concluded that the energy transferred from the knee to the ankle via the gastrocnemius is important for jumping performance.

In the physical model discussed previously and in the examples of Lombard, it was assumed that two-joint muscles are inextensible during the course of motion. In reality, two-joint muscle can stretch and thus dissipate some of the body's energy during flexion and can also transport energy in the distal-to-proximal direction (as illustrated in figure 6.9, *a* and *b*). Two-joint muscles can also shorten during leg extension (these cases are illustrated in figures 6.7 and 6.9, *b* and *c*) and contribute to energy generation while transporting energy in the proximal-to-distal direction.

6.3.1.2 Methods of Energy Transfer Estimation

The mechanical energy considerations we have discussed can be used to quantify the tendon action of two-joint muscles by determining how much and in which direction energy is transported during motion. There are two somewhat similar methods that can be used for this purpose.

6.3.1.2.1 Energy Generated by Joint Moment and Muscles at a Joint

Representative publications: Morrison 1970; Prilutsky and Zatsiorsky 1994; Carroll et al. 2008

The model in figure 6.9 and the analysis of energy transfer through GA and RF indicate that power developed by a joint moment (the product in equations 6.4-6.6),

$$P_j = M_j \cdot \dot{\theta}_j, \qquad [6.11]$$

where P_j, M_j, and $\dot{\theta}_j$ are joint power, resultant moment, and joint angular velocity, respectively, does not have to be equal to the sum of powers developed by each muscle crossing that joint:

$$P_m = \sum_{i=1}^{N} P_i = \sum_{i=1}^{N} F_i \cdot V_i, \qquad [6.12]$$

where P_m is the total power of muscles crossing a given joint, F_i and V_i are force and velocity of muscle i crossing the joint, and N is the number of muscles crossing the joint. If all muscles across a joint are one-joint, it follows from the principle of virtual work (see equation 5.15) that $P_j = P_m$. However, if at least 1 two-joint muscle crosses the joint, the powers P_j and P_m may be unequal (compare, e.g., equations 6.4-6.6 and 6.10). Thus, the difference between the joint moment power and the total power of muscles crossing the joint,

$$P_t = P_j - P_m, \qquad [6.13]$$

indicates the rate at which energy is transmitted through two-joint muscles from or to the joint. It is important to note that the sum of joint powers across all joints must be equal to the sum of muscle powers across all muscles of the system,

$$P_{tot} = \sum_{j=1}^{N_j} P_j = \sum_{i=1}^{N_i} P_i, \qquad [6.14]$$

where N_j and N_i are the number of joints and muscles in the system, respectively.

Equation 6.13 can be used to determine direction and rate of energy transfer, which depend on signs of the powers comprising the equation. There are five possible variants of energy transfer by two-joint muscles. They are described in table 6.2 in *Kinetics of Human Motion*. We repeat this table here for convenience.

Some of the variants of energy transfer by two-joint muscles were considered in detail earlier (figure 6.9, *a* and *b*). For instance, variant 4 ($P_t < 0$, $P_j < 0$, $P_m \leq 0$, see equation 6.13) corresponds to energy transfer from the ankle by two-joint GA at the rate of $-P_t = -F_{GA} \cdot V_{GA,k}$, whereas the ankle moment absorbs energy at a rate of $-P_j = -F_{GA} \cdot V_{GA,a}$ and GA absorbs energy at a rate of $P_{GA} = (F_{GA} \cdot V_{GA,k} \cdot -F_{GA} \cdot V_{GA,a}) = 0$ (see equations 6.7 and 6.10). Variant 5 (b) ($P_t = 0$, $P_j < 0$, $P_m \leq 0$) corresponds to the knee joint receiving energy delivered by GA from ankle at the rate of $P_t = F_{GA} \cdot V_{GA,k}$, energy is transferred from the knee to the hip by RF at the rate of $-P_t = -F_{RF} \cdot V_{RF,k}$, and the muscles crossing the knee generate energy at the rate of $P_m = P_{GA} + P_{RF} = 0$. When variant 3 ($P_t > 0$, $P_j \leq 0$, $P_m < 0$) occurs, the hip joint is receiving energy delivered by RF at the rate of $P_t = F_{RF} \cdot V_{RF,k}$ and is absorbing energy at the rate of $P_h = (F_{RF} \cdot V_{RF,h} - F_{GL} \cdot V_{GL,h})$ (equation 6.9), whereas muscles around the hip absorb energy at a rate of $P_m = P_{RF} + P_{GL} < 0$.

Table 6.2   Mechanical Energy Transfer By Two-Joint Muscles

				Energy transfer		
Variant of energy transfer	Joint power, $P_j(t)$	Summed muscle power, $P_m(t)$	Difference, $P_t(t)$	Rate	Direction	Muscles of the jth joint generate *(gen)* or absorb *(abs)* power at a rate of
1	>0	≥0	>0	$|P_t(t)|$	To joint *j*; from joints $(j-1)$, $(j+1)$, or both	$|P_j(t)| - |P_t(t)|$ *gen*
2	≥0	>0	<0	$-|P_t(t)|$	From joint *j*; to joints $(j-1)$, $(j+1)$, or both	$|P_j(t)| + |P_t(t)|$, *gen*
3	≤0	<0	>0	$|P_t(t)|$	To joint *j*; from joints $(j-1)$, $(j+1)$, or both	$-[|P_j(t)| + |P_t(t)|]$, *abs*
4	<0	≤0	<0	$-|P_t(t)|$	From joint *j*; to joints $(j-1)$, $(j+1)$, or both	$-[|P_j(t)| - |P_t(t)|]$, *abs*
5	≥0 <0	≥0 <0	=0	(a) $|P_t(t)| = 0$ or (b) $|P_t(t)|$ and $-|P_t(t)|$	No energy transfer or transfer to joint *j* from one of the joints and from joint *j* to another joint	$|P_j(t)|$, *gen* or $-|P_j(t)|$, *abs*

Adapted from *Journal of Biomechanics* 27(1), B.I. Prilutsky and V.M. Zatsiorsky, "Tendon action of two-joint muscles: Transfer of mechanical energy between joints during jumping, landing, and running," 25-34, copyright 1994, with permission from Elsevier.

Variants of energy transfer 1 ($P_t > 0$, $P_j > 0$, $P_m \geq 0$) and 2 ($P_t < 0$, $P_j \geq 0$, $P_m > 0$) occur during leg extensions (figure 6.9, *b* and *c*). The reader is left to derive the corresponding equations for these variants.

Estimates of the rate (P_t) and amount of energy transfer by two-joint muscles (the latter is the time integral of P_t) have been obtained for human landing from a drop jump, for a squat vertical jump, and jogging.

■ ■ ■ *FROM THE LITERATURE* ■ ■ ■

Two-Joint Leg Muscles Transfer Mechanical Energy in the Distal-to-Proximal Direction During the Yield Phase of Landing and Running and in the Proximal-to-Distal Direction During the Push-Off Phase of a Squat Jump or Running

Source: Prilutsky, B.I., and V.M. Zatsiorsky. 1994. Tendon action of two-joint muscles: transfer of mechanical energy between joints during jumping, landing, and running. *J Biomech* 27:25-34

Equation 6.13 and table 6.2 were used to determine the directions of energy transfer between the leg joints by two-joint muscles during three tasks: landing after a drop jump, a squat jump, and jogging. Powers P_i, P_j, and P_m (see equation 6.13 and table 6.2) were computed based on a musculoskeletal model of the leg and inverse dynamics analysis. According to the results, the mechanical energy was transferred from distal to proximal joints (variants 3 and 4 of the energy transfer; table 6.2) during the yield phases of landing and jogging. The amount of energy transfer was up to 50% of the negative work done at the proximal joints (knee and hip). During the push-off phases of a squat jump and jogging, two-joint muscles transferred energy from proximal to distal joints (variants 1 and 2 of energy transfer; table 6.2) and the amount of this energy transfer approached the energy generated by the hip moment in the squat jump. It was concluded that during locomotion, the tendon action of two-joint muscles compensates for the insufficient capacity of distal muscles to do work by distributing mechanical energy along the leg. During the push-off phase, large proximal muscles help extend the distal joints by transferring a part of the generated mechanical energy. During the shock-absorbing phase, the proximal muscles help distal muscles dissipate the mechanical energy of the body.

6.3.1.2.2 Work Done by a Two-Joint Muscle at the Adjacent Joint

Representative publications: Elftman 1939a, 1939b; Bobbert et al. 1986b; Jacobs et al. 1996; Prilutsky et al. 1996a; Voronov 2004

The energy transfer estimation method shown here does not distinguish the contribution of individual two-joint muscles in this process. Energy transferred by an individual two-joint muscle can be obtained based on the relation between total muscle power (P_i) and the power at each joint it spans (distal and proximal, $P_{i,d}$ and $P_{i,p}$), which follows from the principle of virtual work (see equation 5.15):

$$P_i = F_i \cdot V_i = M_d \cdot \dot{\theta}_d + M_p \cdot \dot{\theta}_p = P_{i,d} + P_{i,p}. \qquad [6.15]$$

It follows that the power produced by the muscle moment at the distal (or proximal) joint depends on both the total muscle power and the power produced by the muscle at the proximal (distal) joint. Several variants of energy transfer by a two-joint muscle can be distinguished based on the signs and values of the terms in equation 6.15) (table 6.3).

Table 6.3   Selected Variants of Mechanical Energy Transfer By a Single Two-Joint Muscle

Variant of energy transfer	Power developed by the muscle, P_i	Power developed by the muscle at the proximal joint, $P_{i,p}$	Power developed by the muscle at the distal joint, $P_{i,d}$	Direction and rate of energy transfer by the muscle
1	<0	>0	<0	Energy transfer from distal joint at rate $-P_{i,d}$. Muscle absorbs energy at rate $-P_i$. Energy delivered to proximal joint at rate $P_{i,p}$.
2	>0	>0	<0	Energy transfer from distal joint at rate $-P_{i,d}$. Muscle generates energy at rate P_i. Energy delivered to proximal joint at rate $P_{i,p}$.
3	<0	<0	>0	Energy transfer from proximal joint at rate $-P_{i,p}$. Muscle absorbs energy at rate $-P_i$. Energy delivered to distal joint at rate $P_{i,d}$.
4	>0	<0	>0	Energy transfer from proximal joint at rate $-P_{i,p}$. Muscle generates energy at rate P_i. Energy delivered to distal joint at rate $P_{i,d}$.

Note: P_i is power developed by the muscle; $P_{i,p}$ and $P_{i,d}$ are powers developed by the muscle at the proximal and distal joints, respectively.

Adapted from *Journal of Biomechanics* 29(4), B.I. Prilutsky, W. Herzog, and T. Leonard, "Transfer of mechanical energy between ankle and knee joints by gastrocnemius and plantaris muscles during cat locomotion," 391-403, copyright 1996, with permission from Elsevier.

As can be noticed, variants 1 and 2 (energy transfer in the distal-to-proximal direction) can occur during a shock absorbing phase of locomotion (see, for example, figure 6.9, *a* and *b*). Some two-joint muscles can also transfer energy from distal to proximal joints during the push-off phase of leg extension, as suggested by Lombard and demonstrated in figure 6.10. Specifically, two-joint hamstrings deliver energy to the hip joint during jumping and sprint push-off in the amount of 7% and 11% of the work done at the hip. Variants 3 and 4 typically correspond to a push-off phase and energy transfer in the proximal-to-distal direction (see figure 6.9, *b* and *c,* and figure 6.10).

Mechanical energy transferred from one segment (or a joint) to another does not necessarily reach its final destination during a given motion. For instance, the mechanical energy generated by proximal muscles may be partly transferred by two-joint muscles to the distal ankle joint and increase the rate of ankle energy generation. The additional ankle energy then contributes to acceleration of other leg segments via multiple mechanical pathways, including joint forces and moments (see chapter 6 in *Kinetics of Human Motion*).

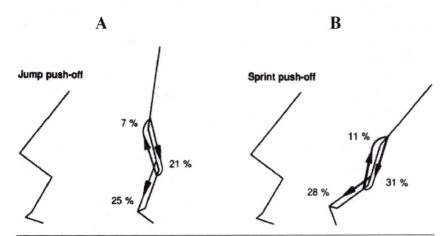

Figure 6.10   The relative contribution of energy transfer by two-joint muscles to work done by leg joint moments during jump push-off *(a)* and sprint push-off *(b)*. Arrows along the two-joint gastrocnemius, rectus femoris, and hamstrings muscles specify direction of energy transfer. The values of energy transfer during the last 90 ms of push-off are expressed in percentage of work done at the joint. Powers generated by the muscles at the joints were calculated based on motion recordings and musculoskeletal modeling.

Reprinted from *Journal of Biomechanics* 29(4), R. Jacobs, M.F. Bobbert, and G.J. van Ingen Schenau, "Mechanical output from individual muscles during explosive leg extensions: The role of biarticular muscle," 513-523, copyright 1996, with permission from Elsevier.

6.3.1.3 Tendon Action and Jumping Performance

Representative publications: van Ingen Schenau 1989; Pandy and Zajac 1991; van Soest et al. 1993

W.P. Lombard and a number of other researchers have speculated that the presence of two-joint muscles and their tendon action might improve the performance of propulsive movements in frogs and humans. Experimental studies to test these hypotheses are virtually impossible. Even though it may be possible to convert a two-joint muscle into a one-joint muscle using tendon transfer, the two-joint muscle would be much shorter and thus its force generation abilities would be compromised. Muscle shortening velocity and tendon slack length would also change. Therefore, the only realistic way to test the performance of the musculoskeletal systems with different designs is to use mathematical, physical, or computer simulation models.

A model called Jumping Jack was developed to demonstrate how energy generated at the knee can be more fully exploited to increase jumping performance (figure 6.11). The trunk of the model can move only in the vertical direction along a vertical rail. A one-joint knee extensor is represented by a spring that can store potential energy when the trunk is pushed downward. A wire of adjustable length represents the gastrocnemius. The length determines the knee angle at which the wire engages and starts ankle extension following model release. It appears that there is an optimal length of the wire at which the Jumping Jack achieves maximum jumping height. If the wire engages too early, the model loses contact with the ground before the knee spring completely releases its potential energy; that is, the knee is not fully extended at takeoff. If the wire is too long, ankle extension is delayed and a substantial amount of knee potential energy is converted into rotational energy of the shank and thigh and dissipated as heat (in human performers, joint angular velocity decreases before the extremity assumes a fully extended position—see section **3.1.1.1.7** in *Kinematics of Human Motion*). At the optimal wire length, an optimal amount of knee spring energy is used to propel the trunk upwards.

Similar results were obtained using computer simulation models. When the two-joint gastrocnemius muscle is substituted with two equivalent one-joint muscles (ankle extensor and knee flexor with the same moment arms as gastrocnemius), coupling between the knee and ankle joints is removed and the maximum jump height decreases. These results can be explained by the fact that a system without a two-joint gastrocnemius would have to deactivate one-joint knee extensor muscles much earlier before the takeoff to prevent overextension of the knee. As a consequence, knee extensors do a decreased amount of positive work. When the gastrocnemius is present, it

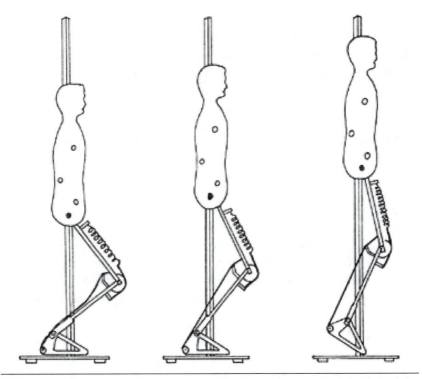

Figure 6.11 Jumping Jack, a physical model demonstrating the role of two-joint gastrocnemius in jumping performance (see text for further explanations).

Reprinted from *Human Movement Science* 8(4), G.J. van Ingen Schenau, "From rotation to translation: Constraints on multi-joint movements and the unique action of bi-articular muscles," 301-337, copyright 1989, with permission from Elsevier.

slows knee extension, thereby allowing the knee extensors to generate more energy, part of which is transmitted by the gastrocnemius to the calcaneus to increase the work done at the ankle.

6.3.2 Saving Mechanical Energy by Two-Joint Muscles

Representative publications: Elftman 1940; Aleshinsky 1986a, 1986b; Wells 1988; Gielen and van Ingen Schenau 1992; Prilutsky et al. 1996b; Alexander 1997; de Lussanet and Alexander 1997

The presence of two-joint muscles in the human body may affect the total amount of mechanical energy that muscles generate and absorb during movements. Consider 2 three-link chains representing the pelvis, thigh, and shank with hip and knee joints and depicted during late swing in spring running (figure 6.12). One chain is equipped with 2 one-joint muscles, hip extensor and knee flexor, and the other chain with an equivalent two-joint muscle (i.e., equivalent in the sense that the moment arms are the same). The muscles produce hip extensor and knee flexor moments proportional to the radii of the arcs in the figure, and the knee and hip extend at 18 rad/s and 6 rad/s, respectively. As a result, power developed at the hip joint is positive, 1,128 W, and at the knee joint is negative, –1,888 W.

With only one-joint muscles, the positive hip joint power corresponds to hip extensor power ($P_{he} = F_{he}V_{he} = M_h\dot{\theta}_h > 0$, where F_{he} and V_{he} are muscle force and velocity and M_h and $\dot{\theta}_h$ are hip joint moment and angular velocity, respectively) and the knee power corresponds to power absorbed by the knee flexor ($P_{kf} = F_{kf}V_{kf} = M_k\dot{\theta}_k < 0$). Because the muscles are independent actuators (sources of mechanical energy), the energy absorbed by the knee flexor cannot be used by the hip extensor to do positive work and extend the hip. These types of actuators are called *nonintercompensated* sources of mechanical energy (see section **6.3.2** in *Kinetic of Human Motion*). The total energy expended by the hip extensor and knee flexor muscles of chain

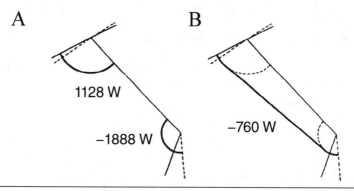

Figure 6.12   Diagrams illustrating power produced at the hip and knee joints by 2 one-joint muscles *(a)* and an equivalent two-joint muscle *(b)*. Solid and dashed lines indicate earlier and subsequent segment positions 27 ms apart (the knee and hip are extending at 18 rad/s and 6 rad/s) during late swing of sprint running. The muscles develop hip extensor and knee flexor moments proportional to the radii of the arcs. The numbers in *(a)* correspond to power produced by the joint moments (in watts) which are the same as power produced by the one-joint knee flexor and hip extensor muscles. In *(b)*, the power developed by an equivalent two-joint muscle is lower although the power developed by the muscle at the knee and hip is the same as in *(a)*. The joint moments and powers were calculated using inverse dynamics analysis.

Adapted, by permission, from H. Elftman, 1940, "The work done by muscles in running," *American Journal of Physiology* 129(3): 672-684.

1 can be estimated using a measure called mechanical energy expenditure, MEE_1 (explained in chapter 6 of *Kinematics of Human Motion*):

$$MEE_1 = \int_{t_o}^{t_1} \left(|P_{he}| + |P_{kf}| \right) dt. \qquad [6.16]$$

Note that the absolute values of power reflect the fact that the two muscles are nonintercompensated sources of energy and the muscle that is being stretched spends energy to do negative work (to resist the stretching).

Although power generated at the joints of the two chains is the same, power and energy expended by the two-joint muscle of the second chain are lower because hip and knee powers are developed by the same two-joint muscle $(P_{he,kf} = M_h \dot{\theta}_h + M_k \dot{\theta}_k = F_{he} V_{he} + F_{kf} V_{kf})$. Thus, the moments developed by this muscle at the hip and knee are *intercompensated* sources of energy. Therefore, mechanical energy expenditure of the second chain, MEE_2, is lower in the depicted example and calculated as

$$MEE_2 = \int_{t_o}^{t_1} P_{he,kf}\, dt = \int_{t_o}^{t_1} \left(P_{he} + P_{kf} \right) dt. \qquad [6.17]$$

Note that $MEE_1 > MEE_2$. The reduction of mechanical power and energy generated or absorbed by the chain with the two-joint muscle in figure 6.12 occurs because (a) the two-joint muscle does not generate energy (does no positive work) and (b) the rate of energy absorption by the muscle is reduced compared with the one-joint muscles. Notice that knee extension causes the distal attachment point of the two-joint muscle to move away from the muscle proximal attachment on the pelvis and permits the active muscle to pull on the pelvis. This is another demonstration of the tendon action of a two-joint muscle, which allows the hip joint to receive energy at the rate of 1,128 W generated by an external moment at the knee (variant of energy transfer 1, table 6.3).

The example in figure 6.12 illustrates that a two-joint muscle can reduce the amount of mechanical energy necessary for performing a given motion. This reduction occurs if both of the following conditions are met:

- The two-joint muscle produces force.
- Powers developed by the muscle at the two joints have opposite signs.

In human motion, these conditions are often met. For example, during walking, the gastrocnemius, rectus femoris, and hamstrings (all two-joint) develop oppositely signed power at the joints they cross in several phases of the gait cycle (figure 6.13). This is especially apparent for rectus femoris

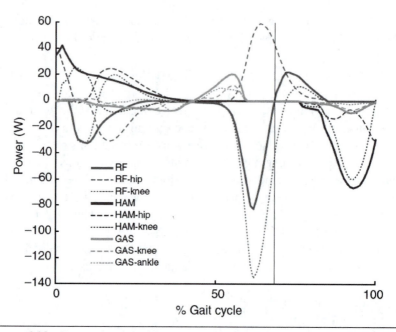

Figure 6.13   Power developed by the two-joint rectus femoris (RF), hamstrings (HA), and gastrocnemius (GAM) during walking. Power values are obtained as a result of forward dynamic simulations of a musculoskeletal model of the leg (methods of forward dynamics are explained in chapter 8). Thick lines are powers of the muscles (the product of force and velocity of the muscle–tendon complex); dashed lines are powers developed by the two-joint muscles at the proximal joints; dotted lines are powers developed by the muscles at the distal joints.

Adapted, by permission, from K. Sasaki, R.R. Neptune, and S.A. Kautz, 2009, "The relationships between muscle, external, internal and joint mechanical work during normal walking," *The Journal of Experimental Biology* 212(5): 738-744. Website: jeb.biologists.org.

in late stance when it demonstrates positive power at the hip and negative power at the knee, whereas power produced by the muscle itself has much lower peak values.

Figures 6.12 and 6.13 demonstrate examples of the energy-saving mechanism whereby two-joint muscles reduce mechanical energy demands on other muscles by using work done on two-joint muscles at one joint for doing work on the adjacent joint. For example, during late swing of running, the hamstring uses the energy of the rotating shank to extend the hip (figure 6.12).

6.4 SUMMARY

Multijoint muscles (or polyarticular muscles) cross two, three, or more joints and have up to three functions at each spanned joint. A special case of multijoint muscles is two-joint muscles that cross two joints. Anatomically, two-joint muscles are long, spanning the full length of the segment that lies between muscle attachments. Some, but not all, two-joint muscles achieve long lengths by having a long tendon and relatively short muscle fibers with a large angle of pennation. This design permits a large physiological cross-sectional area and thus high force potential of the muscles, which, in the case of distal muscles, permits sufficient moment development despite relatively small moment arms at the distal joints (small mechanical advantage). These long tendons allow for substantial storage and release of elastic energy, creating favorable conditions for high movement efficiency. Furthermore, proximal location of short bellies of the distal two-joint and multijoint muscles reduces the moment of inertia of the distal segment, thereby decreasing the burden on proximal muscles to develop limb acceleration.

Most two-joint muscles in the proximal limb segments, like semitendinosus, gracilis, and sartorius in the leg and biceps brachii in the arm, have long muscle fibers, small pennation angles, and relatively short tendons. Because two-joint muscles have large moment arms, and because opposite joint angle changes can stretch the muscle from both ends, it is conceivable that these long fibers may be an adaptation to prevent overstretch. Although certain muscle fibers span the whole muscle belly length, running from the proximal to distal tendons and reaching lengths of 40 to 60 cm, as in sartorius, the majority of these muscles contain a large proportion of shorter nonspanning fibers that taper at one or both ends. As a consequence, lateral force transmission through the endomysial and perimysial connective tissue plays a dominant role in these muscles. It promotes mechanical stability inside a muscle by preventing excessive shortening of activated muscle regions and overstretching of passive or less active regions.

Several approaches have been used to classify the functional role of muscles, including two-joint muscles, and these can be classed as kinematic or kinetic. A standard kinetics method involves determination of the moment of the muscle force and its components with respect to the joint axes at a given joint posture ($M_{OX}^m, M_{OY}^m, M_{OZ}^m$); determination of the resultant joint moment and its three components $M_{OX}^R, M_{OY}^R, M_{OZ}^R$; and comparison of the directions of the muscle joint moment and the resultant joint moment about each joint axis. If the directions of the muscle and resultant moment components about an axis are the same (i.e., their corresponding scalar products are positive, $M_j^m \cdot M_j^R > 0$), the muscle is defined as a joint agonist (for the given axis and given motion). If the directions are opposite ($M_j^m \cdot M_j^R < 0$), the muscle is called a joint antagonist for a given task. According to this definition a muscle can be a joint agonist with respect to one joint axis and a joint antagonist with respect to another. Note that this definition of muscle agonist–antagonist functions is different from the traditional anatomical definition according to which agonist muscles lie on the same side of a bone, act on the same joint, and move the joint in the same direction, whereas antagonist muscles act on the same joint but move it in the opposite direction.

It is rather common for a two-joint muscle to have an agonistic action at one joint and an antagonistic action at the other, especially in the sagittal plane. This implies that the same muscle can act against itself by unavoidably producing an antagonistic action at one of the joints. Two-joint muscles can also contribute to the resultant moments at both joints as joint agonists (which intuitively seems the most efficient use of two-joint muscles) or as joint antagonists at both joints (apparently a more unfavorable use of two-joint muscles).

In movements such as rising from a chair or jumping, the anatomical antagonists, two-joint hamstrings and rectus femoris, are simultaneously active; not only does each produce antagonistic action at one of the joints, but they also oppose each other (the so-called Lombard's paradox). In such cases, which of the two muscles is an agonist and which is an antagonist? The solution can be obtained using a kinematic method for classifying muscle action.

The kinematic procedures proposed to determine functional roles of two-joint muscles in multijoint systems are based on analysis of muscle length changes during motion. It is postulated that if an active two-joint muscle shortens during propulsive motion, the muscle acts as a task agonist and assists the ongoing motion. The rationale is that when a muscle shortens and produces force, it is supplying mechanical energy to the system by generating positive power, $P_m = F_m \cdot V_m$ (where P_m, F_m, and V_m are muscle power, force, and velocity). If an active two-joint muscle lengthens, the muscle resists motion and absorbs a portion of the body's mechanical energy (the developed muscle power is negative). In this case the two-joint muscle is defined as a task antagonist.

Three potential advantages of two-joint muscles over one-joint muscles have been suggested: (1) two-joint muscles distribute mechanical energy along the limb to allow large proximal muscles to generate and absorb more mechanical energy, (2) they improve performance of explosive movements, and (3) they decrease mechanical energy expenditure (increase efficiency). The three advantages are achieved by the tendon action of two-joint muscles. This term refers to tow-rope–like behavior in which the muscle transmits tension and energy from the one segment to the other.

The rate at which mechanical energy is transported to or from a joint P_t can be determined as the difference between the joint moment power P_j and the total power of muscles crossing the joint P_m:

$$P_t = P_j - P_m. \qquad [6.13]$$

For instance, if $P_t < 0$, $P_j < 0$, $P_m \leq 0$, energy is transferred from joint j to joints $(j-1)$, $(j+1)$, or both by two-joint muscles at the rate of $-P_t$, whereas muscles crossing joint j absorb energy at the rate of $-P_m$ (or $P_m = 0$), that is, at a rate lower than the rate at which energy is absorbed at the joint, $-P_j$. If $P_t > 0$, $P_j > 0$, $P_m \geq 0$, the energy is transferred by two-joint muscles to joint j at the rate of P_t, whereas muscles crossing joint j generate energy at the rate of P_m (or $P_m = 0$), that is, at a rate lower than the rate of energy generation at the joint. During the yield phase of landing and jogging, the energy is transferred from distal to proximal joints. During the push phase of jogging or vertical jumping, two-joint muscles transfer energy from proximal to distal joints.

Energy transferred by an individual two-joint muscle can be obtained based on the relation between total power generated by a muscle (P_i) and the power of the muscle generated at each joint it spans (distal and proximal, $P_{i,d}$ and $P_{i,p}$), which follows from the principle of virtual work (see equation 5.15):

$$P_i = F_i \cdot V_i = M_d \cdot \dot{\theta}_d + M_p \cdot \dot{\theta}_p = P_{i,d} + P_{i,p}, \qquad [6.15]$$

where M_d and M_p are moments developed by the two-joint muscle at the distal and proximal joints, and $\dot{\theta}_d$ and $\dot{\theta}_p$ are angular velocities at the distal and proximal joints, respectively. It follows from the equation that the power produced by the muscle moment at the distal (or proximal) joint depends on both power produced by the muscle and power produced by the muscle at the proximal (or distal) joint.

Several variants of energy transfer by a two-joint muscle can be distinguished based on the signs and values of the terms in equation 6.15. For instance, if $P_i < 0$, $P_{i,p} > 0$, $P_{i,d} < 0$, work done on distal joint and absorbed at the rate $-P_{i,d}$ appears at the proximal joint as positive work; thus energy is transferred from distal to proximal joint at a rate of $-P_{i,d}$, muscle absorbs energy at a rate of $-P_i$, and energy is delivered to proximal joint at a rate of $P_{i,p}$.

Thus, the energy transfer between proximal and distal joints through two-joint muscles enables larger proximal muscles to more fully use their energy absorption and generation potential by absorbing and generating energy at the distal joints.

The second advantage of the tendon action of two-joint muscles (besides efficient use of larger proximal muscles for absorption and generation of energy) may be to increase the maximum mechanical performance of explosive movements such as jumping and sprint push-off.

The third possible advantage of two-joint muscles may be that they can reduce the amount of mechanical energy necessary for motion. This occurs if the following conditions are met:

- The two-joint muscle produces force.
- Powers developed by the two-joint muscle at the two joints have opposite signs.

These conditions are often met during locomotion.

6.5 QUESTIONS FOR REVIEW

1. Define multijoint and two-joint muscles. What are the major differences between these and one-joint muscles?

2. Name the two-joint muscles in the lower and upper extremities.

3. Define a moment arm vector of a two-joint muscle in the sagittal plane. Show this vector graphically.

4. List anatomical and morphological features of distal two-joint and multijoint muscles. Speculate on possible advantages and disadvantages of their design.

5. List anatomical and morphological features of proximal limb two-joint and multijoint muscles. Speculate on possible advantages and disadvantages of their design.

6. What are spanning and nonspanning muscle fibers? Name the longest human muscle and give a range of its fiber lengths. Given that long spanning fibers cannot be activated simultaneously along the muscle's length, speculate on how these fibers develop force and transmit it to tendons and bones.

7. Explain a standard kinetic method for classifying a muscle's functional role at a joint.

8. List four possible functional roles of a two-joint muscle at its two joints.

9. Give examples of movements for each of the functional roles in item 8.

10. Formulate Lombard's paradox. Can it be resolved using a standard kinetic method of classification of muscle function?

11. How is the functional role of a two-joint muscle determined using kinematic classification?

12. Give examples of movements at two joints spanned by a two-joint muscle in which

    a. two-joint muscle does not change its length,

    b. two-joint muscle is stretched, and

    c. two-joint muscle shortens.

13. Formulate the conditions at which the velocities of the hamstrings and rectus femoris are zero during leg extensions.

14. Plot muscle isometric constraints in the plane of knee and hip joint velocities. Define regions on this plane in which

    a. both hamstrings and rectus femoris muscles shorten,

    b. both muscles lengthen,

    c. rectus femoris shortens while hamstrings lengthen, and

    d. hamstrings shortens and rectus femoris lengthens.

15. Explain Lombard's paradox.

16. Define the tendon action of two-joint muscles.

17. Give examples of tendon action in which a two-joint muscle

    a. transfers mechanical energy but does not do mechanical work,

    b. receives mechanical energy at one attachment point but absorbs it and does not transfer energy to the other segment, and

    c. receives mechanical energy at one attachment point, generates energy, and transfers it to the other segment.

18. Describe two methods for estimating the rate, direction, and amount of two-joint muscle energy transfer.

19. In which direction do two-joint muscles transfer energy during the shock-absorbing phase of motion? Give movement examples.

20. In which direction do two-joint muscles transfer energy during push-off? Give movement examples.

21. How can tendon action at a two-joint muscle improve jumping performance?

22. Explain how two-joint muscles can reduce mechanical energy necessary for movement. Give examples of movement phases during which this reduction takes place.

23. List conditions for which such mechanical energy reduction (see item 22) can take place.

6.6 LITERATURE LIST

Aleshinsky, S.Y. 1986a. An energy "sources" and "fractions" approach to the mechanical energy expenditure problem: I. Basic concepts, description of the model, analysis of a one-link system movement. *J Biomech* 19(4):287-293.

Aleshinsky, S.Y. 1986b. An energy "sources" and "fractions" approach to the mechanical energy expenditure problem: V. The mechanical energy expenditure reduction during motion of the multi-link system. *J Biomech* 19(4):311-315.

Alexander, R.M. 1997. A minimum energy cost hypothesis for human arm trajectories. *Biol Cybern* 76(2):97-105.

Andrews, J.G. 1985. A general method for determining the functional role of a muscle. *J Biomech Eng* 107(4):348-353.

Andrews, J.G. 1987. The functional roles of the hamstrings and quadriceps during cycling: Lombard's paradox revisited. *J Biomech* 20(6):565-575.

Andrews, J.G., and J.G. Hay. 1983. Biomechanical considerations in the modeling of muscle function. *Acta Morphol Neerl Scand* 21(3):199-223.

Basmajian, J.V. 1957. Electromyography of two-joint muscles. *Anat Rec* 129(3):371-380.

Bobbert, M.F., P.A. Huijing, and G.J. van Ingen Schenau. 1986a. An estimation of power output and work done by the human triceps surae muscle-tendon complex in jumping. *J Biomech* 19(11):899-906.

Bobbert, M.F., P.A. Huijing, and G.J. van Ingen Schenau. 1986b. A model of the human triceps surae muscle-tendon complex applied to jumping. *J Biomech* 19(11):887-898.

Bobbert, M.F., and G.J. van Ingen Schenau. 1988. Coordination in vertical jumping. *J Biomech* 21(3):249-262.

Bock, W.J. 1968. Mechanics of one- and two-joint muscles. *American Museum Novitates* 2319:1-45.

Bolhuis, B.M., C.C. Gielen, and G.J. van Ingen Schenau. 1998. Activation patterns of mono- and bi-articular arm muscles as a function of force and movement direction of the wrist in humans. *J Physiol* 508(Pt 1):313-324.

Carlsoo, S., and S. Molbech. 1966. The functions of certain two-joint muscles in a closed muscular chain. *Acta Morphol Neerl Scand* 6(4):377-386.

Carroll, A.M., D.V. Lee, and A.A. Biewener. 2008. Differential muscle function between muscle synergists: Long and lateral heads of the triceps in jumping and landing goats (Capra hircus). *J Appl Physiol* 105(4):1262-1273.

Cleland, J. 1867. On the actions of muscles passing over more than one joint. *J Anat Physiol* 1:85-93.

de Looze, M.P., H.M. Toussaint, J.H. van Dieen, and H.C. Kemper. 1993. Joint moments and muscle activity in the lower extremities and lower back in lifting and lowering tasks. *J Biomech* 26(9):1067-1076.

de Lussanet, M.H.E., and R.M. Alexander. 1997. A simple model for fast planar arm movements; optimizing mechanical and moment-arms of uniarticular and biarticular arm muscles. *J Theor Biol* 184:187-201.

Doorenbosch, C.A., J. Harlaar, M.E. Roebroeck, and G.J. Lankhorst. 1994. Two strategies of transferring from sit-to-stand; the activation of monoarticular and biarticular muscles. *J Biomech* 27(11):1299-1307.

Dostal, W.F., and J.G. Andrews. 1981. A three-dimensional biomechanical model of hip musculature. *J Biomech* 14(11):803-812.

Elftman, H. 1939a. Forces and energy changes in the leg during walking. *Am J Appl Physiol* 125:339-356.

Elftman, H. 1939b. The function of muscles in locomotion. *Am J Appl Physiol* 125:357-366.

Elftman, H. 1940. Work done by muscles in running. *Am J Appl Physiol* 129:672-684.

Enklaar, M.J.E. 1954. Antagonisme entre demi-tendineux et demi-membraneux chez l'homme. Paper read at 6ième Congrès de la Société Internationale de Chirurgie orthopédique et de Traumatologie.

Fujiwara, M., and J.V. Basmajian. 1975. Electromyographic study of two-joint muscles. *Am J Phys Med* 54(5):234-242.

Gielen, C.C.A.M., and G.J. Van Ingen Schenau. 1992. The constrained control of force and position by multilink manipulators. *IEEE Trans Syst Man Cybern* 22(5):1214-1219.

Gregersen, C.S., N.A. Silverton, and D.R. Carrier. 1998. External work and potential for elastic storage at the limb joints of running dogs. *J Exp Biol* 201(Pt 23):3197-3210.

Gregoire, L., H.E. Veeger, P.A. Huijing, and G.J. van Ingen Schenau. 1984. Role of mono- and biarticular muscles in explosive movements. *Int J Sports Med* 5(6):301-305.

Harris, A.J., M.J. Duxson, J.E. Butler, P.W. Hodges, J.L. Taylor, and S.C. Gandevia. 2005. Muscle fiber and motor unit behavior in the longest human skeletal muscle. *J Neurosci* 25(37):8528-8533.

Heron, M.I., and F.J. Richmond. 1993. In-series fiber architecture in long human muscles. *J Morphol* 216(1):35-45.

Jacobs, R., M.F. Bobbert, and G.J. van Ingen Schenau. 1996. Mechanical output from individual muscles during explosive leg extensions: The role of biarticular muscles. *J Biomech* 29(4):513-523.

Jacobs, R., and G.J. van Ingen Schenau. 1992. Control of an external force in leg extensions in humans. *J Physiol* 457:611-626.

Kuo, A. 2001. The action of two-joint muscles: The legacy of W.P. Lombard. In *Classics in Movement Studies,* edited by V.M. Zatsiorsky and M.L. Latash. Champaign, IL: Human Kinetics, pp. 289-316.

Landsmeer, J.M. 1961. Studies in the anatomy of articulation: II. Patterns of movement of bi-muscular, bi-articular systems. *Acta Morphol Neerl Scand* 3:304-321.

Loeb, G.E., and F.J.R. Richmond. 1994. Architectural features of multiarticular muscles. *Human Movement Science* 13(5):545-556.

Lombard, W.P. 1903. The action of two-joint muscles. *American Physical Education Reviews* 9:141-145.

Markee, J.E., J.T. Logue, Jr., M. Williams, W.B. Stanton, R.N. Wrenn, and L.B. Walker. 1955. Two-joint muscles of the thigh. *J Bone Joint Surg Am* 37-A(1):125-142.

Molbech, S. 1965. On the paradoxical effect of some two-joint muscles. *Acta Morphol Neerl Scand* 6:171-178.

Morrison, J.B. 1970. The mechanics of muscle function in locomotion. *J Biomech* 3(4):431-451.

Murray, W.M., T.S. Buchanan, and S.L. Delp. 2000. The isometric functional capacity of muscles that cross the elbow. *J Biomech* 33(8):943-952.

Pandy, M.G., and F.E. Zajac. 1991. Optimal muscular coordination strategies for jumping. *J Biomech* 24(1):1-10.

Peters, S.E., and C. Rick. 1977. The actions of three hamstring muscles of the cat: A mechanical analysis. *J Morphol* 152(3):315-328.

Prilutsky, B.I. 2000a. Coordination of two- and one-joint muscles: Functional consequences and implications for motor control. *Motor Control* 4(1):1-44.

Prilutsky, B.I. 2000b. Muscle coordination: The discussion continues. *Motor Control* 4(1):97-116.

Prilutsky, B.I., R.J. Gregor, and M.M. Ryan. 1998a. Coordination of two-joint rectus femoris and hamstrings during the swing phase of human walking and running. *Exp Brain Res* 120(4):479-486.

Prilutsky, B.I., W. Herzog, and T. Leonard. 1996a. Transfer of mechanical energy between ankle and knee joints by gastrocnemius and plantaris muscles during cat locomotion. *J Biomech* 29(4):391-403.

Prilutsky, B.I., T. Isaka, A.M. Albrecht, and R.J. Gregor. 1998b. Is coordination of two-joint leg muscles during load lifting consistent with the strategy of minimum fatigue? *J Biomech* 31(11):1025-1034.

Prilutsky, B.I., L.N. Petrova, and L.M. Raitsin. 1996b. Comparison of mechanical energy expenditure of joint moments and muscle forces during human locomotion. *J Biomech* 29(4):405-415.

Prilutsky, B.I., and V.M. Zatsiorsky. 1994. Tendon action of two-joint muscles: Transfer of mechanical energy between joints during jumping, landing, and running. *J Biomech* 27(1):25-34.

Rozendal, R.H. 1994. Human poly-articular muscles: An anatomical comment. *Hum Mov Sci* 13(5):557-568.

Sasaki, K., R.R. Neptune, and S.A. Kautz. 2009. The relationships between muscle, external, internal and joint mechanical work during normal walking. *J Exp Biol* 212(Pt 5):738-744.

Spoor, C.W., and J.L. van Leeuwen. 1992. Knee muscle moment arms from MRI and from tendon travel. *J Biomech* 25(2):201-206.

van Ingen Schenau, G.J. 1989. From rotation to translation: Constraints on multi-joint movements and the unique action of bi-articular muscles. *Hum Mov Sci* 8:301-337.

van Ingen Schenau, G.J., and M.F. Bobbert. 1993. The global design of the hindlimb in quadrupeds. *Acta Anat (Basel)* 146(2-3):103-108.

van Ingen Schenau, G.J., M.F. Bobbert, and R.H. Rozendal. 1987. The unique action of bi-articular muscles in complex movements. *J Anat* 155:1-5.

van Soest, A.J., A.L. Schwab, M.F. Bobbert, and G.J. van Ingen Schenau. 1993. The influence of the biarticularity of the gastrocnemius muscle on vertical-jumping achievement. *J Biomech* 26(1):1-8.

Visser, J.J., J. Hoogkamer, M.F. Bobbert, and P.A. Huijing. 1990. Length and moment arm of human leg muscles as a function of knee and hip-joint angles. *Eur J Appl Physiol Occup Physiol* 61(5-6):453-460.

Voronov, A.V. 2004. The role of monoarticular and biarticular muscles of the lower limb in ground locomotion (in Russian). *Fiziol Cheloveka* 30(4):114-123.

Ward, S.R., C.M. Eng, L.H. Smallwood, and R.L. Lieber. 2009. Are current measurements of lower extremity muscle architecture accurate? *Clin Orthop Relat Res* 467(4):1074-1082.

Wells, R., and N. Evans. 1987. Functions and recruitment patterns of one-joint and 2-joint muscles under isometric and walking conditions. *Hum Mov Sci* 6(4):349-372.

Wells, R.P. 1988. Mechanical energy costs of human movement: An approach to evaluating the transfer possibilities of two-joint muscles. *J Biomech* 21(11):955-964.

Yang, D., S.F. Morris, and L. Sigurdson. 1998. The sartorius muscle: Anatomic considerations for reconstructive surgeons. *Surg Radiol Anat* 20(5):307-310.

7

ECCENTRIC MUSCLE ACTION IN HUMAN MOTION

Eccentric muscle actions are innate to many human movements. In walking, for instance, during early support the support leg flexes, the support leg's extensors are active, and the body's mechanical energy decreases; during late swing the swing leg's knee flexors are active to dissipate the shank's energy and prevent excessive shank extension.

As discussed in chapter 4, active muscle stretch (eccentric action) modifies both muscle force development and energy production during subsequent muscle action. These changes can be explained by muscle mechanical properties. Although these properties are important in human movements, additional factors such as involvement of multiple muscles, unequal muscle length changes, effects of joint motion on two-joint muscles, and muscle coordination play a large role.

This chapter considers how eccentric actions in human motion affect muscle performance. First, it discusses how the amount and intensity of eccentric muscle action can be quantified during natural motion (section **7.1**) and how much eccentric action is performed in selected activities (section **7.2**). It subsequently demonstrates (section **7.3**) that many of muscle's in vitro stretch-related mechanical properties (e.g., dynamic and residual force enhancements) are also manifested in vivo. However, additional in vivo factors must be considered for a more complete understanding of eccentric muscle action. Reflexes, for example, and the stretch reflex in particular, can greatly affect eccentric muscle behavior. Other neural factors like muscle activation intensity affect a muscle's response to stretch. Section **7.4** addresses muscle activity during eccentric actions including issues such as surface electromyography (EMG) (**7.4.1**), motor unit activity (**7.4.2**), and electromechanical delay (**7.4.3**). Section **7.5** deals with the physiological costs of eccentric action including oxygen consumption (**7.5.1**), fatigue and perceived exertion (**7.5.2**), and muscle soreness (**7.5.3**).

This chapter also considers how stretching a muscle immediately before shortening (the stretch–shortening cycle, SSC), can be beneficial (section **7.6**). The benefit manifests in increased mechanical output and decreased metabolic cost as well as in reduced mechanical impact through energy absorption in stretching muscles. Enhancement of positive work and power production (**7.6.1**), mechanisms of the performance enhancement in the SSC (**7.6.2**), and efficiency of work in SSC (**7.6.3**) are described.

7.1 JOINT POWER AND WORK AS MEASURES OF ECCENTRIC MUSCLE ACTION

The theory of negative power and work in multilink chains was described in chapter 6 of *Kinetics of Human Motion,* but this assumed revolute joints powered by joint moment (torque) actuators (the terms joint moment and joint torque have the same meaning). The goal of this section is to refresh and augment the above theory for musclelike actuators.

7.1.1 Negative Power and Work at a Joint

In section **6.2.2.1** of *Kinetics of Human Motion,* joint power was defined as the scalar product of joint moment (**M**) and relative angular velocity ($\dot{\boldsymbol{\theta}}$) at a joint, both in the global reference system (see equation 6.15 in *Kinetics of Human Motion*):

$$P(\text{watts}) = \mathbf{M} \cdot \dot{\boldsymbol{\theta}} = \mathbf{M}\,(\dot{\boldsymbol{\theta}}_d - \dot{\boldsymbol{\theta}}_p), \qquad [7.1a]$$

where $\dot{\boldsymbol{\theta}}$ is the vector of the angular velocity of the body link rotation and subscripts d and p refer to the distal and proximal links adjacent to the joint, respectively. In two dimensions, **M** and $\dot{\boldsymbol{\theta}}$ can be considered scalars, and the power equation is

$$P = M \cdot \dot{\theta}. \qquad [7.1b]$$

Joint power is negative when M and $\dot{\theta}$ are in opposite directions ($-M \cdot \dot{\theta} < 0$ or $M \cdot (-\dot{\theta}) < 0$). Physically, negative joint power indicates that mechanical energy flows to the joint. The joint structures—muscles and connective tissues—absorb mechanical energy and either store it as elastic deformation energy or dissipate it into heat (however, see section **4.2.2**, where conversion to chemical energy is discussed).

Negative joint power can be partially attributed to forcible muscle stretch (i.e., to eccentric muscle action). Several issues should be taken into account:

1. Possible contribution of passive tissue resistance (e.g., passive muscles, ligaments, and joint capsule; passive mechanical resistance at joints is described in section **5.4.1.2** of *Kinetics of Human Motion*). Passive tissues do not spend energy. Mechanical energy flows to the muscles and passive tissues; external forces do work on them.

2. When negative power is seen at a joint, active single-joint muscles are forcibly stretched and act eccentrically. However, two-joint muscle behavior depends on neighboring joint movements and thus cannot be predicted from single joint behavior. It is possible that the length of a two-joint muscle stays the same or even decreases (i.e., the muscle acts concentrically) when joint power is negative. This fact is the basis of the energy transfer between nonadjacent segments, as discussed in section **2.2.2.3** of *Kinetics of Human Motion* and chapter 6 here.

3. For an active muscle working eccentrically, there are two flows of energy—to and from the muscle. The external force does work on the muscle, and the muscle spends energy to provide resistance against the external force.

4. Although negative joint power and work are unambiguous measures of the intensity and amount of eccentric action at the joint level, the muscle level is different. Negative joint power and work can be produced not only while muscle fibers act eccentrically but also when they act isometrically or even concentrically. This happens because of elastic stretch and subsequent recoil of the tendon and aponeurosis. We explain this phenomenon later in the chapter when we discuss negative work and the SSC in selected activities.

▪ ▪ ▪ *FROM THE LITERATURE* ▪ ▪ ▪

Energy of Rotating Shank Is Used to Operate a Cell Phone, Reducing the Negative Work at the Knee Joint

Source: Donelan, J.M., Q. Li, V. Naing, J.A. Hoffer, D.J. Weber, and A.D. Kuo. 2008. Biomechanical energy harvesting: generating electricity during walking with minimal user effort. *Science* 319:807-810

The authors developed a device called a *biomechanical energy harvester* that can absorb a portion of the shank energy during late swing, thus reducing the negative work done by the hamstrings. The harvester converts the absorbed mechanical energy into electrical current that can simultaneously power 10 typical cellular phones.

Negative joint work is computed as the time integral of the joint power for periods of negative joint power. If there are intervals during which the power is negative (external forces do work on the muscles and other joint structures) as well as intervals during which the power is positive (muscles serving the joint do work on the environment), the integral of joint power over the entire time period may not represent the work done at the joint (see section **6.3.2.3.2** on *compensation during time [recuperation]* in *Kinetics of Human Motion*).

7.1.2 Total Negative Power and Work in Several Joints

For kinematic chains with several joints, two main situations may occur: (a) the joint power is positive or negative at all joints, and (b) the joint power is positive at some joints and negative at others. In the first case, the power values summed over all joints can be used to calculate the total power; in the second the outcome of total power calculations using the summed joint power and its interpretation depend on whether the sources of energy are *intercompensated* (the intercompensation of sources is discussed in section **6.3.2.3.1** in *Kinetics of Human Motion*). Recall that single joint muscles are nonintercompensated sources of energy, whereas two-joint muscles are intercompensated.

7.1.3 Negative Power of Center of Mass Motion

The work and power associated with center of mass (CoM) motion is discussed in section **6.3.3.3.2** of *Kinetics of Human Motion*. When the speed and vertical coordinates of the CoM decrease, the kinetic and potential energy associated with its motion also decreases. The rate of decrease, that is, negative power, can be determined as the derivative of the CoM energy with respect to time. This energy does not include the energy for body rotation and movement of the body parts with respect to the CoM.

7.1.4 Two Ways of Mechanical Energy Dissipation: Softness of Landing

Representative publication: Zatsiorsky and Prilutsky 1987

In many movements there are phases in which the total mechanical energy of the body decreases. Part or all of this energy is dissipated by muscles acting eccentrically, whereas any remaining part is dissipated in the passive structures of the body, such as heel pads, joint cartilage, intervertebral discs, and bones, as well as in shoe soles and the ground (if they are soft). Imagine a sudden fall after tripping. The energy of the body in this situation is dissipated by passive body structures, which often leads to bone fractures and damage of internal organs due to high-impact accelerations. The more the energy dissipated by joint moments, the less the energy dissipated by passive body structures.

The following index of softness of landing (ISL) has been suggested to describe how much mechanical energy is dissipated by joint moments, that is, by muscles and tendons:

$$ISL = \frac{\text{Total negative work of joint moments}}{\text{Decrease of total energy of the body}}. \qquad [7.2]$$

The index has values between 0 and 1. If it is 0, the mechanical energy of the body is dissipated passively without eccentric muscle action (negative work and power of joint moments). When $ISL = 1$, all of the body mechanical energy is dissipated in muscles and tendons as a result of negative work of joint moments.

■ ■ ■ *FROM THE LITERATURE* ■ ■ ■

Softness of Landing Is Correlated With the Rate of Initial Vertical Ground Reaction Force Increase

Source: Zatsiorsky, V.M., and B.I. Prilutsky. 1987. Soft and stiff landing. In *Biomechanics X-B*, edited by B. Jonsson. Champaign, IL: Human Kinetics pp. 739-743

An index (ISL, the ratio between negative work of joint moments and decrease of total energy of the body) was introduced to quantify softness of landing and to relate it to vertical ground reaction forces. One subject performed landing after jumping down from heights of 20 and 50 cm. The subject was asked to land in different trials with varying degree of yielding (softness). Joint moments, joint work, and total energy change of the body were determined using an inverse dynamics analysis. The ISL varied between 0.25 (maximally stiff landing) and 0.99 (maximally soft landing). Strong negative correlation was found between the ISL and the rate of initial vertical ground force increase at contact.

In maximally soft landings, the total negative work of joint moments and the reduction in total body energy are essentially equal; that is, in soft landing nearly all body energy is dissipated actively by joint moments.

7.2 NEGATIVE WORK IN SELECTED ACTIVITIES

Most human movements include phases in which the total mechanical energy of the body or some of its segments decreases. The energy decrease can be caused by external forces (e.g., air or water resistance) or by muscle forces and passive anatomical structures such as ligaments and cartilage. In some activities, such as swimming, rowing, high-resistance ergometer cycling, and high-speed road cycling, body energy is dissipated mostly by external forces, and muscles do little negative work—eccentric muscle action is negligible.

■ ■ ■ FROM THE LITERATURE ■ ■ ■

Little Joint Negative Work Is Done During Ergometer Cycling

Source: Ericson, M.O., A. Bratt, R. Nisell, U.P. Arborelius, and J. Ekholm. 1986. Power output and work in different muscle groups during ergometer cycling. *Eur J Appl Physiol Occup Physiol* 55(3):229-235

The aim of this study was to calculate the instantaneous power output at the hip, knee, and ankle joints during ergometer cycling. Six healthy subjects pedaled a weight-braked bicycle ergometer at 120 W and 60 revolutions per minute. The subjects were filmed and pedal reaction forces were recorded. The work at the hip, knee, and ankle joints was calculated using inverse dynamics. The mean peak positive power output was 74.4 W for the hip extensors, 18.0 W for the hip flexors, 110.1 W for the knee extensors, 30.0 W for the knee flexors, and 59.4 W for the ankle plantar flexors. At the ankle joint, energy absorption through eccentric plantar flexor action was observed, with a mean peak power of only –11.4 W and negative work of 3.4 J for each limb and complete pedal revolution. The energy production for the major muscle groups was computed; the contributions to the total positive work were hip extensors 27%, hip flexors 4%, knee extensors 39%, knee flexors 10%, and ankle plantar flexors 20%.

In many activities, however, muscles do substantial negative work. In this section we analyze some of these activities: walking, descending stairs, running, and landing. It is assumed here that the work done by the joint moments is the most accurate estimate of muscle work in these activities. However, muscle work may differ from joint moment work as a result of coactivation of antagonistic muscles and specific functions of two-joint muscles, as discussed in chapter 6.

Let us now consider negative work done during selected human activities in more detail.

7.2.1 Walking

Representative publications: Elftman 1939; Winter 1983a; Eng and Winter 1995; Prilutsky and Zatsiorsky 1992; Prilutsky 2000; Hof et al. 2002; Devita et al. 2007; Sawicki et al. 2009

During walking at constant speed, absolute values of total negative and positive work done in the major joints of the body are very close to each other. Slightly higher values of positive joint moment work are caused by some dissipation of the body mechanical energy by passive tissues, as explained previously. In the cycle of walking at speeds of 1.6 to 2.4 m/s, estimates of the total negative work summed across the three orthogonal planes and major joints (three joints for each lower and upper extremity, and also trunk and head-trunk articulations) range between 125 and 190 J. Most of the negative work is done (or energy absorbed) by the joints of lower extremity, 75% to 90% of the total negative work. On the order of 75% to 90% of negative work of the lower extremities is done in the sagittal plane.

There are several phases of the walking cycle in which joints of the lower extremity absorb mechanical energy (figure 7.1).

At the ankle joint, after touchdown during approximately first 10% of the cycle, the dorsiflexors act eccentrically to decelerate foot plantar flexion (this phase is typically absent during barefoot walking). During racewalking, this phase is greatly enhanced and has been implicated in anterior lower leg pain, a common problem among racewalkers. This syndrome is partly caused by high values of negative power and work produced by the ankle dorsiflexors.

When the distal portion of the foot touches the ground, the ankle plantar flexors start acting eccentrically and absorb energy during 10% to 40% of the walking cycle (figure 7.1), just before the ankle plantar flexors begin to generate energy in late stance (40%-60%, figure 7.1). During the phase

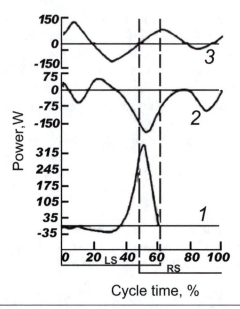

Cycle time, %

Figure 7.1 Power at the ankle, knee, and hip joints of the left leg during walking at 1.82 m/s. LS and RS are stance phases of the left and right legs, respectively. At the ankle joint, the period of positive power production is much shorter than the period of mechanical power absorption (mechanical energy is accumulated over a longer period of time than is energy release). This fact represents the so-called *catapult muscle action* (discussed in more detail in section **7.6**).

Adapted from B.I. Prilutsky and V.M. Zatsiorsky, 1992, "Mechanical energy expenditure and efficiency of walking and running," *Human Physiology* (in Russian, *Fiziologia Cheloveka*) 18(3): 118-127.

of negative ankle plantar flexor power, the muscle–tendon units (MTUs), fascicles, and tendinous structures of soleus and gastrocnemius muscles act eccentrically; that is, their length increases while the muscles are active. Some authors have reported small fascicle length changes during this phase. Also during this phase (10%-40% of cycle), the Achilles tendon and aponeuroses stretch and absorb mechanical energy as demonstrated by in vivo ultrasound measurements. The resultant moments of ankle plantar flexors do up to 5 times less negative work than they do positive work during push-off in the subsequent 40% to 60% of the cycle. Part of this negative work is absorbed in the stretched tendinous structures of the ankle extensors and is released during the energy production phase (40%-60% of cycle time, figure 7.1).

During stance the knee extensors mostly absorb energy (figure 7.1). The negative work in the knee during walking at different speeds has been reported to exceed positive work five- to six-fold.

Hip flexors and extensors do approximately the same amount of negative and positive work during the walking cycle. The hip flexor muscles decelerate the thigh extension during approximately the last third of stance (figure 7.1, 25%-50% of cycle time).

7.2.2 Stair Descent and Ascent

Representative publications: McFadyen and Winter 1988; Chleboun et al. 2008; Spanjaard et al. 2009

During stair descent, work done by moments at the knee and ankle is mostly negative, whereas very little negative or positive work is done at the hip joint (figure 7.2). The ankle plantar flexors absorb energy during approximately the

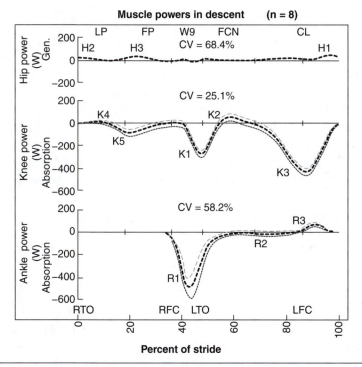

Figure 7.2 Mean and standard deviations of muscle powers at the ankle, knee, and hip joints in the sagittal plane during stair descent. RTO, right toe-off; RFC, right foot contact; LTO, left toe-off; LFC, left foot contact. Step dimensions are 22 cm (height) by 28 cm (depth); the total slope of the stairs is 37°.

Reprinted from *Journal of Biomechanics* 21(9), B.J. McFadyen and D.A. Winter, "An integrated biomechanical analysis of normal stair ascent and descent," 733-744, copyright 1988, with permission from Elsevier.

first third of stance (weight acceptance phase). The knee extensors also absorb energy of the body from about midstance to the beginning of swing (controlled lowering phase). Thus, during stair descent, most of the work done by joint moments is negative, and the ankle plantar flexors and knee extensors absorb most of the energy. The opposite is true for stair ascent—the knee extensors and ankle plantar flexors do most of the positive work, whereas all three leg joints absorb little energy. Hence during stair descent and ascent, muscles do most work which is negative and positive, respectively.

During stair descent and ascent, the work estimated from the CoM's total energy change is very close to the joint moment work because power in different joints is of the same sign. That is, in movements where power signs are the same across joints, the joint moment work is similar to the body's total mechanical energy change if little or no passive dissipation of energy occurs (see chapter 6 in *Kinetics of Human Motion*).

Although values of work done by knee joint moments during stair descent and ascent are mostly negative or positive, vastii fascicles operate mostly isometrically (ascending) or concentrically (descending) during stance. During descending (slope –10%) and ascending stairs (slope +10%), fascicles of gastrocnemius shorten throughout most of stance, indicating concentric gastrocnemius action despite the ankle moment's mostly negative work. This implies that the ankle moment's negative work is absorbed in tendinous structures of the ankle plantar flexors. (Note the difference between the work of the ankle joint moment—mostly negative—and the work of fascicles of the gastrocnemius that serves the joint, which is mostly positive.)

7.2.3 Level, Downhill, and Uphill Running

Representative publications: Aleshinsky 1978; Winter 1983b; Zatsiorsky et al. 1982; Prilutsky and Zatsiorsky 1992; Stefanyshyn and Nigg 1997; Hof et al. 2002; Lichtwark and Wilson 2006; Ishikawa and Komi 2007; Devita et al. 2008

When running at constant and relatively low speeds, the total joint moment negative work (summed across the major body joints) is slightly smaller than the positive work as a result of passive energy dissipation in the body. This situation is similar to walking. During sprint running, however, the difference between total positive and negative work increases to overcome aerodynamic drag. The total negative work of joint moments per cycle ranges from 200 to 800 J for speeds of 3.3 to 6.0 m/s and for individuals with body masses of about 70 kg.

During the first half of stance, sagittal plane ankle and knee moments (ankle plantar flexors and knee extensors) absorb energy, but during the second half of stance they play the opposite role, generating energy (figure 7.3). Running downhill at an 8% grade increases the negative work done by the ankle plantar flexors and knee extensors twofold. The changes in the hip joint power profile, however, do not follow a stereotypical pattern during stance; the pattern appears to depend on speed. Typically, the hip flexors absorb energy in late stance to decelerate hip extension. Metatarsophalangeal joint moments (plantar flexors) mostly absorb energy during stance (not shown in the figure).

The facts that the energy generation phase of the ankle and knee joints in the stance of running follows immediately after the energy absorption phase (figure 7.3) and that absolute values of the negative and positive work done by the ankle plantar flexors and knee extensors during the stance are similar

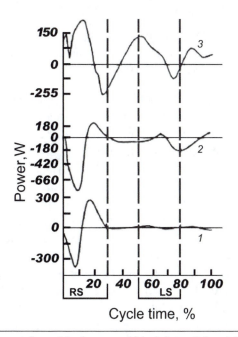

Figure 7.3 Power at the ankle, knee, and hip joints of the right leg during running at 3.30 m/s. RS and LS are stance phases of the right and left legs, respectively.

Adapted from B.I. Prilutsky and V.M. Zatsiorsky, 1992, "Mechanical energy expenditure and efficiency of walking and running," *Human Physiology* (in Russian, *Fiziologia Cheloveka*) 18(3): 118-127.

support the notion that storage and release of the elastic potential energy (i.e., energy compensation during time; see section **6.3.2.3.2** in *Kinetics of Human Motion*) play a large role in running.

During negative work production at the ankle in early stance, soleus and gastrocnemius fascicles may or may not be stretched when the corresponding MTUs are stretched. The outcome depends on running speed and slope (the faster speed and greater slope, the more stretch may occur in the fascicles of the ankle plantar flexors).

■ ■ ■ *FROM THE LITERATURE* ■ ■ ■

More Positive Work of Joint Moments Is Done During Upslope (+10°) Running Than Negative Work in Downslope (–10°) Running Despite Similar Changes in Total Energy of the Body

Source: Devita, P., L. Janshen, P. Rider, S. Solnik, and T. Hortobágyi. 2008. Muscle work is biased toward energy generation over dissipation in non-level running. *J Biomech* 41(16):3354-3359

The authors tested the hypothesis that skeletal muscles generate more mechanical energy in gait tasks that raise the CoM compared with the mechanical energy they dissipate in gait tasks that lower the CoM despite equivalent changes in total mechanical energy. Subjects ran on a 10° decline and incline surface at a constant average velocity. The authors computed work done by joint moments of the leg and the external work of the body in three dimensions using inverse dynamics. During incline and decline running, both negative work and positive work were observed at each joint. During decline running, joint moments did a greater amount of total negative joint work compared with incline running (–1.06 vs. –0.77 J/kg·m, a 38% difference), but during incline running the total joint positive work was 265% greater than that during decline running (1.43 vs. 0.54 J/kg·m). The differences (38% vs. 265%) were observed despite similar changes in the total energy of the body in the two running conditions. The differences were most likely due to a larger dissipation of mechanical energy by passive tissues in decline running.

In swing phase, the knee joint moments mostly absorb energy—the knee extensors decelerate knee flexion in the first half of swing and the knee flexors decelerate knee extension in the second half of swing. During the first part of swing, the hip flexors accelerate hip flexion, and in the second half of swing, the hip extensors first decelerate hip flexion and then accelerate hip extension.

7.2.4 Landing

Representative publications: Zatsiorsky and Prilutsky 1987; Prilutsky and Zatsiorsky 1994; Devita and Skelly 1992; McNitt-Gray 1993; Prilutsky 1990; Galindo et al. 2009

In landing after a drop jump or dismount in gymnastics, the goals are to stop downward body movement, to keep balance during and after landing, and to avoid pain associated with a hard landing. A performer needs to control the energy dissipation by skillfully regulating resistance to the impact. This resistance modulation has been suggested to originate from a preplanned reduction of excitability of short-latency stretch reflexes.

Let us consider examples of soft (ISL = 0.96) and more stiff (ISL = 0.73) barefoot landings on a stiff surface after jumping down from a height of 0.5 m (figure 7.4).

During both landing styles, the decrease in total body energy is accompanied by negative power and work of leg joint moments; that is, the leg extensors primarily do negative work. The amount of the work, the relative contribution of different joints to the total work, and patterns of joint powers depend substantially on whether the landing is soft or stiff. In the example depicted in figure 7.4, the total negative work done by the leg joints during soft landing (shaded area) is 592 J (or 96% of the TE decrease), and the ankle, knee, and hip moments absorb 159, 248, and 185 J, respectively. Thus, nearly the entire decrease in the total energy of the body in soft landing is attributable to negative work of the joint moments. The total negative work done by the leg joints during the same period of stiff landing is 455 J (or 73% of the TE decrease of 620.9 J), and the ankle, knee, and hip moments absorb 20, 228, and 197 J, respectively. Thus, during stiff landing about 30% of body energy is dissipated in the passive anatomical structures.

The softness and stiffness of landing relate to the types of extreme loading injuries that can occur during this activity: those of passive

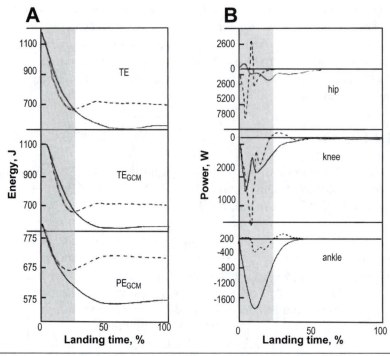

Figure 7.4 Mechanical energy of the body *(a)* and leg joint powers *(b)* during landing after jumping down from a height of 0.5 m. The subject (an experienced athlete, body mass 76 kg) was alternately asked to land as softly as possible and as stiffly as possible. Continuous lines, soft landing (ISL = 0.96); dashed lines, stiff landing (ISL = 0.73). Representative trials of one subject. *(a)* PE_{GCM}, potential energy of the center of mass; TE_{GCM}, total energy of the center of mass; TE, total mechanical energy of the body. Mechanical energy of the body was calculated based on a 14-segment, two-dimensional human body model. The decreases in TE during soft and stiff landings during the period marked by the shaded area are 618 J and 621 J, respectively. *(b)* The joint power during landing.

Reprinted from B.I. Prilutsky, 1990, "Negative mechanical work during human locomotion" (in Russian), PhD thesis, Latvian Research Institute of Orthopedics and Traumatology, Riga.

anatomical tissues (e.g., ligaments, cartilage, intervertebral discs) and those of muscle–tendon complexes (e.g., the patella tendon). If landing is maximally stiff, as when landing on the heels by keeping the legs straight, for example, no joint work will be done and all of the energy of the body will be dissipated in the passive anatomical structures. Such an activity would be very harmful for the body, specifically for the joint cartilages and intervertebral discs.

■ ■ ■ *FROM THE LITERATURE* ■ ■ ■

Landing Strategy and Patellar Tendinopathy

Source: Bisseling, R.W., A.L. Hof, S.W. Bredeweg, J. Zwerver, and T. Mulder. 2007. Relationship between landing strategy and patellar tendinopathy in volleyball. *Br J Sports Med* 41(7):e8

To see whether landing strategy might be a risk factor for the development of patellar tendinopathy (jumper's knee), this study examined whether landing dynamics from drop jumps differed among healthy volleyball players (controls) and volleyball players with jumper's knee. The patients with jumper's knee were divided into an asymptomatic group with a previous jumper's knee and a symptomatic group with a recent jumper's knee. Inverse dynamics analyses were used to estimate lower extremity joint dynamics from 30, 50, and 70 cm drop jumps in the three groups. The authors concluded that whereas subjects with recent jumper's knee used a soft landing technique to avoid high patellar tendon loading, subjects with previous jumper's knee used a stiffer landing strategy.

Based on the preceding discussion of landing softness, we can hypothesize that athletes with previous patellar tendon injuries tend to preferentially dissipate more energy in the passive body structures.

7.3 JOINT MOMENTS DURING ECCENTRIC ACTIONS

7.3.1 Maximal Joint Moments During Eccentric Actions

Representative publications: Komi 1973; Westing et al. 1990; De Ruiter and De Haan 2001

When people exert single-joint maximal efforts, the moments are largest during eccentric actions, intermediate in the isometric actions, and smaller in concentric actions (figure 7.5). The joint moment–joint angular velocity curves are usually obtained using isokinetic dynamometers, which measure exerted moments at a preset constant joint velocity while the subjects are instructed to flex or extend the joint with maximum effort. In general, the moment–angular velocity relations are similar to the force–velocity relations obtained on single muscles (see chapters 3 and 4).

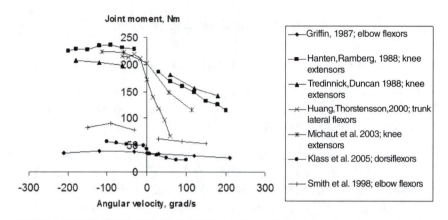

Figure 7.5 Joint moment–joint velocity relation in humans. Positive and negative velocities correspond to concentric and eccentric actions, respectively; zero velocity corresponds to isometric action. Each symbol on the graph corresponds to a peak of joint moment measured on an isokinetic dynamometer at a preset constant joint velocity. Each moment–angular velocity relation was obtained for a single muscle group (see the corresponding symbols): elbow flexors (Griffin 1987; Smith et al. 1998), knee extensors (Hanten and Ramberg 1988; Tredinnick and Duncan 1988; Michaut et al. 2003), ankle dorsiflexors (Klass et al. 2005), and trunk lateral flexors (Huang and Thorstensson 2000).

The maximum eccentric moment produced voluntarily by humans typically exceeds the isometric moment by approximately 10% to 30%, which is much smaller than almost a twofold difference seen in experiments on isolated muscles (see figure 4.3). The smaller enhancement of eccentric moments in vivo during maximal efforts may be partially explained by inhibitory influences imposed by the central nervous system, possibly to prevent muscle tissue damage. When subjects' muscles are electrically stimulated, the difference between maximum eccentric and isometric moments increases and better resembles results of in vitro experiments (figure 7.6a). Even relatively low electrical stimulation (30% of stimulation intensity that elicits maximum voluntary moment) dramatically enhances eccentric moment production.

The magnitude of eccentric moment also depends on muscle temperature and fatigue. This is expected because the dynamic force enhancement during muscle stretch is dependent on properties of engaged crossbridges. When muscle temperature decreases from 36.8 °C to 22.3 °C, the absolute magnitude of eccentric force drops by about 30% at the plateau of the force–velocity relation (figure 7.6b); however, the difference between the eccentric

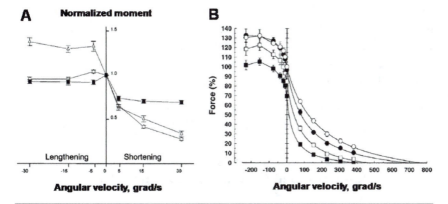

Figure 7.6 Effects of electrical muscle stimulation and temperature on moment–angular velocity relation in humans. *(a)* Plantar flexion ankle moment as a function of ankle angular velocity in human subjects. The moments are normalized to the maximum voluntary isometric moment produced at ankle and knee joint angles of 90° and 120°, respectively. Filled boxes correspond to maximum voluntary action; open circles correspond to submaximal voluntary action (EMG level corresponded to 30% of maximum isometric voluntary effort); and open triangles correspond to submaximal electrical stimulation that elicits 30% of voluntary isometric moment. *(b)* Force–velocity relation of adductor pollicis muscle at different temperatures (open circles, 36.8 °C; filled circles, 32.6 °C; open boxes, 26.6 °C; filled boxes, 22.3 °C). Values of force are normalized to the maximal isometric force at 36.8 °C. The force refers to that exerted by the thumb during percutaneous supramaximal electrical stimulation of the ulnar nerve. The angular velocity refers to the rate of change in angle between the thumb and the index finger. Muscle temperature was manipulated by immersing the subjects' hand and forearm in a water bath of different temperature for 20 min.

(a) With kind permission from Springer Science+Business Media: *European Journal of Applied Physiology,* "Tension regulation during lengthening and shortening actions of the human soleus muscle," 81(5), 2000, 375-383, G.J. Pinniger and J.R. Steele, fig. 4.

(b) Reprinted, by permission, from C.J. De Ruiter and A. De Haan, 2001, "Similar effects of cooling and fatigue on eccentric and concentric force-velocity relationships in human muscle," *Journal of Applied Physiology* 90(6): 2109-2116.

and isometric forces becomes greater at lower temperatures. For example, at 36.8 °C, eccentric force exceeds isometric by about 30%, whereas at 22.3 °C, eccentric force exceeds isometric by about 60% (figure 7.6*b*).

Because maximum isometric force and maximum shortening velocities are substantially reduced at lower temperatures, the greater enhancement of eccentric force in this situation—60%—could be related to a slower rate of crossbridge cycling. Muscle fatigue also slows down crossbridge cycling rate, which in turn leads both to a greater enhancement of eccentric force with respect to isometric force and to a declining maximum isometric force and shortening velocity. Perhaps similar mechanisms of slowing crossbridge cycling can explain the fact that slow-twitch muscles also have a greater increase in eccentric force compared with isometric force than do fast-twitch muscles.

■ ■ ■ *FROM THE LITERATURE* ■ ■ ■

Eccentric Force Enhancement Is Greater in Slow-Twitch Than in Fast-Twitch Muscles

Source: Mutungi, G., and K.W. Ranatunga. 2001. The effects of ramp stretches on active contractions in intact mammalian fast and slow muscle fibres. *J Muscle Res Cell Motil* 22(2):175-184

The effects of a ramp stretch (amplitude < 6% muscle fiber length $[L_0]$, speed < 13 L_0/s) on twitch tension and twitch tension redevelopment were examined in intact mammalian (rat) fast and slow muscle fiber bundles. The experiments were done in vitro at 20 °C and at an initial sarcomere length of 2.68 μm. In both fiber types, a stretch applied during the rising phase of the twitch response (including the time of stimulation) increased the redeveloped twitch tension (15%-35%). A stretch applied before the stimulus had little or no effect on the twitch myogram in fast muscle fibers, but it increased the twitch tension (approximately 5%) in slow muscle fibers. A similar stretch had little or no effect on tetanic tension in either muscle fiber type. In general, the results indicate that the contractile-activation mechanism may be stretch sensitive, and this is particularly pronounced in slow muscle fibers.

7.3.2 Force Changes During and After Stretch

Representative publications: De Ruiter et al. 2000; Onambele et al. 2004

The eccentric muscle moment–angular velocity relation discussed in the previous section is obtained by recording peak muscle moment or force during stretching an active muscle group or a single muscle. This peak corresponds to the dynamic force enhancement that was discussed in chapter 4 (**4.1.1**). The behavior of muscle moment or force during stretch of human muscles in vivo is similar to that of isolated muscles or even isolated fibers whose response to stretch is not influenced by such serial elastic structures as tendons or aponeuroses. Although reflexes affect muscle responses to stretch in vivo conditions (see section **7.3.2.2**), many features of the force response to stretch described for isolated muscles are observed in healthy humans.

7.3.2.1 Dynamic Force Enhancement

During stretch of human muscles with constant moderate speeds, the stretch-induced force increase has two distinct phases (figure 7.7a). In the first phase force rises quickly as a linear function of length change (elastic response). After a break point, the second phase starts in which force continues to increase but with progressively slower rate (velocity-dependent response) until the force reaches its peak at the end of stretch.

These two phases constitute dynamic force enhancement (figure 7.7). After stretch, the force of activated muscles decreases in a similar fashion: first quickly with the rate similar to that of initial force increase and then more slowly until force reaches a plateau. The force at plateau exceeds the isometric force at the same muscle length and activation and represents the residual force enhancement (see **4.1.2**; also discussed in **7.3.3**).

Muscle force changes during and after stretch can be described as the sum of three components (figure 7.7b):

Component A (which combines effects of the first two phases of force rising) is stretch-velocity dependent.

Component B is muscle-length dependent but this dependency is not related to crossbridge function and is likely determined by structural proteins like titin.

Component C is also length dependent and has active (muscle force–length relation) and passive (perhaps sarcomere structural proteins) contributions.

Adding these three components together results in a force trace that is qualitatively similar to that obtained experimentally (compare figures 7.7a and 7.7b, total force trace).

The dynamic force enhancement in human muscles (evaluated as the difference between the peak of muscle force at the end of stretch and the plateau force, $[F_{peak} - F_{after}]$, figure 7.7a) depends strongly on the initial muscle force (F_{before}, figure 7.7, a and c) and velocity of stretch (figure 7.7d). Because the isometric muscle force is a function of the number of crossbridges formed, and the dependence of muscle force on stretch velocity was linked to the turnover speed of crossbridges, it is likely that dynamic force enhancement is determined to a large extent by behavior of crossbridges during stretch (see **4.1.1** for details). Beside the crossbridges, the tendon and aponeurosis

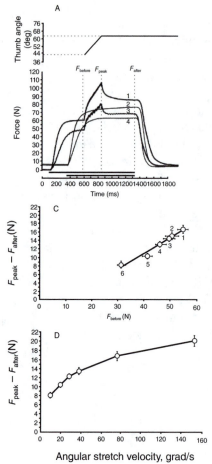

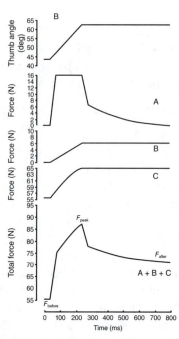

Figure 7.7 Force enhancement during and after stretch of human adductor pollicis muscle. *(a)* Angular displacement (top) and force data (bottom) for one subject. Stimulation (indicated by the bars above the time axis) was either at 80 Hz (traces 1 and 2) or 30 Hz (traces 3 and 4). Vertical dotted lines indicate where the forces were measured, before (F_{before}), at the peak (F_{peak}), and 500 ms following the stretch (F_{after}). The thick traces are the force responses before, during, and after stretches from 44° to 63° thumb angle at 76.4°/s. Note that with both the 80 and 30 Hz stimulation, the isometric force after the stretch (F_{after}) was enhanced compared with the isometric forces at the 63° thumb angle without a preceding stretch (thin traces). *(b)* A diagram illustrating the contribution of hypothetical components to stretch-induced force enhancement. Component A is the force response of crossbridges to constant velocity stretch. Component B is the length-dependent element. Component C follows the total (active + passive) force–thumb angle relation of the muscle. The difference between the peak force (F_{peak}) and F_{after} is used in panels *(c)* and *(d)* to quantify the dynamic force enhancement shown in panel *(a)*. *(c)* Effect of the isometric force level before stretch (F_{before}) on the dynamic force enhancement ($F_{peak} - F_{after}$). Stretch velocity 76.4°/s. 1, stretches from 44° to 63°, 80 Hz stimulation; 2, from 44° to 63°, 80 Hz, after recovery; 3, from 36° to 55°, 80 Hz; 4, from 36° to 55°, 80 Hz, after recovery; 5, from 44° to 63°, 30 Hz; 6, from 44° to 63°, 30 Hz, after recovery. *(d)* Effect of velocity of muscle stretch on the dynamic force enhancement ($F_{peak} - F_{after}$). Stimulation 80 Hz.

From C.J. De Ruiter, W.J. Didden, D.A. Jones, and A. De Haan, 2000, "The force-velocity relationship of human adductor pollicis muscle during stretch and the effects of fatigue," *Journal of Physiology* 526(3): 671-681. Reprinted by permission of John Wiley & Sons Ltd.

that also constitute the series elastic component of the muscle may contribute to the dynamic force enhancement. As discussed in sections **3.3.2** and **4.1.1**, the series elasticity largely determines the muscle behavior during stretch and the amount of energy stored in the muscle and released during subsequent muscle shortening (see section **7.6** for more details).

7.3.2.2 Short-Range Stiffness

Representative publications: Rack and Westbury 1974, 1984; Nichols and Houk 1976; Morgan 1977; Lin and Rymer 1993; Loram et al. 2007; Cui et al. 2008

The first fast increase in force prior to the break point during stretch has been suggested to originate from deformation of engaged crossbridges. Force is rising in this phase as a linear function of muscle elongation (the elastic response). If the muscle is allowed to shorten within this phase, the force falls in almost the same way as it rises, that is, in proportion to muscle shortening and with similar rate of force change. Thus muscle during this phase behaves like an ideal spring with little net work done (energy absorbed during stretch is used for shortening), and its resistance to an imposed stretch (stiffness) is greater than after the break point. The range of muscle elongation during which the muscle exhibits such behavior is rather small and is determined by the maximal range of deformation of the engaged crossbridges without their forceful detachment (approximately 12-14 nm per half sarcomere depending on a muscle and experimental conditions such as temperature and stretch velocity).

Muscle resistance to stretch during the initial phase of muscle elongation is called *short-range stiffness*. The short-range stiffness (i.e., the ratio of force change and corresponding muscle elongation, $\Delta F/\Delta L$) depends on muscle activity or muscle force before the stretch (for the cat soleus muscle it changes from 0.9 to 2.1 N/mm per newton of the initial force). The dependence of the short-range stiffness on the muscle force is expected given that the muscle resistance to stretch is a result of deformation of the engaged crossbridges. Thus, the larger the muscle force the larger the number of active crossbridges and therefore the greater resistance they provide during deformation. The deformation of crossbridges during this short-magnitude muscle stretch is approximately 12 to 14 nm per half sarcomere.

The short-range stiffness allows instantaneous resistance to sudden small-magnitude external perturbations that occur, for example, as a result of unexpected contacts with external objects. An external perturbation leads to changes in joint angles and, in turn, to muscle stretch. Owing to short-range stiffness, the stretched muscles resist length changes before the fastest reflexes (monosynaptic stretch reflexes) start to operate in about 20 to 40 ms. Thus short-range stiffness can be viewed as the first line of defense against unexpected postural perturbations.

Stretch reflexes, the neural excitation of muscles in response to stretch, provide the second line of defense that allows the muscles to resist stretch beyond the span of short-range stiffness (figure 7.8). Stretch reflexes increase muscle resistance two- to threefold compared with muscles without stretch reflex (see figure 7.8).

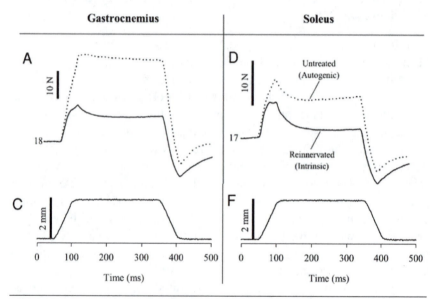

Figure 7.8 Comparisons of responses to a constant velocity stretch (2 mm in 50 ms) of intact muscles and muscles without stretch reflex. Data are shown for cat gastrocnemius (left) and soleus (right) muscles. *(A)* and *(D)* Dotted lines, force responses of intact muscles; solid lines, responses of muscles without stretch reflex. The numbers 18 and 17 at the force traces correspond to muscle initial forces (in newtons). *(C)* and *(F)* Imposed length changes. Muscles were activated physiologically by eliciting the withdrawal reflex of the contralateral hindlimb (cross-extension reflex). Stretch reflex of ipsilateral soleus and gastrocnemius was removed by self-reinnervating soleus and gastrocnemius muscles. Self-reinnervation was performed at least 9 months before the experiments by cutting the nerves, innervating the muscles, and suturing the cut nerve stumps together. Axons of motoneurons grew back and reinnervated the muscles over several months, but stretch reflex did not recover. Note that the intact muscles respond to stretch with high resistance over a longer range of muscle elongation and thus maintain the initial high muscle stiffness.

Reprinted, by permission, from C.M. Huyghues-Despointes, T.C. Cope, and T.R. Nichols, 2003, "Intrinsic properties and reflex compensation in reinnervated triceps surae muscles of the cat: Effect of activation level," *Journal of Neurophysiology* 90(3): 1537-1546.

Measurements of joint or muscle stiffness in humans are more challenging given difficulties in controlling a constant level of muscle activity and variable experimental conditions and assessing internal deformation of contractile and tendon components (a more detailed discussion of the concept of joint stiffness and the difficulties associated with its measurements is provided in *Kinetics of Human Motion,* section **3.5**). Nevertheless, estimations of the short-range stiffness and the range of its operation in humans are generally in agreement with studies on other mammals. In particular, measurements show that short-range stiffness is important in stabilizing human posture against small external perturbations.

7.3.2.3 Decay of Dynamic Force Enhancement

Representative publication: Cavagna 1993

The force that an active muscle develops during a moderate-speed ramp stretch starts to decay as soon as stretch stops, even if the muscle has the same activation level and the same constant length achieved at the end of ramp stretch (see figures 4.2 and 7.7*a*). The force decay normalized to the peak of dynamic force enhancement has been typically described by an equation with two exponential terms corresponding to two phases of decay, fast and slow:

$$(F - F_{after}) / (F_{peak} - F_{after}) = A_f e^{-r_f(t)} + A_s e^{-r_s(t)}, \qquad [7.3]$$

where F_{peak} and F_{after} are the force values corresponding to the peak during stretch and at a steady-state level, respectively (see also figure 7.7*a*); A_f and A_s are the magnitudes of force decay during the fast and slow phases; and r_f and r_s are rate constants.

As was already mentioned, the first fast phase starts immediately after the cessation of stretch and lasts for several dozen milliseconds depending on muscle, temperature, and other factors. During this phase the force drops by 20% to 60%. The magnitude and rate of force decay during the fast phase increase with larger stretch velocity, temperature, and the percentage of fast twitch fibers in the muscle. These facts suggest that the crossbridges are possibly responsible for the fast drop in muscle force immediately after stretch.

The force decay during the second (slow) phase lasts for several hundred milliseconds and does not depend markedly on muscle temperature. The force drop in this phase decreases with stretch velocity (in contrast, the force drop in the first phase increases with stretch velocity).

If following muscle stretch the muscle is allowed to shorten against a load equal to F_o (maximum isometric force at the corresponding length) at different times during force decay, the muscle will shorten and demonstrate enhanced performance. This subject is discussed below in section **7.6** dealing with the SSC.

7.3.3 Residual Force Enhancement in Humans

Representative publications: De Ruiter et al. 2000; Oskouei and Herzog 2005; Hahn et al. 2007; Pinniger and Cresswell 2007

Experiments on fully activated isolated muscles (see section **4.1.2**) demonstrate, following muscle stretch, muscle force changes in three phases. In the first two phases the force decreases, first rapidly and then more slowly. During the third phase the force reaches a steady-state level in which there is a residual force enhancement compared with isometric force at the same length. If similar effects exist in human muscles activated voluntarily, the effects will be beneficial for many occupational and athletic activities because, as we learned from chapter 4, this residual force enhancement can be substantial and does not require extra muscle excitation and energy expenditure.

A residual muscle moment enhancement after muscle stretch in voluntary human movements apparently exists (table 7.1). It has been observed in forearm flexors, small muscles of the hand, and ankle dorsiflexors and plantar flexors (figure 7.9) during a variety of experimental conditions: maximum and submaximum voluntary contractions and maximum and submaximum electrical muscle or nerve stimulations. Residual moment enhancement during voluntary muscle activation can exceed 10% of the isometric moment at the same joint angle 5 to 10 s after the end of muscle stretch. Electrically stimulated muscles demonstrate a greater moment enhancement by 5% to 10% compared with voluntarily contracting muscles.

Table 7.1 Residual Force Enhancement in Human Muscles in Vivo

Muscle	Subjects	Muscle activity	Residual force	Source
Forearm flexors	15 athletes; 15 schoolgirls	MVC	Present	Gulch et al. 1991
Adductor pollicis (thumb adductor)	8 females	Maximum stimulation	Present	De Ruiter et al. 2000
Adductor pollicis (thumb adductor)	12 males; 5 females	30% MVC	Present in 8 of 17 subjects	Oskouei and Herzog, 2005
Knee extensors	15 males	95%-100% MVC	Absent	Hahn et al. 2007

Note: MVC, maximum voluntary contraction. Maximum stimulation denotes electrical stimulation that caused highest isometric muscle moment at the joint.

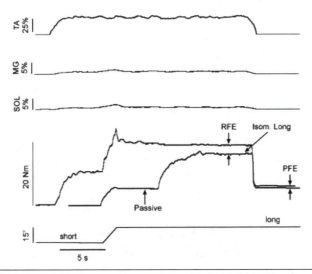

Figure 7.9 Representative traces from submaximal (25% MVC) voluntary eccentric action of ankle dorsiflexors. From top to bottom: Rectified and low-pass filtered EMG activity of tibialis anterior (TA), medial gastrocnemius (MG), and soleus (SO) muscles; ankle dorsiflexion active and passive moments, respectively, before, during, and after ankle plantar flexion; and ankle plantar flexion angle before (short), during, and after (long) angle change. EMG is normalized to MVC. The initial increase in active moment is caused by voluntary muscle activation; the second increase corresponds to the dynamic moment enhancement caused by muscle stretch; the maximal moment is reached at the end of the joint angle ramp and then the moment decreases for about 10 s to a steady-state level. This steady-state moment exceeds the isometric moment at the same joint angle (Isom. Long) and muscle activation; the difference between the two traces is residual force enhancement (RFE). Stretch of passive dorsiflexors also results in moment enhancement (passive force enhancement, PFE), which is 25% of RFE in these experiments.

Reprinted, by permission, from G.J. Pinniger and A.G. Cresswell, 2007, "Residual force enhancement after lengthening is present during submaximal plantar flexion and dorsiflexion actions in humans," *Journal of Applied Physiology* 102(1): 18-25.

As in experiments on isolated muscle, the residual moment enhancement increases with the magnitude of stretch and initial muscle length. In isolated muscles that are kept at short lengths (on the ascending part of the force–length relation), force enhancement depends on muscle extension speed, but at long lengths, at which a reduced number of crossbridges are formed, the residual force enhancement does not depend on stretch velocity. In humans, muscles commonly operate on the ascending limb or plateau of the force–length relation. This explains why stretch velocity affects residual force enhancement. For example, when knee extensors are stretched so fast that the knee moment reaches its peak in the middle of the stretch and then the moment decreases while the stretch continues (an indication of the yield effect, or forceful detachment of crossbridges, see **4.1.2**), residual moment enhancement is not seen (table 7.1, knee extensors).

Although the examples of the dynamic and residual force enhancement in humans described in sections **7.3.2** and **7.3.3** are obtained in specially designed experimental conditions, they indicate that the dynamic and residual force enhancement could increase muscle performance during natural human movements such as locomotion, jumping, and throwing, as discussed in the following sections.

7.4 MUSCLE ACTIVITY DURING ECCENTRIC ACTIONS

Because concentric moments sharply decline with increasing angular velocity whereas eccentric moments either stay the same or slightly increase, the difference in the magnitude between eccentric and concentric moments becomes larger as absolute values of angular velocity increase (see figures 7.5 and 7.6). Note that the muscle moment–joint angular velocity curves are obtained in conditions of identical or similar muscle activity when subjects are instructed to exert maximum voluntary effort or to maintain similar magnitude of EMG activity based on visual feedback while performing eccentric, isometric, or concentric actions.

The difference in joint moment magnitudes between different types of muscle action at similar efforts has important consequences. First, to exert the same value of moment, an individual needs to exert less effort during eccentric action compared with concentric action. Second, to maintain the same moment with increasing absolute values of angular joint velocity, similar muscle excitation is required during eccentric action (because the eccentric moment does not change substantially with velocity after the velocity exceeds a certain relatively low value). In contrast, during concentric action, increasing muscle excitation with velocity is expected to compensate for the declining muscle force with shortening velocity (see figures 7.5 and 7.6). Third, because muscles are less activated while producing the same muscle moment in eccentric compared with concentric action, a larger force per cross-sectional area of the activated muscle fibers is exerted during eccentric activity. This higher force per active muscle area has been implicated in causing damage and delayed muscle soreness (see **7.5.3**).

7.4.1 Surface Electromyographic Activity During Eccentric Actions

Representative publications: Abbott et al. 1952; Asmussen 1953; Bigland and Lippold 1954

In eccentric actions, EMG activity increases with exerted force (figure 7.10*a*) by a smaller amount than in concentric actions. The EMG magnitude does not depend on the joint angular velocity during eccentric actions, whereas EMG in concentric exercises increases with the velocity (figure 7.10*b*). These facts are consistent with and can be explained by the muscle force–velocity (see figure 4.3) and muscle moment–angular velocity relations (see figure 7.5).

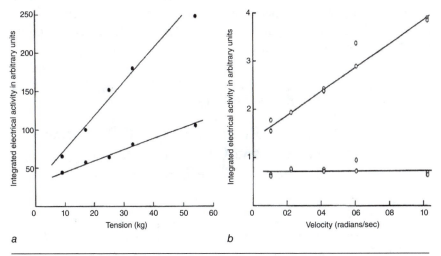

a   b

Figure 7.10   *(a)* The relation between integrated electrical activity and tension in the human calf muscles. Recordings from surface electrodes. Shortening at constant velocity (above) and lengthening at the same constant velocity (below). Each point is the mean of the first 10 observations on one subject. Tension represents weight lifted and is approximately one tenth of the tension calculated in the tendon. *(b)* The relation between integrated electrical activity and velocity of shortening (above) and lengthening (below) at the same tension (3.75 kg). Each point is the mean of the first 10 observations on one subject.

From B. Bigland and O.C.J. Lippold, 1954, "The relation between force, velocity and integrated electrical activity in human muscles," *Journal of Physiology* 123(1): 214-224. Reprinted by permission of John Wiley & Sons Ltd.

7.4.2 Motor Unit Activity During Eccentric Actions

Representative publications: Nardone et al. 1989; Duclay and Martin 2005; Pasquet et al. 2006; Duchateau and Enoka 2008

Motor unit (MU) recruitment during both eccentric and concentric muscle actions follows the size principle (the principle is explained in section **3.1.2**). However, although order of MU recruitment (from small to large motor neurons and, correspondingly, from slow to fast MUs) and derecruitment (from large to small motor neurons) is similar in eccentric and concentric actions with rising and declining muscle force, respectively, recruited MUs do not seem to change their firing rate substantially with rising force in eccentric action, whereas the MU firing rate increases with force in concentric action, especially when the force exceeds a certain threshold value. In other words, during concentric actions of high intensity, muscle force is controlled by two main mechanisms equally—MU recruitment and rate coding—whereas during eccentric actions MU recruitment and derecruitment mechanism appears more prominent.

■ ■ ■ *From the Literature* ■ ■ ■

Order of Motor Unit Recruitment and Derecruitment in Eccentric and Concentric Actions Is Consistent With the Size Principle

Source: Pasquet, B., A. Carpentier, and J. Duchateau. 2006. Specific modulation of motor unit discharge for a similar change in fascicle length during shortening and lengthening contractions in humans. *J Physiol* 577(Pt 2):753-765

The authors compared motor unit recruitment and discharge rate in the human tibialis anterior during shortening and lengthening contractions that involved a similar change in joint moment. The dorsiflexor moment, muscle fascicle length changes, and surface and intramuscular EMGs from the tibialis anterior were recorded in eight subjects. The behavior of the same 63 motor units was compared during submaximal shortening and lengthening contractions performed at a constant velocity (10°/s) with the dorsiflexor muscles over a 20° range of motion around the ankle neutral position.

Motor units that were active during a shortening action were always active during the subsequent lengthening action. Furthermore, additional motor units ($n = 18$) of higher force threshold that were recruited during the shortening action to maintain the required moment were derecruited first during the following lengthening action. Although the change in fascicle length was similar for both shortening and lengthening actions, modulation of discharge rate was different between them.

Compared with discharge rates during initial isometric actions at short (11.9 ± 2.4 Hz) or long (11.7 ± 2.2 Hz) muscle length the discharge rate increased only slightly and stayed nearly constant throughout the lengthening action (12.6 ± 2.0 Hz; $p < .05$), whereas it changed progressively during the shortening action, reaching 14.5 ± 2.5 Hz ($p < .001$) at the end of the movement. The authors concluded that there was a clear difference in motor unit discharge rate modulation, but there was no indication of changes in their recruitment order between shortening and lengthening action when performed with a similar change in muscle fascicle length and moment.

7.4.3 Electromechanical Delay

Representative publications: Cavanagh and Komi 1979; Norman and Komi 1979

The *electromechanical delay* (EMD) is a time interval between the onset of muscle EMG activity and developed force or joint moment (see **3.1.2.2**). The type of muscle action affects the duration of EMD. Because EMD depends heavily on the measurement protocol (no standardized protocol exists), cross-study comparisons of EMD values is not meaningful. However, data from single studies in which the same protocol was used in both eccentric and concentric actions can be used. The data suggest that EMD is slightly shorter in eccentric actions. For example in one of the studies, the EMD determined for the biceps brachii during eccentric action was 38 ± 13 ms (mean $\pm$ standard deviation; at the slow joint angular velocity) and 28 ± 11 ms (at the faster velocity), whereas EMD during concentric action was 41 ± 13 ms and was independent of joint velocity. It is thought that a major portion of EMD is associated with the stretch of the series elastic component (SEC) to the point where muscle force can be detected. Therefore, it seems that conditions for a shorter reaction time and rapid force development are more advantageous during eccentric action.

7.5 PHYSIOLOGICAL COST
OF ECCENTRIC ACTION

When an active muscle exerts a force F and shortens a distance x, the muscle performs work $W^+ = Fx$. If the muscle is stretched a distance x while exerting the same force, it absorbs mechanical energy and it can be said that the muscle does negative work $W^- = -Fx$ (the sign is negative because the force and displacement of its point of application are in opposite directions). For example, when a person walks upstairs, the leg extensors shorten and do positive work against gravity; when she descends, the same muscles are stretched while actively resisting gravity. Common experience suggests that walking downstairs is less effortful than walking upstairs. This claim can be tested by measuring, for instance, the oxygen consumption in both tasks.

7.5.1 Oxygen Consumption During Eccentric and Concentric Exercise

Representative publications: Abbot et al. 1952; Assmussen 1953

Oxygen uptake during negative work production is lower than during positive work of the same magnitude. An elegant demonstration of this fact was obtained in an experiment in which two bicycle ergometers were placed back to back and connected by a chain; when one cyclist (A) pedaled in the usual forward direction, the legs of the other cyclist (B) were driven backward (figure 7.11). The magnitudes of the forces exerted on the pedals and the work (power) of these forces were equal for both cyclists (external work, explained in chapter 6 in *Kinetics of Human Motion*).

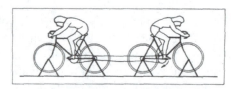

The recorded values of oxygen consumption were sharply different; they increased with power at different rates (figure 7.12). In part, the differences can be explained by the dissimilar magnitudes of the work done by two subjects for the leg movements. Cyclist A spent energy moving the legs of both subjects, whereas cyclist B did not. We invite the reader to perform the following

Figure 7.11 Two bicycle ergometers coupled together in opposition (a schematic). During the experiment, the left cyclist (cyclist A) pedaled at an agreed speed set by a metronome and cyclist B resisted sufficiently to maintain an agreed force reading as displayed on a gauge. Cyclists A and B, by definition, exerted the same force, otherwise the pedaling rate would have accelerated or slowed down.

Adapted from B.C. Abbott, B. Bigland, and J.M. Ritchie, 1952, "The physiological cost of negative work," *Journal of Physiology* 117(3): 380-390.

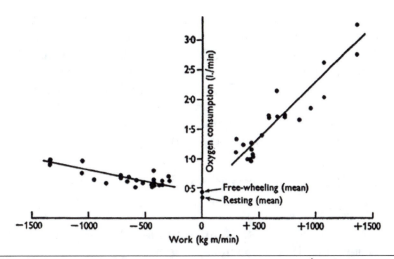

Figure 7.12   Variation in the rate of oxygen consumption ($\dot{V}O_2$) with external mechanical power for pedaling with both concentric and eccentric muscle actions. All measurements were performed at a steady state when oxygen consumption was equal to the oxygen requirement. We recommend comparing this figure with Figure 7.10 where the data on the integrated EMG activity during concentric and eccentric activities are presented. The general patterns in both figures look similar.

From B.C. Abbott, B. Bigland, and J.M. Ritchie, 1952, "The physiological cost of negative work," *Journal of Physiology* 117(3): 380-390. Reprinted by permission of John Wiley & Sons Ltd.

mental experiment. Imagine that subject B is replaced by a mechanical dummy with the same mass-inertial characteristics. Such dummies are used in biomechanical research involving high injury risks, for instance, in studying car accidents. In the conditions of the described experiment, subject A will spend energy for leg movement whereas the dummy will spend nothing. However in the actual (human–human) experiment, subject B spent energy creating a resisting force and dissipating energy of the rotating pedals.

Similar findings were reported for other activities (ascending and descending stairs, lifting and lowering loads). The ratio of oxygen uptake during concentric ($\dot{V}O_{O_2}^+$) and eccentric ($\dot{V}O_{O_2}^-$) exercise at the same absolute values of mechanical power ($\dot{V}O_{O_2}^+ / \dot{V}O_{O_2}^-$) always exceeds 1 and depends on exercise (walking, running, cycling), velocity of movement, and methods of determining $\dot{V}O_2$ (gross oxygen uptake, gross oxygen uptake minus oxygen uptake at rest). For example, the ratio $\dot{V}O_{O_2}^+ / \dot{V}O_{O_2}^-$ during cycling increases with cadence from about 2 at 15 rpm to 5 to 10 at 100 rpm.

The fact that the ratio $\dot{V}O_{O_2}^+ / \dot{V}O_{O_2}^-$ exceeds 1 can be explained by the hypothesis that eccentric actions require fewer active fibers compared with the concentric actions against the same load (see figure 7.10). The same hypothesis can be used to explain the increase of the oxygen uptake ratio with the speed of movement: the difference in the maximum developed force between eccentric and concentric actions increases with the speed of muscle length change (figure 7.5). In addition, a lower oxygen uptake of eccentric actions per unit of muscle activation can also contribute to the high $\dot{V}O_{O_2}^+ / \dot{V}O_{O_2}^-$ ratio. $\dot{V}O_2$ per unit of integrated EMG of working muscles is about 3 times lower in eccentric actions than in concentric actions.

7.5.2 Fatigue and Perceived Exertion During Eccentric Action

Representative publications: Komi and Viitasalo 1977; Knuttgen 1986; Hortobagyi et al. 1996; Pandolf et al. 1978

Two major types of exercise-induced fatigue can be distinguished: (1) metabolic fatigue, which is related to a failure to maintain adenosine triphosphate production rates and to tolerate high accumulation of by-products of metabolic reactions, and (2) nonmetabolic fatigue caused by high internal muscle stress, which is believed to be associated with a disruption of internal muscle structures. As discussed previously, eccentric muscle action is much less metabolically demanding than are concentric and isometric actions, whereas force per unit of cross-sectional area of active muscle fibers is likely higher during eccentric action compared with concentric and isometric actions. Therefore, differences in fatigue between eccentric and other types of muscle action can be expected.

When eccentric and concentric actions are compared at the same intensity level, moderate eccentric muscle action appears to cause substantially lower fatigue (smaller declines in developed force and power) and perceived exertion. Most probably it happens because fewer muscle fibers are activated during eccentric exercise compared with concentric and isometric exercise against the same load (see figure 7.10).

In contrast, when eccentric and concentric actions are compared at the same oxygen consumption level or when eccentric and concentric exercises are performed with maximum effort, muscle fatigue occurs faster in eccentric exercise. Correspondingly, perceived exertion in eccentric exercise in this case is higher than in the corresponding concentric exercise. Note that muscles develop higher forces in maximum eccentric than in maximum concentric exercise (see figure 7.5).

At the point of exhaustion after a 30 min eccentric exercise of high intensity (corresponding to 90% of $\dot{V}O_2max$ in the corresponding concentric exercise), none of the signs of exhaustion typical for concentric exercise (high values of $\dot{V}O_2$ uptake, heart rate, muscle and blood lactate) are present. After 6 weeks of eccentric training with the same intensity, the subjects are able to continue exercise for several hours. It has been thought that inability of untrained subjects to continue eccentric exercise is caused by damage of muscle fibers and inappropriate motor unit recruitment.

7.5.3 Muscle Soreness After Eccentric Exercise

Representative publications: Asmussen 1956; Armstrong 1984; Proske and Morgan 2001; Lieber and Friden 2002

Almost everyone has experienced discomfort and soreness in muscles after a hard workout or atypical physical activity. In these situations, muscle discomfort and soreness do not start immediately after exercise but are delayed for several hours. The intensity of soreness increases within the first 24 hours, reaches a maximum between 24 and 72 hours, and then subsides within a week without any special treatment. Such symptoms are referred to as *delayed-onset muscle soreness* (DOMS). The other general symptoms of DOMS include muscle swelling, increased muscle temperature, reduced joint range of motion, and reduced maximum force production.

Eccentric muscle action is a primary trigger of the events leading to DOMS. This was first demonstrated in the elegant experiments of Erling Asmussen in 1956. Subjects repeatedly raised themselves on a step of 0.5 m by one leg (concentric action) and then stepped down by descending the body by the other leg (eccentric action). In 24 to 48 hours, muscle soreness and increased hardness (i.e., resistance to indentation) of the quadriceps were reported in the leg that performed eccentric action. These first observations have since been confirmed in a variety of other activities involving eccentric and concentric actions: walking and running downhill and uphill, normal cycling and resisting rotating pedals, and lowering and raising loads. In all these activities, tasks involving mostly eccentric actions lead to DOMS, whereas concentric actions with the same magnitude and intensity typically do not.

Why are muscles susceptible to damage and soreness after eccentric but not after concentric action? It is because of the difference in the force–velocity properties between the two types of actions. When a muscle with a given level of excitation is stretched, it develops force that exceeds force developed by the same activated muscle during shortening at the same speed (see figures 7.5 and 7.6). Therefore, to produce the same force during concentric and eccentric actions (as in stepping on and stepping down a step), less muscle excitation and presumably fewer active muscle fibers are required during eccentric action (see figure 7.10). Thus, stress on active muscle fibers (the ratio of developed muscle force per area of active fibers) is higher in eccentric action. A greater stress on muscle fibers is thought to cause mechanical disruption of fiber membranes and perhaps some structural proteins inside muscle cells. Muscle tissue damage after eccentric action has been documented by electron and light microscopy of muscle materials taken from human and animal muscles. Most structural damage occurs in fast twitch fibers. Discussion of the biological mechanisms of muscle damage leading to DOMS is beyond the scope of this textbook.

Among prevention and treatment strategies used to eliminate or reduce DOMS symptoms, only physical activity has demonstrated unequivocal positive effects. People who exercise regularly and perform activities consisting of eccentric action do not typically experience DOMS. Moderate exercise can also reduce the sensation of muscle soreness caused by eccentric action, although this exercise-induced analgesic effect is temporary.

7.6 REVERSIBLE MUSCLE ACTION: STRETCH–SHORTENING CYCLE

In many human movements, muscle shortening is preceded by the stretch of active muscle. An example is leg yield and push-off in stance of running. Stretch immediately followed by shortening is called the *stretch–shortening cycle* (SSC). The term *reversible muscle action* is also used. The SSC is a natural part of some movements, such as running (a spring in the leg; figure 7.13); in other movements, such as throwing, these actions must be learned.

The reversible muscle action can be roughly classified as either springlike or catapult-like. In springlike SSCs, the durations of the eccentric

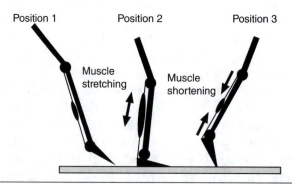

Figure 7.13 Stretch–shortening cycle during the support period in running. The plantar flexors of the foot are stretched during the first part of the support period (from position 1 to position 2) and shorten afterward, from position 2 to position 3.

and concentric phases are approximately equal. In catapult-like SSCs, the eccentric phase is much longer than the concentric phase; the elastic energy is accumulated over a long time period and then released over a short time period (figure 7.14).

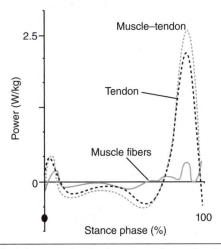

Figure 7.14 Ankle joint muscle–tendon catapult mechanism during walking. Ultrasound imaging data from walking humans show the instantaneous rate of mechanical energy absorption and production for gastrocnemius muscle fibers and Achilles tendon separately (watts per kilogram). The muscle–tendon power (bold gray) is the sum of the gastrocnemius muscle power (dotted gray) and the Achilles tendon power (black dashed). Positive values indicate energy generation, and negative values indicate energy absorption. Note that in this activity, a larger portion of elastic energy is stored in the tendon than in the muscle belly. The muscle fibers contribute very little to total muscle–tendon mechanical power output but the Achilles tendon stores and returns a significant amount of mechanical energy. See also figure 7.1, the power at the ankle joint, in this chapter.

Reprinted, by permission, from G.S. Sawicki, C.L. Lewis, and D.P. Ferris, 2009. "It pays to have a spring in your step," *Exercise and Sport Sciences Reviews* 37(3): 130-138. Adapted, by permission, from M. Ishikawa, P.V. Komi, M.J. Grey, V. Lepola, and G.P. Bruggemann, 2005, "Muscle-tendon interaction and elastic energy usage in human walking," *Journal of Applied Physiology* 99(2): 603-608.

■ ■ ■ *FROM THE LITERATURE* ■ ■ ■

Catapult Action in Jumping

1. Human and Insects Use Different Jumping Techniques

Alexander, R.M. 1995. Leg design and jumping technique for humans, other vertebrates and insects. *Philos Trans R Soc Lond B Biol Sci* 347(1321):235-248

(continued)

From the Literature   *(continued)*

Humans, bush babies, frogs, locusts, fleas, and other animals jump by rapidly extending a pair of legs. Mathematical models were used to investigate the effect of muscle properties and jumping technique on jump height. When ground forces are small multiples of body mass (as for humans), countermovement and catapult jumps are about equally high, and both are much better than squat (noncountermovement) jumps. Vertebrates have not evolved catapult mechanisms and use countermovement jumps instead. When ground forces are large multiples of body mass, catapult jumps (as used by some insects) are much higher than the other jump styles.

2. The Best Insect Jumpers Use Catapult Technique

Source: Burrows, M. 2009. Jumping performance of planthoppers (*Hemiptera, Issidae*). *J Exp Biol* 212(17):2844-2855

Jumping insects were filmed at rates up to 7,500 images per second. In the best jumps, the insect body was accelerated in 0.8 ms to a takeoff velocity of 5.5 m/s, was subjected to acceleration of 719 g, and was displaced a horizontal distance of 1.1 m. The takeoff force was more than 700 times body weight. This performance implies that a catapult mechanism must be used: the muscles contract and store energy in advance of the jump.

7.6.1 Enhancement of Positive Work and Power Production

Representative publications: Cavagna et al. 1968; Gregor et al. 1988; Roberts et al. 1997; Biewener et al. 1998; Stefanyshyn and Nigg 1998; Komi 2000; Hof et al. 2002

All the main features observed during muscle shortening after stretch, as described in section **4.2** for isolated muscles, are also manifested in reversible muscle action in humans. In particular, muscle stretch of an appropriate magnitude and speed performed immediately before the muscle shortens will enhance muscular performance (e.g., positive work and power, muscle shortening velocity, jump height; table 7.2). A simple observation supports this claim: when performing vertical jumps, people can jump higher if they perform countermovement (bend the legs)

Table 7.2   Enhancement of Concentric Performance During the Stretch–Shortening Cycle (Selected Studies)

Subjects	Movement	Eccentric action	Performance index	Enhancement	Authors
$n = 6$ (22-29 y)	Leg extension from a squat position	Countermovement	Mean power	29%	Thys et al. 1972
$n = 3$	Elbow flexion	Countermovement	Positive work per unit of EMG	23%	Cnockaert 1978
	Elbow extension	Countermovement		111%	
$n = 18$ (18-25 y, M)	Push of a pendulum	Countermovement at speed:	Pendulum speed		Bober et al. 1980
		0.91 m/s		0.14 m/s	
		1.37 m/s		0.19 m/s	
		1.82 m/s		0.21 m/s	
		2.27 m/s		0.22 m/s	
		2.72 m/s		0.24 m/s	
$n = 9$ (27 ± 4 y, 7 M, 2 F)	Knee extension	Countermovement at speed $120° \cdot s^{-1}$	Extension moment	10%	Finni et al. 2001

Note: M, males; F, females. Enhancement is computed with respect to concentric performance without preliminary muscle stretch.

immediately prior to leg extension than if they start from a static squat posture.

Because performance enhancement in SSC activities is observed over a broad range of velocities, it can be said that the SSC results in the displacement of almost the entire force–velocity curve. In most SSC activities, the maximum concentric performance occurs at an optimal intensity (magnitude, speed, load) of the preliminary muscle stretch. Performance improves as the preceding eccentric action increases, until the preceding eccentric action reaches a point beyond which the performance starts to decline (figure 7.15).

Even higher values of performance enhancement can be expected in experienced athletes in sport activities involving multijoint movements. For instance, the peak values of positive power during the stance phase of running long jumps reported in the literature can reach 3,000 W, 1,000 W, and 2,500 W for the ankle, knee, and hip joints, respectively. These values, especially the power at the ankle joint, are much higher than the maximum joint power obtained in single-joint tasks. However, the enhancement in the preceding example is caused by two mechanisms, the SSC and energy

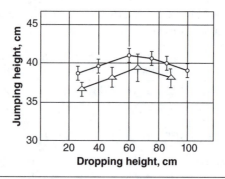

Figure 7.15 Dependence of vertical jump height (ordinate, centimeters) on the height of the preceding drop jump (abscissa). The subjects performed drop jumps from an elevation (abscissa) and then immediately jumped up. Triangles, data are from Komi and Bosco (1978; $n = 18$); circles, data are from Zatsiorsky et al. (1981; $n = 31$). Drop jumps are popular in performance training. This training method has been erroneously called *plyometrics* by some. The term is not appropriate in this case, because reversible, not eccentric, muscle action is the training objective.

Adapted from V.M. Zatsiorsky, A.S. Aruin, and S.N. Seluyanov, 1981, *Biomechanics of human musculoskeletal system* (in Russian) (Moscow: FiS Publishers), 44.

transfer from other joints (for details see chapter 6). For instance, the power production at the ankle joint in excess of 3,000 W is partly (25%) explained by the energy transfer from the proximal joints by two-joint muscles.

In both humans and isolated muscles, the enhanced performance subsides with force decay (described in section **7.3.2.3**), and thus there is a limited window of opportunity for taking advantage of the preceding muscle stretch. For instance, in countermovement jumps the jumpers must perform the leg flexion and extension in a single uninterrupted motion. If they stop after squat and then resume the movement, the advantage of the countermovement disappears (table 7.3).

Table 7.3 Effect of 1-Second Pause at the Bottom Body Position on the Jumping Performance in the Countermovement Jumps, Height of the Jump

Jumping with a pause	Jumping without a pause
49.5 ± 5.9 cm	53.2 ± 6.5 cm

Note: The difference is statistically significant ($P \leq 0.5$).

The data are from Zatsiorsky et al. 1981.

7.6.2 Mechanisms of the Performance Enhancement in the Stretch–Shortening Cycle

Several factors contribute to performance enhancement during the SSC:

• Increased time for force development. During a SSC, the force starts increasing during an eccentric phase of the cycle; therefore, there is more time for the force increase. An additional advantage is provided by the quick extension of the muscle SEC, which allows for manifesting the active state of the muscle externally (this mechanism is described in section **3.1.2**).

• Increased force after stretch due to dynamic force enhancement (described in section **4.1.1**) and low, or even zero, muscle fiber shortening speed at the instant of maximal force production. The peak force production is usually observed at the transition from lengthening to shortening. At this instant, the muscle acts isometrically. According to the force–velocity relations described in section **3.2.2** (figures 3.26 and 3.27), muscles generate larger forces in isometric conditions than in concentric conditions.

• Energy storage and release due to muscle and tendon elasticity. If a tendon or active muscle is stretched, the elastic potential energy is stored within these biological structures (this mechanism is discussed in chapter 2). The deformation energy is recoiled and used to enhance motor output in the concentric phase of the SSC. According to physical principles, the magnitude of the stored energy is proportional to the applied force and the induced deformation. Because muscle and tendon are arranged in series, they are subjected to the same force, and the distribution of the stored energy between them in this case is only a function of their deformation. The deformation, in turn, is a function of muscle or tendon stiffness (or its inverse value, compliance; figure 7.16).

At low muscle forces, the tendon may be in its toe region (explained

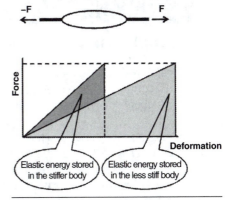

Figure 7.16 Accumulation of elastic potential energy in the muscle and tendons connected in series: a schematic. Equal forces act on the muscle and tendon. A linear relation between the force and deformation (Hooke's law) is assumed. The areas of the triangles represent the stored (or absorbed) mechanical energy. Depending on muscle and tendon stiffness, different amounts of mechanical energy are stored (absorbed).

in section **2.1.1.1.1**, see figure 2.2) and thus may be more compliant than the muscle. In this case the tendon will be deformed more than the muscle and may store a larger amount of the deformation energy. This happens in walking when a larger portion of elastic energy is stored in the Achilles tendon than in the gastrocnemius muscle fibers (see figure 7.14). When the force increases, the tendon is in the linear region of its stress–strain curve (described in section **2.1.1.1.2**). In this region, tendon stiffness is constant, whereas muscle stiffness (i.e., muscle resistance to stretching) is variable and depends on the exerted forces. The greater the muscle tension, the more the muscle resists stretch. If the tension is very large—most probably such muscle forces can be exerted only by well-trained athletes—the stiffness of the muscles may exceed the stiffness of their tendons. That is why elastic energy in elite athletes (e.g., during push-offs) may be stored primarily in tendons rather than in muscles. Animals who are fast runners (e.g., horses) have short, strong muscles and lengthy compliant tendons. Such tendons work as springs; they allow for a large amount of mechanical energy storage and release during each step. However, although the SSC is energetically efficient, the SSC and eccentric action require muscle activation and metabolic energy, as discussed previously (see figure 7.10 and section **7.5.1**).

• • • PHYSIOLOGY REFRESHER • • •

Stretch Reflex and Golgi Tendon Organ Reflex

The stretch reflex, also called *myotatic reflex*, accomplishes a length-feedback control. When a muscle is stretched, the *muscle spindles*—the sensory receptors within the muscle belly—are also stretched and their activity increases. The increased activity of the spindles provides direct excitatory feedback to the *alpha motor neurons* innervating the muscle. As a result, the muscle contracts and its length decreases. The sensitivity of the spindles is regulated by the *gamma activation* of the *intrafusal fibers* that comprise the spindles. The intrafusal fibers are arranged in parallel with the *extrafusal fibers* (standard muscle fibers) that make up the main part of the muscle and are innervated by the alpha motor neurons.

Golgi tendon organ (GTO) *reflex*, also called *inverse myotatic reflex*, accomplishes a force-feedback control. Golgi tendon organs are the sensory receptors located at the muscle–tendon junctions. They detect the muscle force. The GTO reflex is an inhibitory reflex. If the force is excessively large, the GTO reflex causes the muscle to relax (so-called *clasped-knife reaction*). One of the functions of the GTO reflex is to prevent the muscle and tendon injury.

• Neural mechanisms. Muscle forces in the SCC are influenced by the action of two reflexes: stretch reflex and inverse stretch reflex (GTO reflex). These reflexes constitute two feedback systems that assist in

  • keeping the muscle close to a preset length (stretch reflex, length feedback) and

  • preventing unusually high and potentially damaging muscle tension (GTO reflex, force feedback).

• The stretch reflex increases muscle activation. On the contrary, the GTO reflex evokes the inhibition of muscle action. The outcome depends on the interaction of these two reflexes: the positive (excitatory) effect from the stretch reflex and the negative (inhibitory) effect from the GTO reflex (figure 7.17).

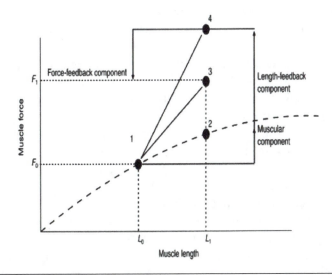

Figure 7.17 Neural mechanisms of enhanced force output in the SSC. As a result of stretch from L_0 to L_1, the muscle force increases from F_0 to F_1. Three functional components responsible for strength enhancement are shown. (2) The muscular component—force increases during lengthening due to muscle and tendon elasticity (stiffness). (4) The length-feedback component—force increases due to facilitatory spindle discharge (myotatic reflex). (3) The force-feedback component—force decreases during high tension due to feedback originating in Golgi tendon organs. The length-feedback component increases muscle stiffness (resistance to the lengthening) whereas the force-feedback component decreases it. The final outcome is the line from 1 to 3. The slope of this line defines the apparent muscle stiffness (the concept of apparent stiffness in biomechanics is discussed in section **3.1** in *Kinetics of Human Motion*).

Adapted, by permission, from P.V. Komi, 1986, "Training of muscle strength and power: Interaction of neuromotoric, hypertrophic and mechanical factors," *International Journal of Sport Medicine* 7(Suppl): 15.

As an example let us apply the theory explained in figure 7.17 to the results of the drop jump experiments presented in figure 7.15. During landing, a stretch applied to a leg extensor produces (via myotatic reflex) an additional excitation in that muscle; simultaneously, high muscle tension elicits a GTO reflex in the same muscle, thus inhibiting its activity. At low drop heights (<60 cm), the effect of the stretch reflex dominates and the jumping performance increases with the drop height. At large drop heights (>65 cm), the inhibitory action of the GTO reflex starts dominating and the height of the jump starts decreasing with the rise of drop height. At very high drop heights (not shown in the figure), the athletes are unable to perform the takeoff in one uninterrupted movement: the leg extensors are inhibited, and the athletes get "stuck" to the floor and can perform the takeoff only after a pause. As a result of specific training (e.g., in triple jumping), the GTO reflex can be inhibited (an "inhibition of inhibition" takes place) and athletes can sustain very high landing forces without a decrease in muscular force. The dropping height can then be increased.

The preceding description simplifies the physiological control mechanisms acting during the SSC. For instance, it was not mentioned that the GTO reflex inhibits muscle activity through a decrease in the excitability of spinal motor neurons.

The relative contribution to power enhancement of these mechanisms (accumulation and release of elastic potential energy, stretch reflex) is not known precisely. Most probably the contribution is different in different tasks and in different performers.

Stretch reflex–induced positive power enhancement requires additional motor unit recruitment and, hence, increased metabolic energy consumption. In contrast, performance enhancement due to recuperated elastic potential energy is not associated with increased metabolic cost. Therefore, energy consumption can be used to probe the mechanisms of performance enhancement in SSC activities.

••• PHYSIOLOGY REFRESHER •••

H-Reflex

The *H-reflex*, also called *Hoffmann reflex* (suggested by P. Hoffmann in 1910), is a monosynaptic reflex induced by an electrical stimulation of a peripheral muscle nerve containing large diameter group Ia afferents originating from the muscle spindles. The reflex elicits EMG and muscle force responses. The H-reflex is used to test the excitability of the motor neurons in the spinal cord.

■ ■ ■ *FROM THE LITERATURE* ■ ■ ■

Reduction of Short-Latency Reflex Excitability in the Drop Jump

Source: Leukel, C., W. Taube, M. Gruber, M. Hodapp, and A. Gollhofer. 2008. Influence of falling height on the excitability of the soleus H-reflex during drop-jumps. *Acta Physiol (Oxf)* 192(4):569-576

In drop-jumps, the short latency component of the stretch reflex (SLR) was shown to increase with falling height. However, in jumps from excessive heights, the SLR was diminished. The present study assessed the spinal excitability of the soleus Ia afferent pathway at SLR during jumps from low height (LH, 31 cm) and excessive height (EH, 76 cm). In 20 healthy subjects (age 25 ± 3 years), H-reflexes were timed to occur at the peak of the SLR during drop-jumps from LH and EH. H-reflexes were significantly smaller at EH than at LH ($p < .05$). Differences in the H-reflex between EH and LH indicate that spinal mechanisms are involved in the modulation of the SLR. A decreased excitability of the H-reflex pathway at EH compared with LH is argued to serve as a prevention strategy to protect the musculotendinous system from potential injuries caused by the high load.

7.6.3 Efficiency of Positive Work in Stretch–Shortening Cycle

Representative publications: Thys et al. 1972; Aruin et al. 1979; van Ingen Schenau et al. 1997; Prilutsky 1997; Woledge 1997; Umberger and Martin 2007

Because the SSC involves elastic energy storage during stretch in the muscles and tendinous structures and its subsequent release during shortening, the positive work during the SSC is done with relatively smaller metabolic energy expenditure than during movements without the SSC. The ratio between the work output and the metabolic energy expenditure for this task (input) is called *efficiency*. In humans, efficiency computations can involve a variety of parameters and depend on adopted definitions of work output and metabolic energy expenditure. Note that muscle work, that is, the work of individual muscles and their total sum, cannot be directly determined using present experimental techniques (for a discussion of the relevant mechanics theory, see chapter 6 in *Kinetics of Human Motion*). Hence, the work must be estimated, and this is done in different ways, for example, work of the forces exerted on the environment (external work), work of joint moments, work expended to move the CoM of the body (e.g., for walking upstairs this work is computed as the product of the body weight and the body lifting distance). For the same task, the values of these different types of work can be quite different. There are similarly several ways to characterize metabolic energy spent during human movements. As a result many indices of mechanical efficiency can be computed. Some commonly used indices are described in figure 7.18.

If mechanical positive work W done by a performer is somehow estimated (as during ergometer cycling against resistance, stair ascent, or steep uphill walking), the definitions of efficiency in figure 7.18 seem straightforward

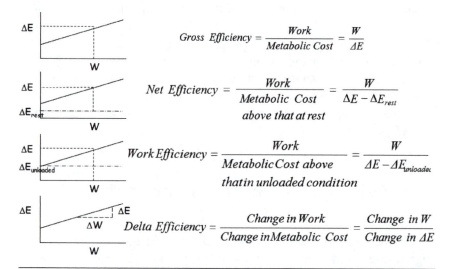

$$Gross \ Efficiency = \frac{Work}{Metabolic \ Cost} = \frac{W}{\Delta E}$$

$$Net \ Efficiency = \frac{Work}{\substack{Metabolic \ Cost \\ above \ that \ at \ rest}} = \frac{W}{\Delta E - \Delta E_{rest}}$$

$$Work \ Efficiency = \frac{Work}{\substack{Metabolic \ Cost \ above \\ that \ in \ unloaded \ condition}} = \frac{W}{\Delta E - \Delta E_{unloaded}}$$

$$Delta \ Efficiency = \frac{Change \ in \ Work}{Change \ in \ Metabolic \ Cost} = \frac{Change \ in \ W}{Change \ in \ \Delta E}$$

Figure 7.18 Different definitions of efficiency of human motion derived from the metabolic cost–external mechanical work relation (left panel). External work is the work of a force exerted on the environment, for example, the work of a force applied on the pedals during pedaling on a bicycle ergometer. When one is computing gross efficiency, no adjustments are made for metabolic energy spent at rest. Computation of the net efficiency involves subtraction of the metabolic energy spent at rest, whereas for the work efficiency the metabolic energy spent for unloaded movement (e.g., cycling without resistance) is subtracted. To determine the delta efficiency, the differences (delta) in work magnitude in two tasks are compared with the differences in the metabolic energy expenditures during these tasks.

and yield values between 20% and 25%. Measurements of positive work and heat production performed on amphibian and mammalian muscles in physiological conditions give similar results. The situation becomes more complicated when one tries to evaluate mechanical efficiency of movements with the SSC. In such movements (e.g., level walking and running with constant speeds), there are phases of eccentric and concentric action with about equal negative and positive power and work production. As a result, the total net work *Wnet* and therefore mechanical efficiency per SSC are close to zero.

Let us consider the human body as a closed thermodynamic system; that is, only energy transfer (and no mass transfer) takes place between the body and the environment. Energy is transported between the body and the environment in forms of work and heat. Work done by the body on the environment is positive and work done on the body is negative (figure 7.19). The conservation-of-energy equation for the body can be written as

$$\Delta E = Wnet + Q, \qquad [7.4a]$$

where Q is energy transported from the body due to heat transfer; $Wnet = Wpos - Wneg$, where $Wpos$ is positive work done by the body (or energy transported from the body as the result of work done on the environment) and $-Wneg$ is negative work (i.e., the work of the external forces

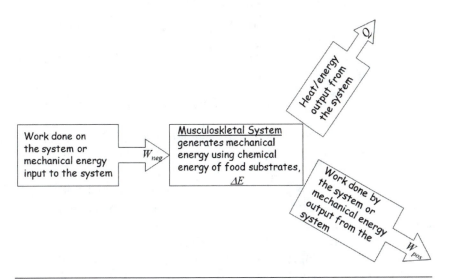

Figure 7.19 Definition of efficiency of positive work for movements with SSC. See text for explanations.

on the body or the mechanical energy transported to the body in a form of work $Wneg > 0$); and ΔE is chemical energy released (which is usually assessed by measuring the total metabolic energy spent). Equation 7.4a can be rewritten as

$$\Delta E + Wneg = Wpos + Q. \qquad [7.4b]$$

The performance of thermodynamic systems is often estimated as the ratio of the desired effect of the system to the energy input required to produce the desired effect. The desired effect of muscle contractions in the SSC is, typically, the generation of mechanical energy (production of positive work). The energy input to the system is the sum of chemical energy released and energy transported to the system as a result of work done on the system ($\Delta E + Wneg$). Efficiency of positive work can then be defined as

$$e_p = Wpos \, / \, (\Delta E + Wneg), \qquad [7.5]$$

where e_p can have different values depending on how W_{pos}, W_{neg}, and ΔE are measured, but it must satisfy the inequality $0 \le e_p < 1$. This definition of efficiency is consistent with the following notions:

- Energy cannot be created or destroyed but only transformed from one form to another, for example, from chemical energy to mechanical work and heat (the first law of thermodynamics).

- The human body performing motion always wastes some of its energy resources to heat ($Q > 0$), which cannot be reused to do mechanical work, so the body cannot produce movement indefinitely (the second law of thermodynamics).

Also, e_p does not equal zero in the SSC with equivalent values of negative and positive work.

It should be noted that ΔE is usually determined as the difference between the metabolic energy expenditure during exercise and a certain base level (at rest or in unloaded movement). Unfortunately, the base level cannot be determined in an indisputable way (e.g., should it be the metabolic rate at rest in bed or at working posture?). Also, different decisions can be made about including or not including in the computations the mechanical work spent for breathing and heart function. Hence, the absolute values of ΔE should be taken with the grain of salt. However, if the same procedures are used for computation of ΔE and e_p in different activities, comparison of the obtained e_p values may have some merit.

The efficiency of movements in which positive work is done immediately after an appropriate amount of preliminary muscle stretch—in level running, countermovement jumping, and squatting—is higher compared with the efficiency obtained during uphill walking and running or during cycling, where muscles do little or no negative work. An increase in mechanical efficiency in different movements with the SSC was reported to range between 20% and 50%.

Simultaneous in vivo measurements of forces and fiber length changes of selected distal muscles during running in birds, wallabies, quadruped animals, and humans show that substantial fraction of the positive work results from the release of strain energy from stretched tendons or aponeuroses, which explains higher mechanical efficiency of positive work in movements with SSC.

Recently developed methods for determining oxygen consumption of individual muscles in vivo will further advance our understanding of mechanical efficiency of the SSC.

■ ■ ■ *FROM THE LITERATURE* ■ ■ ■

Stretch–Shortening Cycle
Increases Efficiency of Movement

Source: Aruin, A.S., N.I. Volkov, V.M. Zatsiorsky, L.M. Raitsin, and E.A. Shirkovets. 1977. Effect of the elastic forces of muscles on the efficiency of muscular work. *Hum Physiol* 3:420-426

Five subjects performed two series of 4 min vertical jumps: with and without pause between the countermovement flexion and subsequent extension. The jump height corresponded to 50% of individual maximum. Mechanical work was determined from jump height, and metabolic energy expenditure was determined from oxygen consumption. The coefficients of efficiency (the ratio between positive mechanical work and energy expenditure) in jumps with and without pause were 9.4% and 21.7%, respectively. These results support the hypothesis that during the concentric part of the SSC, a part of the positive work is supplied from elastic recoil of stretched musculotendinous structures.

■ ■ ■ *FROM THE LITERATURE* ■ ■ ■

Mechanical Efficiency of an Individual Muscle Increases With Running Speed

Source: Rubenson, J., and R.L. Marsh. 2009. Mechanical efficiency of limb swing during walking and running in guinea fowl *(Numida meleagris). J Appl Physiol* 106(5):1618-1630

The goal of the study was to examine the mechanical determinants of the energy cost of limb swing during locomotion. By combining inverse dynamic modeling and muscle-specific energetics from blood flow measurements, the authors used the guinea fowl *(Numida meleagris)* as a model to explore whether mechanical work at the joints explains limb-swing energy use. Overall efficiency of the limb swing increased markedly from walking (3%) to fast running (17%) and was well below the usually accepted maximum efficiency of muscle, except at the fastest speeds recorded. The estimated efficiency of a single muscle used during ankle flexion (tibialis cranialis) paralleled that of the total limb-swing efficiency (3% walking, 15% fast running). These findings do not support the hypothesis that joint work is the major determinant of limb-swing energy use across the animal's speed range and warn against making simple predictions of energy use based on joint mechanical work. To understand limb-swing energy use, mechanical functions other than accelerating the limb segments need to be explored, including isometric force production and muscle work arising from active and passive antagonist muscle forces.

7.7 SUMMARY

Eccentric muscle actions are innate parts of many human movements. The intensity and amount of eccentric action in vivo can be quantified by computing negative joint power and work. Joint power is defined as the scalar product of the vectors of joint moment and relative angular velocity at a joint. In two dimensions, the joint moment M and joint angular velocity $\dot{\theta}$ are scalars and the power equation is

$$P = M \cdot \dot{\theta}. \qquad [7.1b]$$

The joint power is negative when M and are in opposite directions ($-M \cdot \dot{\theta} < 0$ or $M \cdot (-\dot{\theta}) < 0$). Physically, negative joint powers indicate that mechanical energy flows to the joint. The joint structures—the muscles and connective tissues—absorb mechanical energy and either store it as elastic deformation energy or dissipate it into heat. Negative joint work is computed as the joint power time integral for periods of negative joint power.

In many human movements, there are phases in which the total mechanical energy of the body or some of its segments decreases. The energy decrease can be caused by external forces (e.g., air or water resistance) or by forces developed by muscles and passive anatomical structures such as ligaments and cartilage. In some activities, such as swimming, rowing, ergometer cycling against high resistance, and road cycling with high speeds, the body's mechanical energy is dissipated mostly by external forces, and muscles do little negative work—eccentric muscle action is negligible. In many activities, however, muscles do substantial amount of negative work. These activities include level and downslope walking and running, stair descent, and landing after a drop jump. During level walking and running, leg joints do about the same amounts of positive and negative work.

The behavior of muscle moments and forces during in vivo stretch is similar to that of isolated muscles. In general, the moment–angular velocity relations are similar to the force–velocity relations obtained on single muscles; however, the maximum eccentric moment produced voluntarily by humans typically exceeds the isometric moment by only 10% to 30% as opposed to an almost twofold difference seen in experiments on isolated muscles. This result may be partially explained by inhibitory influences imposed by the central nervous system. When subjects' muscles are electrically stimulated, the difference between maximum eccentric and isometric moments increases and better resembles the results of in vitro experiments.

During moderate speed muscle stretch in vivo, force increase has two distinct phases, as also observed in isolated muscle experiments. In the first phase, force increases quickly as a linear function of length change (elastic response). After a break point, the second phase starts in which force continues to increase but with progressively slower rate (velocity-dependent response) until the force reaches its peak at the end of stretch. These two phases constitute dynamic force enhancement. After the stretch, the force of activated muscles decreases in a similar fashion: first quickly, with a rate similar to that of initial force increase, and then more slowly until a plateau is reached. The plateau force exceeds the isometric force at the same muscle length and activation and represents residual force enhancement.

The initial fast increase in force prior to the break point during stretch has been suggested to originate from deformation of engaged crossbridges. Force rises in this phase as a linear function of muscle elongation (the elastic response). The range of elongation during which the muscle exhibits such behavior is rather small and is determined by the maximal range of deformation of the engaged crossbridges without their forceful detachment (approximately 12-14 nm per half sarcomere). Muscle resistance to stretch during this initial phase of muscle elongation is called *short-range stiffness*. Because of short-range stiffness, the stretched muscles resist length changes before the fastest reflexes (monosynaptic stretch reflexes) start to operate in about 20 to 40 ms. Thus short-range stiffness can be viewed as the first line of defense against unexpected postural perturbations.

The difference in the magnitude of joint moments between eccentric and concentric muscle actions at similar efforts has important consequences. First, to exert the same value of moment, a lesser effort and smaller muscle activation are required during eccentric action compared with concentric. Second, to maintain the same moment with increasing absolute values of angular joint velocity, similar muscle excitation is required during eccentric action (because the eccentric moment does not change substantially with velocity after the velocity exceeds a certain relatively low value). In contrast, during concentric action, increasing muscle excitation with velocity leads to an increase in muscle activation to compensate for the declining muscle force with shortening velocity. Third, because muscles are less activated while producing the same moment in eccentric compared with concentric action, a larger force per cross-sectional area of the active muscle fibers is exerted during eccentric activity. This higher force per active muscle area has been implicated in causing damage and delayed muscle soreness.

Motor unit recruitment during both eccentric and concentric actions follows the size principle. However, in eccentric action, recruited motor units do not change their firing rate substantially with increasing force.

The electromechanical delay, the time interval between the onset of EMG and developed force or joint moment, is shorter (up to 65%) during eccentric than concentric actions, and thus conditions for a shorter reaction time and rapid force development are more advantageous during eccentric action.

The differences in muscle force–velocity relations and in joint moment–joint angular velocity relations between eccentric and concentric actions determine the differences in physiological cost of activities with a preferential use of eccentric or concentric muscle actions. For example, oxygen uptake during negative work production $\dot{V}O_{O_2}^-$ is lower than during positive work of the same magnitude $\dot{V}O_{O_2}^+$. The ratio of oxygen uptake during concentric and eccentric exercise at the same absolute values of mechanical power ($\dot{V}O_{O_2}^+ / \dot{V}O_{O_2}^-$) always exceeds 1 and depends on the exercise (e.g., walking, running, cycling), movement velocity, and $\dot{V}O_2$ measurement method (e.g., gross oxygen uptake, gross oxygen uptake minus oxygen uptake at rest). When eccentric and concentric actions are compared at the same intensity, moderate eccentric muscle action appears to cause substantially lower fatigue (smaller declines in force and power) and perceived exertion.

In many human movements, muscle shortening is preceded by stretch. An example is leg yield and push-off in stance of running. Stretch immediately followed by shortening is called the *stretch–shortening cycle* (SSC). The term *reversible muscle action* is also used. The SSC permits the concentric phase of movement to be performed with a greater power and efficiency because of dynamic and residual force enhancement. Force enhancement enables the muscle to do more mechanical work, leads to a greater stretch of the series elastic component, and does not require additional metabolic energy expenditure.

Reversible muscle action can be roughly classified as a spring-like and a catapult-like. In spring-like SSCs, the durations of the eccentric and concentric phases are approximately equal. In catapult-like SSCs, the eccentric phase is much longer than the concentric phase; elastic energy is accumulated over a long time period and then released over a short period of time.

Several factors contribute to SSC performance enhancement:

- Increased time for force development. During an SSC, the force starts increasing during the eccentric phase of the cycle; therefore, there is more time for the force increase.
- Dynamic force enhancement after stretch and low, or even zero, speed of shortening of muscle fibers at the instant of maximal force production.
- Muscle and tendon elasticity. If a tendon or active muscle is stretched, the elastic potential energy is stored within these biological structures. The deformation energy is recoiled and used to enhance motor output in the concentric phase of the SSC.
- Muscle forces in the SCC are influenced by the action of two reflexes: stretch reflex and inverse stretch reflex (force-sensitive reflex, Golgi tendon organ [GTO] reflex). The stretch reflex increases muscle activation. The GTO reflex contrastingly evokes the inhibition of muscle activation. The outcome depends on the combined effect of these two reflexes: the positive (excitatory) effect from the stretch reflex and the negative (inhibitory) effect from the Golgi tendon organ reflex.

Because the SSC involves storage of elastic energy during stretch in the muscles and tendinous structures and its subsequent release during shortening, positive work is done with relatively smaller metabolic energy expenditure than during movements without SSCs. The ratio between the work output and the metabolic energy expenditure for this task (input) is called *efficiency*. The work used in the computations of efficiency is calculated in different ways, for example, as work of joint moments or work expended to move the CoM of the body. For the same task, the values of work can be quite different depending on the way it is calculated. There are also several ways to characterize metabolic energy spent during human movements. As a result, many indices of mechanical efficiency can be computed: gross efficiency, net efficiency, work efficiency, delta efficiency (see figure 7.18). If mechanical positive work W done by a performer is somehow estimated during such activities as ergometer cycling against resistance, stair ascent, or steep uphill walking, the preceding definitions of efficiency seem straightforward and yield values of between 20% and 25%. When the total net work $Wnet$ and therefore mechanical efficiency in SSCs are close to zero (as in level walking and running with constant speeds), the efficiency of positive work during the SSC can be defined as

$$e_p = Wpos / (\Delta E + Wneg),\qquad [7.5]$$

where $Wpos$ is positive work done during the SSC, ΔE is chemical energy released by the human body (which is usually assessed by measuring the total metabolic energy), and $Wneg$ is work done on the body by the environment (or the mechanical energy transported to the body in a form of work). The denominator ($\Delta E + Wneg$) is the maximum attainable positive work during the SSC. The efficiency of movements where positive work is done immediately after an appropriate amount of preliminary muscle stretch—in level running, countermovement jumping, and squatting—is higher compared with the efficiency obtained during uphill walking and running or during cycling where muscles do little or no negative work.

7.8 QUESTIONS FOR REVIEW

1. Define the terms *concentric, isometric,* and *eccentric muscle action* in human motion.

2. What are the measures of the amount and intensity of eccentric action in human motion?

3. List human movements in which muscles mostly operate eccentrically.

4. Give examples of movements in which little or no eccentric action occur.

5. Plot a joint moment–joint velocity relation including stretching and shortening velocities. Is it similar to the force–velocity relation of an isolated muscle? What are the differences?

6. Does dynamic force enhancement occur in human movements? If yes, list several examples. If no, explain why not.

7. How does dynamic force enhancement depend on stretch speed, fatigue, temperature, and percentage of slow-twitch fibers in the muscle?

8. Define short-range stiffness. Explain its functional significance.

9. Explain mechanisms of short-range stiffness.

10. Does residual force enhancement exist in human movements?

11. Why, with similar loads, is EMG magnitude lower in eccentric than concentric action?

12. Does motor unit recruitment order differ in eccentric and concentric actions?

13. Define the stretch–shortening cycle.

14. Name possible reasons for positive work and power enhancement during the stretch–shortening cycle.

15. Define efficiency of positive work.

16. Why is efficiency of positive work higher in movements with the stretch–shortening cycle than in movements without it?

17. Why is running efficiency thought to be higher than walking efficiency?

18. Why is electromechanical delay shorter during eccentric action than during concentric action?

19. Do muscles fatigue faster during eccentric or concentric exercise?

20. Explain the oxygen consumption differences between eccentric and concentric actions.

21. Explain why the oxygen consumption differences increase with motion speed.

22. List the general symptoms associated with delayed onset muscle soreness

23. Why are muscles more susceptible to damage after eccentric than concentric exercise?

7.9 LITERATURE LIST

Abbott, B.C., B. Bigland, and J.M. Ritchie. 1952. The physiological cost of negative work. *J Physiol* 117(3):380-390.

Aleshinsky, S.Y. 1978. Mechanical and mathematical modeling of 3D human movements. In *Biodynamics of Sports Skills* (in Russian), edited by V.M. Zatsiorsky. Moscow: Central Institute of Physical Culture, pp. 54-117.

Alexander, R.M. 1995. Leg design and jumping technique for humans, other vertebrates and insects. *Philos Trans R Soc Lond B Biol Sci* 347(1321):235-248.

Armstrong, R.B. 1984. Mechanisms of exercise-induced delayed onset muscular soreness: A brief review. *Med Sci Sports Exerc* 16(6):529-538.

Aruin, A.S., B.I. Prilutskii, L.M. Raitsin, and I.A. Savel'ev. 1979. Biomechanical properties of muscles and efficiency of movement. *Hum Physiol* 5(4):426-434.

Aruin, A.S., N.I. Volkov, V.M. Zatsiorsky, L.M. Raitsin, and E.A. Shirkovets. 1977. Effect of the elastic forces of muscles on the efficiency of muscular work. *Human Physiology* 3:420-426.

Asmussen, E. 1953. Positive and negative muscular work. *Acta Physiol Scand* 28(4):364-382.

Asmussen, E. 1956. Observations on experimental muscular soreness. *Acta Rheumatol Scand* 2(2):109-116.

Biewener, A.A., D.D. Konieczynski, and R.V. Baudinette. 1998. In vivo muscle force-length behavior during steady-speed hopping in tammar wallabies. *J Exp Biol* 201(Pt 11):1681-1694.

Bigland, B., and O.C. Lippold. 1954. The relation between force, velocity and integrated electrical activity in human muscles. *J Physiol* 123(1):214-224.

Bisseling, R.W., A.L. Hof, S.W. Bredeweg, J. Zwerver, and T. Mulder. 2007. Relationship between landing strategy and patellar tendinopathy in volleyball. *Br J Sports Med* 41(7):e8.

Bober, T., E. Jaskolski, and Z. Nowacki. 1980. Study of eccentric-concentric contraction of the upper extremity muscles. *J Biomech* 13(2):135-138.

Burrows, M. 2009. Jumping performance of planthoppers (Hemiptera, Issidae). *J Exp Biol* 212(17):2844-2855.

Cavagna, G.A. 1993. Effect of temperature and velocity of stretching on stress relaxation of contracting frog muscle fibres. *J Physiol* 462:161-173.

Cavagna, G.A., B. Dusman, and R. Margaria. 1968. Positive work done by a previously stretched muscle. *J Appl Physiol* 24(1):21-32.

Cavanagh, P.R., and P.V. Komi. 1979. Electromechanical delay in human skeletal muscle under concentric and eccentric contractions. *Eur J Appl Physiol Occup Physiol* 42(3):159-163.

Chleboun, G.S., S.T. Harrigal, J.Z. Odenthal, L.A. Shula-Blanchard, and J.N. Steed. 2008. Vastus lateralis fascicle length changes during stair ascent and descent. *J Orthop Sports Phys Ther* 38(10):624-631.

Cnockaert, J.C. 1978. Comparison of the potential elastic energy stored and used by two antagonistic muscular groups. *Eur J Appl Physiol Occup Physiol* 39(3):181-189.

Cui, L., E.J. Perreault, H. Maas, and T.G. Sandercock. 2008. Modeling short-range stiffness of feline lower hindlimb muscles. *J Biomech* 41(9):1945-1952.

De Ruiter, C.J., W.J. Didden, D.A. Jones, and A.D. Haan. 2000. The force-velocity relationship of human adductor pollicis muscle during stretch and the effects of fatigue. *J Physiol* 526:671-668.

De Ruiter, C.J., and A. De Haan. 2001. Similar effects of cooling and fatigue on eccentric and concentric force-velocity relationships in human muscle. *J Appl Physiol* 90(6):2109-2116.

Devita, P., J. Helseth, and T. Hortobagyi. 2007. Muscles do more positive than negative work in human locomotion. *J Exp Biol* 210(Pt 19):3361-3373.

Devita, P., L. Janshen, P. Rider, S. Solnik, and T. Hortobagyi. 2008. Muscle work is biased toward energy generation over dissipation in non-level running. *J Biomech* 41(16):3354-3359.

Devita, P., and W.A. Skelly. 1992. Effect of landing stiffness on joint kinetics and energetics in the lower extremity. *Med Sci Sports Exerc* 24(1):108-115.

Donelan, J.M., Q. Li, V. Naing, J.A. Hoffer, D.J. Weber, and A.D. Kuo. 2008. Biomechanical energy harvesting: Generating electricity during walking with minimal user effort. *Science* 319:807-810.

Duchateau, J., and R.M. Enoka. 2008. Neural control of shortening and lengthening contractions: Influence of task constraints. *J Physiol* 586(Pt 24):5853-5864.

Duclay, J., and A. Martin. 2005. Evoked H-reflex and V-wave responses during maximal isometric, concentric, and eccentric muscle contraction. *J Neurophysiol* 94(5):3555-3562.

Elftman, H. 1939. Forces and energy changes in the leg during walking. *Am J Physiol* 125:339-356.

Eng, J.J., and D.A. Winter. 1995. Kinetic analysis of the lower limbs during walking: What information can be gained from a three-dimensional model? *J Biomech* 28(6):753-758.

Ericson, M.O., A. Bratt, R. Nisell, U.P. Arborelius, and J. Ekholm. 1986. Power output and work in different muscle groups during ergometer cycling. *Eur J Appl Physiol Occup Physiol* 55(3):229-235.

Finni, T., S. Ikegawa, and P.V. Komi. 2001. Concentric force enhancement during human movement. *Acta Physiol Scand* 173(4):369-377.

Galindo, A., J. Barthelemy, M. Ishikawa, P. Chavet, V. Martin, J. Avela, P.V. Komi, and C. Nicol. 2009. Neuromuscular control in landing from supra-maximal dropping height. *J Appl Physiol* 106(2):539-547.

Gregor, R.J., R.R. Roy, W.C. Whiting, R.G. Lovely, J.A. Hodgson, and V.R. Edgerton. 1988. Mechanical output of the cat soleus during treadmill locomotion: In vivo vs in situ characteristics. *J Biomech* 21(9):721-732.

Griffin, J.W. 1987. Differences in elbow flexion torque measured concentrically, eccentrically, and isometrically. *Phys Ther* 67(8):1205-1208.

Gulch, R.W., P. Fuchs, A. Geist, M. Eisold, and H.C. Heitkamp. 1991. Eccentric and posteccentric contractile behaviour of skeletal muscle: A comparative study in frog single fibres and in humans. *Eur J Appl Physiol Occup Physiol* 63(5):323-329.

Hahn, D., W. Seiberl, and A. Schwirtz. 2007. Force enhancement during and following muscle stretch of maximal voluntarily activated human quadriceps femoris. *Eur J Appl Physiol* 100(6):701-709.

Hanten, W.P., and C.L. Ramberg. 1988. Effect of stabilization on maximal isokinetic torque of the quadriceps femoris muscle during concentric and eccentric contractions. *Phys Ther* 68(2):219-222.

Hof, A.L., J.P. Van Zandwijk, and M.F. Bobbert. 2002. Mechanics of human triceps surae muscle in walking, running and jumping. *Acta Physiol Scand* 174(1):17-30.

Hortobagyi, T., J. Tracy, G. Hamilton, and J. Lambert. 1996. Fatigue effects on muscle excitability. *Int J Sports Med* 17(6):409-414.

Houk, J.C. 1974. Feedback control of muscle: A synthesis of the peripheral mechanisms. In *Medical Physiology*, edited by W.B. Mointcastle. St. Louis, MO: Mosby, pp. 668-677.

Huang, Q.M., and A. Thorstensson. 2000. Trunk muscle strength in eccentric and concentric lateral flexion. *Eur J Appl Physiol* 83(6):573-577.

Huyghues-Despointes, C.M., T.C. Cope, and T.R. Nichols. 2003. Intrinsic properties and reflex compensation in reinnervated triceps surae muscles of the cat: Effect of activation level. *J Neurophysiol* 90(3):1537-1546.

Ishikawa, M., and P.V. Komi. 2007. The role of the stretch reflex in the gastrocnemius muscle during human locomotion at various speeds. *J Appl Physiol* 103(3):1030-1036.

Ishikawa, M., P.V. Komi, M.J. Grey, V. Lepola, and G.P. Bruggemann. 2005. Muscle-tendon interaction and elastic energy usage in human walking. *J Appl Physiol* 99(2):603-608.

Klass, M., S. Baudry, and J. Duchateau. 2005. Aging does not affect voluntary activation of the ankle dorsiflexors during isometric, concentric, and eccentric contractions. *J Appl Physiol* 99(1):31-38.

Knuttgen, H.G. 1986. Human performance in high-intensity exercise with concentric and eccentric muscle contractions. *Int J Sports Med* 7(Suppl 1):6-9.

Komi, P.V. 1973. Measurement of the force-velocity relationship in human muscle under concentric and eccentric contractions. In *Medicine in Sport. Biomechanics III*, edited by S. Cerquiglini, A. Venerando, and J. Wartenweiler. Basel, Switzerland: Karger, pp. 224-229.

Komi, P.V. 2000. Stretch-shortening cycle: A powerful model to study normal and fatigued muscle. *J Biomech* 33(10):1197-1206.

Komi, P.V., and C. Bosco. 1978. Utilization of stored elastic energy in leg extensor muscles by men and women. *Med Sci Sports* 10(4):261-265.

Komi, P.V., and J.T. Viitasalo. 1977. Changes in motor unit activity and metabolism in human skeletal muscle during and after repeated eccentric and concentric contractions. *Acta Physiol Scand* 100(2):246-254.

Leukel, C., W. Taube, M. Gruber, M. Hodapp, and A. Gollhofer. 2008. Influence of falling height on the excitability of the soleus H-reflex during drop-jumps. *Acta Physiol (Oxf)* 192(4):569-576.

Lichtwark, G.A., and A.M. Wilson. 2006. Interactions between the human gastrocnemius muscle and the Achilles tendon during incline, level and decline locomotion. *J Exp Biol* 209(Pt 21):4379-4388.

Lieber, R.L., and J. Friden. 2002. Morphologic and mechanical basis of delayed-onset muscle soreness. *J Am Acad Orthop Surg* 10(1):67-73.

Lin, D.C., and W.Z. Rymer. 1993. Mechanical properties of cat soleus muscle elicited by sequential ramp stretches: Implications for control of muscle. *J Neurophysiol* 70(3):997-1008.

Loram, I.D., C.N. Maganaris, and M. Lakie. 2007. The passive, human calf muscles in relation to standing: The short range stiffness lies in the contractile component. *J Physiol* 584(Pt 2):677-692.

McFadyen, B.J., and D.A. Winter. 1988. An integrated biomechanical analysis of normal stair ascent and descent. *J Biomech* 21(9):733-744.

McNitt-Gray, J.L. 1993. Kinetics of the lower extremities during drop landings from three heights. *J Biomech* 26(9):1037-1046.

Michaut, A., M. Pousson, G. Millet, J. Belleville, and J. Van Hoecke. 2003. Maximal voluntary eccentric, isometric and concentric torque recovery following a concentric isokinetic exercise. *Int J Sports Med* 24(1):51-56.

Morgan, D.L. 1977. Separation of active and passive components of short-range stiffness of muscle. *Am J Physiol* 232(1):C45-49.

Mutungi, G., and K.W. Ranatunga. 2001. The effects of ramp stretches on active contractions in intact mammalian fast and slow muscle fibres. *J Muscle Res Cell Motil* 22(2):175-184.

Nardone, A., C. Romano, and M. Schieppati. 1989. Selective recruitment of high-threshold human motor units during voluntary isotonic lengthening of active muscles. *J Physiol* 409:451-471.

Nichols, T.R., and J.C. Houk. 1976. Improvement in linearity and regulation of stiffness that results from actions of stretch reflex. *J Neurophysiol* 39(1):119-142.

Norman, R.W., and P.V. Komi. 1979. Electromechanical delay in skeletal muscle under normal movement conditions. *Acta Physiol Scand* 106(3):241-248.

Onambele, G.N., S.A. Bruce, and R.C. Woledge. 2004. Effects of voluntary activation level on force exerted by human adductor pollicis muscle during rapid stretches. *Pflugers Arch* 448(4):457-461.

Oskouei, A.E., and W. Herzog. 2005. Observations on force enhancement in submaximal voluntary contractions of human adductor pollicis muscle. *J Appl Physiol* 98(6):2087-2095.

Pandolf, K.B., E. Kamon, and B.J. Noble. 1978. Perceived exertion and physiological responses during negative and positive work in climbing a laddermill. *J Sports Med Phys Fitness* 18(3):227-236.

Pasquet, B., A. Carpentier, and J. Duchateau. 2006. Specific modulation of motor unit discharge for a similar change in fascicle length during shortening and lengthening contractions in humans. *J Physiol* 577(Pt 2):753-765.

Pinniger, G.J., and A.G. Cresswell. 2007. Residual force enhancement after lengthening is present during submaximal plantar flexion and dorsiflexion actions in humans. *J Appl Physiol* 102(1):18-25.

Pinniger, G.J., J.R. Steele, A. Thorstensson, and A.G. Cresswell. 2000. Tension regulation during lengthening and shortening actions of the human soleus muscle. *Eur J Appl Physiol* 81(5):375-383.

Prilutsky, B.I. 1990. *Negative Mechanical Work During Human Locomotion* (in Russian, PhD thesis). Riga, Latvia: Latvian Research Institute of Orthopaedics and Traumatology.

Prilutsky, B.I. 1997. Work, energy expenditure and efficiency of the stretch-shortening cycle. *J Appl Biomech* 13(4):466-471.

Prilutsky, B.I. 2000. Eccentric muscle action in sport and exercise. In *Encyclopedia of Sports Medicine. Biomechanics in Sport,* edited by V.M. Zatsiorsky. Oxford, UK: Blackwell Science, pp. 56-86.

Prilutsky, B.I., and V.M. Zatsiorsky. 1992. Mechanical energy expenditure and efficiency of walking and running (in Russian). *Hum Physiol* 18:118-127.

Prilutsky, B.I., and V.M. Zatsiorsky. 1994. Tendon action of two-joint muscles—Transfer of mechanical energy between joints during jumping, landing, and running. *J Biomech* 27(1):25-34.

Proske, U., and D.L. Morgan. 2001. Muscle damage from eccentric exercise: Mechanism, mechanical signs, adaptation and clinical applications. *J Physiol* 537(Pt 2):333-345.

Rack, P.M., and D.R. Westbury. 1974. The short range stiffness of active mammalian muscle and its effect on mechanical properties. *J Physiol* 240(2):331-350.

Rack, P.M., and D.R. Westbury. 1984. Elastic properties of the cat soleus tendon and their functional importance. *J Physiol* 347:479-495.

Roberts, T.J., R.L. Marsh, P.G. Weyand, and C.R. Taylor. 1997. Muscular force in running turkeys: The economy of minimizing work. *Science* 275(5303):1113-1115.

Rubenson, J., and R.L. Marsh. 2009. Mechanical efficiency of limb swing during walking and running in guinea fowl (Numida meleagris). *J Appl Physiol* 106(5):1618-1630.

Sawicki, G.S., C.L. Lewis, and D.P. Ferris. 2009. It pays to have a spring in your step. *Exerc Sport Sci Rev* 37(3):130-138.

Smith, D.B., T.J. Housh, G.O. Johnson, T.K. Evetovich, K.T. Ebersole, and S.R. Perry. 1998. Mechanomyographic and electromyographic responses to eccentric and concentric isokinetic muscle actions of the biceps brachii. *Muscle Nerve* 21(11):1438-1444.

Spanjaard, M., N.D. Reeves, J.H. van Dieen, V. Baltzopoulos, and C.N. Maganaris. 2009. Influence of gait velocity on gastrocnemius muscle fascicle behaviour during stair negotiation. *J Electromyogr Kinesiol* 19(2):304-313.

Stefanyshyn, D.J., and B.M. Nigg. 1997. Mechanical energy contribution of the metatarsophalangeal joint to running and sprinting. *J Biomech* 30(11-12):1081-1085.

Stefanyshyn, D.J., and B.M. Nigg. 1998. Contribution of the lower extremity joints to mechanical energy in running vertical jumps and running long jumps. *J Sports Sci* 16(2):177-186.

Thys, H., T. Faraggiana, and R. Margaria. 1972. Utilization of muscle elasticity in exercise. *J Appl Physiol* 32(4):491-494.

Tredinnick, T.J., and P.W. Duncan. 1988. Reliability of measurements of concentric and eccentric isokinetic loading. *Phys Ther* 68(5):656-659.

Umberger, B.R., and P.E. Martin. 2007. Mechanical power and efficiency of level walking with different stride rates. *J Exp Biol* 210(Pt 18):3255-3265.

van Ingen Schenau, G.J.V., M.F. Bobbert, and d.A. Haan. 1997. Does elastic energy enhance work and efficiency in the stretch-shortening cycle? *J Appl Biomech* 13(4):389-415.

Westing, S.H., J.Y. Seger, and A. Thorstensson. 1990. Effects of electrical stimulation on eccentric and concentric torque-velocity relationships during knee extension in man. *Acta Physiol Scand* 140(1):17-22.

Winter, D.A. 1983a. Energy generation and absorption at the ankle and knee during fast, natural, and slow cadences. *Clin Orthop Relat Res* (175):147-154.

Winter, D.A. 1983b. Moments of force and mechanical power in jogging. *J Biomech* 16(1):91-97.

Woledge, R.C. 1997. Efficiency definitions relevant to the study of SSC. *J Appl Biomech* 13(4):476-478.

Zatsiorsky, V.M., S.Y. Aleshinsky, and N.A. Yakunin. 1982. *Biomechanical Basis of Endurance* (in Russian). Moscow: FiS. The book is also available in German: Saziorski, W.M., S. Aljeschinski, and N.A. Jakunin. 1986. *Biomechanische Grundlagen der Ausdauer.* Berlin: Sportverlag.

Zatsiorsky, V.M., A.S. Aruin, and V. Seluyanov. 1981. *Biomechanics of Human Musculo-Skeletal System* (in Russian). Moscow: Fizkultura i Sport Publishers. The book is also available in German: Saziorski, W.M., A.S. Aruin, and W.N. Selujanow. 1984. *Biomechanik des Menschlichen Bewegungsapparates.* Berlin: Sportverlag.

Zatsiorsky, V.M., and W.J. Kraemer. 2006. *Science and Practice of Strength Training.* 2nd ed. Champaign, IL: Human Kinetics.

Zatsiorsky, V.M., and B.I. Prilutsky. 1987. Soft and stiff landing. In *International Series on Biomechanics, Biomechanics X-B*, edited by B. Jonsson. Champaign, IL: Human Kinetics, pp. 739-743.

MUSCLE COORDINATION IN HUMAN MOTION

This last chapter of the book discusses one of the long-standing problems of biomechanics and motor control—the problem of motor redundancy. The problem is also known as the degree-of-freedom problem and was originally formulated by Nikolai A. Bernstein (1896-1966), the Russian physiologist, biomechanist, and psychologist. The essence of this problem is that because of the enormous number of degrees of freedom of the musculoskeletal system (see the definition of degrees of freedom in section **2.2.1** of *Kinematics of Human Motion*), there are many (often an infinite number) of ways to perform a given motor task. For example, if a person needs to reach a target, she can move her hand along different trajectories, with different joint angle patterns, different velocities, and different combinations of active muscles. The questions are (1) how muscle combinations are selected for this task, (2) whether the selected muscles (i.e., muscle coordination) are the same in different subjects performing this task, and, if yes, (3) why? In this chapter muscle coordination is defined as the distribution of muscle activation or force among individual muscles to produce a given motor task. There are other concepts of motor coordination, for example, bimanual coordination and hand and eye coordination, that are not considered in this chapter. The problem of why and how specific muscle combinations are used for given motor tasks is often called the distribution problem.

We limit the present discussion to the biomechanical aspects of coordination, that is, to discussing main biomechanical findings and their methods of analysis. The neural processes behind these experimental facts, which are still mainly unknown, are beyond the scope of this book. We refrain from contriving hypotheses about the functioning of the central neural controller, the neural structures immediately involved in control of human movements.

The chapter considers musculoskeletal redundancy and its different types (sections **8.1-8.3**). Kinematic redundancy and invariant kinematic features of human motion are considered first (**8.1.1-8.1.5**). Although kinematics of human motion are not by themselves the topic of this book, the observed kinematic features are a consequence of certain muscle actions and may serve as a prerequisite for understanding muscle coordination. Kinetic invariant characteristics of human motion are described in sections **8.2.1** through **8.2.3**. Muscle redundancy and invariant features of muscle activity are discussed in section **8.3**. Static and dynamic distribution problems are formulated in sections **8.4.1** and **8.4.2**. The methods of inverse optimization are briefly reviewed in section **8.4.3**; in some cases inverse optimization may answer the question why specific muscle coordination is used in a given task. Understanding sections **8.4.1** through **8.4.6** requires some basic knowledge of the optimization methods.

8.1 KINEMATIC REDUNDANCY AND KINEMATIC INVARIANT CHARACTERISTICS OF LIMB MOVEMENTS

Representative publications: Bernstein 1947, 1967; Gel'fand et al. 1971; Morecki et al. 1984; Flash and Hogan 1985; Turvey 1990; Neilson 1993; Soechting et al. 1995; Bullock and Grossberg 1988; Lebedev et al. 2001; Haruno and Wolpert 2005; Dounskaia 2007; Latash et al. 2007

The previous chapters of this book presented the complex muscle architecture, muscle and tendon mechanical properties, muscle actions at the joints, and muscle functional roles in different movements. Given such complexity of the muscle and skeletal design and even greater complexity of the neural control system (which is not considered in this book), it is not surprising that movement scientists have struggled to explain how the motor control system produces a vast repertoire of movements. To understand how the actions of multiple muscles are organized to produce purposeful motor behaviors, it is necessary to consider the biomechanics of the musculoskeletal system, because no one-to-one correspondence between a neural command and the resulting movement exists. Indeed, motor responses to the same neural commands are different depending on movement history (e.g., force depression or enhancement after muscle shortening or stretch, see sections **3.3.1** and **4.1**, respectively), the current state of the musculoskeletal system (e.g., current joint angles and corresponding muscle moment arms, (see **5.3**), muscle lengths and velocities (sections **3.2.1**, **3.2.2**, and **4.1.1.1**), intersegmental dynamics (see **5.4.3** in *Kinetics of Human Motion*), and the current conditions of the changing external and internal environment (e.g., external load, errors in neural commands).

Another factor that needs to be considered in an attempt to understand the production of purposeful movements is the mechanically redundant nature of the musculoskeletal system. At least two types of mechanical redundancy can be distinguished: *kinematic redundancy* and *muscle redundancy*. These two types of mechanical redundancy lead to multiple ways of performing given motor tasks.

A kinematically redundant system can be defined as system that has more kinematic degrees of freedom than strictly necessary for motion. For example, if one needs to touch a point x_d in the horizontal plane xy with his fingertip from an initial position x_0 (figure 8.1), the task of computing an initial and final arm posture, fingertip trajectory, and the corresponding joint angle changes is difficult. This is because the fingertip positions $x_0 = (y_0, x_0)$ and $x_d = (y_d, x_d)$ do not have corresponding unique sets of the arm joint angle values, $q_0 = (q_{01}, \ldots, q_{04})$ and $q_d = (q_{d1}, \ldots, q_{d4})$; rather, there are an infinite number of arm configurations consistent with finger positions x_0 and x_d. Also, there is an infinite number of fingertip trajectories in the plane xy available to accomplish this reaching task even though we consider only flexion–extension degrees of freedom at the shoulder, elbow, wrist, and metacarpophalangeal joints. Even if a finger trajectory in the plane is specified a priori (e.g., along a straight line connecting the initial and final points), there are still an infinite number of ways to move along that trajectory in terms of velocity profile of the fingertip—it can be moved faster in the beginning and slower at the end, or it can maintain a relatively constant velocity during most of the movement.

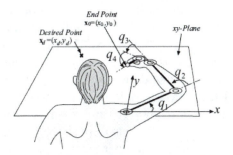

Figure 8.1   Diagram demonstrating kinematic redundancy of the human arm in the horizontal plane. The fingertip position x_0 can be maintained with an infinite number of combinations of joint angles $q_0 = (q_1, \ldots, q_4)$. The same is true for the final fingertip position x_d. There are also an infinite number of trajectories of the fingertip from point x_0 to x_d, and for each such trajectory there are an infinite number of fingertip velocity profiles and the corresponding joint angle trajectories.

From S. Arimoto, M. Sekimoto, H. Hashiguchi, and R. Ozawa, 2005, "Natural resolution of ill-posedness of inverse kinematics for redundant robots: A challenge to Bernstein's degrees-of-freedom problem," *Advanced Robotics* 19(4): 401-434. Reprinted by permission of Koninklijke Brill NV.

Furthermore, the same fingertip trajectory can be produced with an infinite number of possible joint angle trajectories that are consistent with a straight line trajectory of the fingertip. The problem of specifying joint angles for moving the endpoint of a multijoint kinematic chain along a given path is called the *inverse kinematics problem*. The inverse kinematics problem is a topic of extensive research in robotics and is described in detail in section **2.2** of *Kinematics of Human Motion*. To find a unique solution to this problem (unique combination of joint angles), additional assumptions about or constraints on the movement must be made. For instance, a specific combination of joint angle time histories for reaching movements can be selected using, for example, an assumption that movement is performed with the minimum mechanical work done.

▪ ▪ ▪ FROM THE LITERATURE ▪ ▪ ▪

Minimizing the Amount of Work Predicts the Final Arm Postures From Starting Arm Configurations During Reaching in 3-D

Source: Soechting, J.F., C.A. Buneo, U. Herrmann, and M. Flanders. 1995. Moving effortlessly in three dimensions: does Donders' law apply to arm movement? *J Neurosci* 15:6271-6280

The authors tested the hypothesis that for every location of the hand in space there is a unique posture of the arm as defined by shoulder and elbow angles (Donders' law as applied to the arm—the law is described in sections **1.3.2** and **1.3.3** in *Kinematics of Human Motion*). Human subjects made pointing movements to a number of target locations in three dimensions starting from a wide range of initial hand locations while joint angles were recorded. It was found that, in general, the posture of the arm at a given hand location depended on the starting hand location. Therefore, Donders' law was violated in the tested experimental conditions. Also, it was shown impossible to predict the final posture of the arm based only on the initial posture of the arm. However, final arm postures could be predicted assuming that the final posture minimizes the amount of work that must be done to transport the arm from the starting location.

In reality, the kinematic task of reaching by the arm to a given point in space is much more complex than depicted in figure 8.1 because the human arm has many more degrees of freedom (see for details section **2.2.2** in *Kinematics of Human Motion*). How does the motor control system specify a specific trajectory of the arm endpoint, its velocity profile, and the corresponding joint angles? This question and the quest for the answers constitute the essence of the so-called *degree-of-freedom problem* formulated by N.A. Bernstein. In the context of this book, the degree-of-freedom problem can be reformulated in terms of combinations of muscles that need to be activated in order to execute given tasks.

Are observed movement and muscle activity patterns during the performance of a given motor task in the same conditions (1) similar over time for a single individual learning to perform this multijoint task and (2) similar across different healthy people after the task has been mastered? The answer to the former question is negative—movement and muscle coordination are changing during acquisition of a novel task. If a person learns a novel and difficult motor task, she might try to limit excessive kinematic degrees of freedom by coactivating the anatomical antagonists and thus stiffening the joints. This process is called *freezing degrees of freedom* and can simplify the control of a multijoint system. As performance of the task is mastered over time, less and less coactivation of anatomical antagonists is often required and more degrees of freedom may be getting involved with the task performance (*freeing degrees of freedom*). Thus, muscle activity does change during learning of a novel motor task, and so do kinematic movement characteristics.

▪ ▪ ▪ *FROM THE LITERATURE* ▪ ▪ ▪

Hand Path Becomes Straighter and Coactivation of Antagonist Muscles Decreases After Practicing Reaching Movements in Novel Dynamic Environment

Source: Darainy, M., and D.J. Ostry. 2008. Muscle cocontraction following dynamics learning. *Exp Brain Res* 190:153-163

The authors quantified coactivation of shoulder and elbow muscles before, during, and after practicing arm reaching movements to targets

(continued)

From the Literature   *(continued)*

in the horizontal plane while a hand-velocity–dependent force perturbed the arm. It was found that during training, the hand path deviation from the straight line decreased and muscle coactivation became smaller. However, even following substantial training and reaching asymptotic levels of task performance, a relatively small coactivation of antagonist muscles remained.

The intriguing and unexplained finding is that after learning the same novel task, different people typically demonstrate on average very similar movement characteristics and muscle activity patterns, which are called *invariant characteristics* of skilled performance. So the answer to question 2 asked earlier, "Do different healthy individuals perform the same motor skills in a similar way?" appears to be yes. Let us consider invariant movement characteristics in more detail.

Four major kinematic invariant characteristics of skilled arm reaching movements are typically distinguished: (1) straight-line limb endpoint trajectory, (2) bell-shaped velocity profile, (3) power law for curved movements, and (4) Fitts' law.

8.1.1 Straight-Line Limb Endpoint Trajectory

Representative publications: Morasso 1981; Flash and Hogan 1985; Lackner and Dizio 1994; Shadmehr and Mussa-Ivaldi 1994; Thelen et al. 1996; Sainburg et al. 1999; Biess et al. 2007

When we reach to grab a sandwich or a cup of water, our hand moves approximately along a straight line if, of course, there are no objects obstructing this motion. This fact is not trivial considering the number of other available hand trajectories (see figure 8.1). The straight-line feature of hand trajectory seems to hold for planar movements in the horizontal (figure 8.2, *a* and *b*) and sagittal plane and during reaching movements to different locations of three-dimensional space. Although the hand trajectory remains nearly straight during arm point-to-point movements to different spatial locations, the corresponding joint angle time profiles have highly complex shapes and change substantially with changing movement direction.

■ ■ ■ *From the Literature* ■ ■ ■

Pointing to Different Targets Preserves Straight-Line Trajectory of the Hand but Not Joint Angles

Source: Morasso, P. 1981. Spatial control of arm movements. *Exp Brain Res* 42:223-227

The author instructed human subjects to point a hand to different visual targets that randomly appeared in a horizontal plane within arm's reach. The wrist joint was constrained so the arm could be moved by the two joints, shoulder and elbow. The hand displacement and the joint angles were recorded over the course of each pointing movement. The hand trajectory was nearly linear with a specific bell-shaped velocity profile (see section **8.1.2**); these two features of hand movement did not depend on target location. Joint angle trajectories, in contrast, had complex shapes and changed dramatically from target to target (the mechanics of this fact are explained in *Kinematics of Human Motion,* section **3.1.1.1.2**). The author proposed that the central command for these movements is formulated in terms of trajectories of the hand in space.

Straight-line hand trajectory during point-to-point arm reaching movements is acquired starting from a very young age. Infants learn to reach at about 4 months of age. At that time their hand trajectories are jerky and not straight. These movements become straighter and smoother over the first year of life.

In adulthood, this kinematic feature of reaching (trajectory straightness) is very robust. If during reaching movement the arm is unexpectedly perturbed by external forces, the limb endpoint trajectory is no longer straight (figure 8.2c). However, after performing reaching movements in this novel force environment many times with the same instructions to reach the targets fast and accurately, the hand trajectory becomes straighter again (figure 8.2d).

The straightening of hand trajectory means in particular that the subject anticipates and counteracts the externally imposed forces after learning. This type of control is called *feed-forward control*. When control is feed-forward, (1) the corresponding muscles are activated in anticipation of the external forces, and (2) sudden removal of the forces causes deviation of

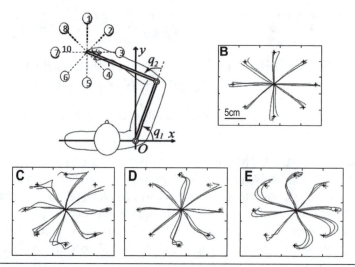

Figure 8.2 Target reaching by a two-joint arm in a novel force field. *(a)* Experimental setup of arm reaching experiments. The subject is sitting in a chair in front of horizontal screen above the arm. The arm is resting on a two-joint arm of robotic exoskeleton called Kinarm (for detail description of Kinarm, see Scott 1999). The subject is instructed to reach as fast and as accurately as possible with the index fingertip to one of eight targets appearing randomly on the horizontal screen in front of the subject. The targets are circles 1 cm in diameter and arranged radially around the start position 10 cm away from it. After reaching all eight targets in a random order (one trial), the next trial is performed. *(b)* Index fingertip trajectories in trials 18, 19, and 20 in a zero-force field (i.e., no external moments produced by the robot are applied to the arm joints). Note nearly straight fingertip trajectories. *(c)* Index fingertip trajectories in trials 1, 2, and 3 in a novel viscous (velocity-dependent) force field. Note the curved trajectories of the fingertip. The novel force field is imposed on the arm endpoint through the Kinarm external moments applied at the shoulder and elbow joints. The force imposed on the subject's arm is a function of the endpoint velocity: $\mathbf{F} = \mathbf{B}\mathbf{v}$, where $\mathbf{F} = \begin{bmatrix} F_x & F_y \end{bmatrix}^T$ is the force vector, $\mathbf{V} = \begin{bmatrix} V_x & V_y \end{bmatrix}^T$ is the hand velocity vector, and $\mathbf{B}$ is a constant matrix representing viscosity of the imposed external environment in endpoint coordinates:

$$\mathbf{B} = \begin{bmatrix} 0 & -10 \\ 10 & 0 \end{bmatrix} (\text{N} \cdot \sec / \text{m}) \qquad [8.1]$$

(d) Index fingertip trajectories in trials 18, 19, and 20 in the same viscous force field (equation 8.1). Note that the fingertip trajectories become straighter after 17 practice trials. *(e)* Index fingertip trajectories in 3 zero-force field trials randomly offered between trials 21 and 30 of force field trials (catch trials). Note the opposite deviation of the fingertip trajectories compared with the first trials performed in the viscous force field.

Reprinted, by permission, from B. Prilutsky, A.N. Klishko, B. Farrell, L. Farley, L. Phillips, and C.L. Bottasso, 2009, Movement coordination in skilled tasks: Insights from optimization. In *Advances in neuromuscular physiology of motor skills and muscle fatigue,* edited by M. Shinohara (Kerala, India: Research Signpost), 142.

hand trajectory in the opposite direction (figure 8.2*e)*. Another conclusion that could be made when observing the straightening of hand trajectory after practicing in a novel force field is that perhaps the control system plans reaching movements in terms of displacement of the limb endpoint in external space (sometimes called *extrinsic* coordinates) rather than in joint (i.e., *intrinsic* coordinates).

8.1.2 Bell-Shaped Velocity Profile

Representative publications: Flash and Hogan 1985; Novak et al. 2003; Richardson and Flash 2002; Dounskaia et al. 2005

Specifying a straight-line hand trajectory to reach a target in space does not completely eliminate kinematic redundancy. The same hand trajectory can be executed with different patterns of joint angles. The problem of specifying joint angles from known endpoint path, the inverse kinematics problem, was briefly discussed earlier (see also section **2.2** in *Kinematics of Human Motion*). Another source of multiple solutions to the problem of moving the hand along a straight line is the hand velocity. It turns out that during reaching to targets, hand velocity often has a *bell-shaped profile* (figure 8.3*a*).

Although a general shape of the velocity profile of limb endpoints is rather reproducible in reaching tasks, as well as in stepping during skilled locomotion, under certain circumstances (task novelty or additional requirements for speed and accuracy of performance) the velocity profile becomes asymmetrical and not as smooth when approaching the target. This fact can be clearly seen in figure 8.3*b* showing multipeak velocity profiles when the subject performs arm reaching movements in a novel force environment. The changes in hand velocity profiles caused by the additional task requirements or changes in the mechanical environment have been associated with the necessity to correct the distorted ongoing hand movement based on visual information and information from arm receptors signaling arm position and exerted muscle forces (called *proprioceptive information*). Practice of novel, difficult tasks improves the symmetry and smoothness of the endpoint velocity profiles, which become bell-shaped again (figure 8.3*c*). Note that subjects in these types of studies are not explicitly asked to control their hand velocity. Nevertheless, hand trajectory and velocity profile tend to change with practice toward their invariant features described here.

A bell-shaped velocity profile of the endpoint limb trajectory is an outcome of a more general kinematic invariant of endpoint limb movements that may be called *maximal movement smoothness*. This property of many

438 *Biomechanics of Skeletal Muscles*

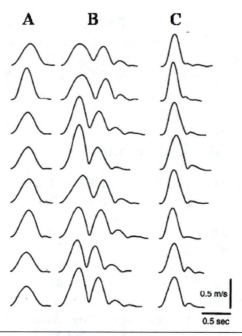

Figure 8.3 Hand velocity–time profiles before and after practicing reaching arm movements to eight targets (as shown in figure 8.2a) in a novel force field. Planar arm-reaching movements were made by the subject while grasping the handle of the robotic arm (manipulandum). The velocity-dependent force field (equation 8.1) with viscosity matrix **B** in endpoint coordinates

$$\mathbf{B} = \begin{bmatrix} -10.1 & -11.2 \\ -11.2 & 11.1 \end{bmatrix} (\text{N sec/m}) \qquad [8.2]$$

was generated. Traces of hand velocities from top to bottom are shown for targets 3, 2, 1, 8, 7, 6, 5, and 4 (see figure 8.2). *(a)* Hand velocities in a zero field before exposure to the novel force field. *(b)* Hand velocities during initial exposure to the force field with viscosity matrix (8.2). *(c)* Hand velocities after 1,000 reaching movements in the force field.

Reprinted, by permission, from R. Shadmehr and F.A. Mussa-Ivaldi, 1994, "Adaptive representation of dynamics during learning of a motor task," *Journal of Neuroscience* 14(5): 3208-3224. Permission conveyed through the Copyright Clearance Center.

skilled movements can be defined and quantified in a variety of ways. A convenient mathematical description of movement smoothness permitting its experimental evaluation is the magnitude of the third time derivative of endpoint displacement (termed *jerk*) in a laboratory-fixed coordinate system:

$$J = \frac{1}{2} \int_0^{t_f} \left(\left(\frac{d^3x}{dt^3} \right)^2 + \left(\frac{d^3y}{dt^3} \right)^2 \right) dt, \qquad [8.3]$$

where J is the time integral of the square of the magnitude of jerk, x and y are the time-varying endpoint position coordinates, and t_f is the movement time. Mathematical minimization of J for a reaching task (i.e., finding such changes in x and y coordinates with time that give the minimum value of J consistent with the initial and final positions of the arm, arm geometry, and movement dynamics) predicts a straight line trajectory of the hand and a bell-shaped velocity profile.

Jerk minimization, which is equivalent to smoothness maximization (see equation 8.3), also predicts the general features of hand path and velocity profile in curved movements of the hand that arise when obstacles need to be avoided during reaching movements or when one needs to trace a nonlinear path or perform drawing movements. Similar to the examples considered previously, practice in fast obstacle avoidance tasks by the hand or the foot leads to smoother movements of the limb endpoint in accordance with the principle of maximum movement smoothness (equation 8.3).

▪ ▪ ▪ *FROM THE LITERATURE* ▪ ▪ ▪

Practice of Fast Arm Movements Around an Obstacle Increases Movement Smoothness

Source: Schneider, K., and R.F. Zernicke. 1989. Jerk-cost modulations during the practice of rapid arm movements. *Biol Cybern* 60:221-230

The authors tested the hypothesis that practice of fast arm movements around an obstacle leads to a decrease in the value of jerk (i.e., to an increase in movement smoothness, as defined by equation 8.3). Four male subjects moved the nondominant arm between an upper target and a lower target while circumnavigating a barrier that extended outward from the plane of the targets. The motion was not restrained except that the subjects had to perform fast pointing movements between the targets, avoid the barrier, and minimize movement time. Arm movements were recorded using high-speed cine film, and the total jerk-cost was obtained for all arm segments. Each subject performed 100 practice trials. It was found that during practice, total movement time decreased, hand paths became more parabolic in shape, and significant changes occurred in hand acceleration magnitude, direction, and timing. Jerk-cost at similar speeds of arm movements was significantly less after than before practice. The decrease in jerk-cost indicated an increased smoothness of the practiced movements.

8.1.3 Power Law

Representative publications: Viviani and Terzuolo 1982; Lacquaniti et al. 1983; Schaal and Sternad 2001; Hicheur et al. 2007

Minimization of jerk (function 8.3) also predicts other experimentally observed stereotyped features of hand trajectories. One such feature is the coupling between hand velocity and trajectory curvature during drawing or curved movements: the smaller the trajectory curvature (i.e., the larger the radius), the larger the endpoint tangential velocity. To understand this coupling let us consider figure 8.4.

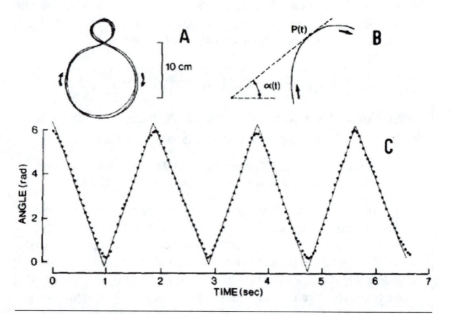

Figure 8.4 *(a)* Illustration of coupling between movement velocity and curvature while drawing a closed pattern of two loops with small and large diameter. *(b)* Definition of the angle α(t) that the tangent to the trajectory forms with an arbitrary reference (the horizontal line). *(c)* The time course of the angle α(t) recorded while drawing the pattern shown in *a* (dotted line). The increasing and decreasing portions of the plot correspond to the larger and smaller loops of pattern, respectively. A linear interpolation of the plot (continuous line) fits the experimental data with accuracy of 20%.

Reprinted from *Acta Psychologica* 54(1-3), F. Lacquaniti, C. Terzuolo, and P. Viviani, "The law relating the kinematic and figural aspects of drawing movements," 115-130, copyright 1983, with permission from Elsevier.

Figure 8.4 shows results of experiments in which the subject is drawing a closed figure consisting of two loops with small and large diameter. When angle α(t), defining the tangent to the trajectory with respect to the horizontal line, is plotted as a function of time, the portions of the angle plots corresponding to drawing a large loop (ascending plot portion) and a small loop (descending portion) have the same absolute slope. This fact means that the time of execution of the loops is rather independent of their size. In other words, the rate of change of angle α(t) (or angular velocity) tends to remain constant when the radius of movement curvature changes (the *isogony principle*—equal angular distances are covered in equal times). The principle typically holds for certain segments of curved movements characterized by a distinct value of the angular velocity.

According to numerous experimental studies, the coupling between angular velocity $\dot{\alpha}$ and curvature of the movement path segment can be expressed as

$$\dot{\alpha} = k \cdot C^{2/3}, \qquad [8.4]$$

where $\dot{\alpha}$ is angular velocity of the tangent to the trajectory, C is curvature of the trajectory, and k is a constant corresponding to a given segment of a curved movement. The equation 8.4 is known as the *two-thirds power law*. The *one-third power law* can be derived from equation 8.4 if angular velocity $\dot{\alpha}$ is substituted with the limb endpoint tangential velocity v:

$$v = k_1 \cdot r \cdot C^{1/3}, \qquad [8.5]$$

where r is the radius of curvature C and k_1 is a coefficient.

The invariant characteristics of hand trajectory discussed previously, including the power law (equations 8.4 and 8.5), have been observed not only in hand movements but also in whole-body trajectories when human subjects walk along prescribed curved lines or when they select walking paths themselves.

■ ■ ■ *FROM THE LITERATURE* ■ ■ ■

The Two-Thirds Power Law Holds for Path Trajectories of Walking Humans

Source: Vieilledent, S., Y. Kerlirzin, S. Dalbera, and A. Berthoz. 2001. Relationship between velocity and curvature of a human locomotor trajectory. *Neurosci Lett* 305:65-69

(continued)

The authors examined whether velocity and curvature during human walking along curved locomotor paths are related in accordance with the two-thirds power law. They recorded the path of subjects walking along circular or elliptical shapes drawn on the ground. The results showed that human subjects adapt their locomotor velocity to the radius of curvature of the path they are following in accordance with the prediction of the power law.

The power law and isogony principle can be derived from the following simple assumptions. The first assumption is that a complex continuous movement is performed as a sequence of linear path segments as shown in figure 8.5. Second, during planning and execution of a curved movement, the performer tries to maintain error of approximation of the curved path by a sequence of line segments within a predetermined limit ε. If one assumes that transitions from one segment to the next occur whenever the mismatch between intended (by linear approximation) and actual (curved) trajectory exceeds a given threshold ε, then the planed linear segment length must be made smaller for smaller curvature radius. In other words, the ratio between the segment length ΔS and the curvature radius r is approximately proportional to a function of the mismatch threshold:

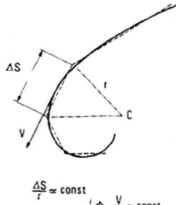

Figure 8.5 Derivation of the *isogony principle*. See text for explanations.

Reprinted from *Neuroscience* 7(2), P. Viviani and C. Terzuolo, "Trajectory determines movement dynamics," 431-437, copyright 1983, with permission from Elsevier.

$$\Delta S\,/\,r \approx f(\varepsilon) = const. \qquad [8.6]$$

If now one assumes that movements along each segment are performed in approximately the same time, then the ratio between the tangential velocity v and segment length ΔS is again proportional to a function of ε:

$$v\,/\,\Delta S \approx g\,(\varepsilon) = const. \qquad [8.7]$$

The isogony principle follows from equations 8.6 and 8.7:

$$v\,/\,r = const. \qquad [8.8]$$

Affine Transformations and Affine Geometry

In *Kinematics of Human Motion,* coordinate transformations involving rotation and translation were described (equation 1.14). The equation can be written as $y = [R]x + b$, where y, x, and b are vectors; and $[R]$ is a rotation matrix whose elements are direction cosines. Vector x designates the initial coordinates and vector y the coordinates after the transformation (i.e., after rotation and translation). If instead of direction cosines other real numbers were used in $[R]$, such that matrix $[R]$ remains nonsingular, the transformation would result not only in rotation and translation but also in scaling, that is, in the change of distances between coordinate points. Vector x will be stretched (or compressed) in the horizontal and vertical directions. The described transformation is an example of an *affine transformation.* Affine transformations (*affinities*) are linear transformations given by a matrix multiplication and a vector addition. In affine transformations the following properties hold: (a) collinearity—any three points that lie on a line will lie on the line after the transformation, and (b) constant distance ratio—for any three collinear points p_1, p_2, p_3, the ratio $| p_2 - p_1 | / | p_3 - p_2 |$ is constant.

Affine geometry is the field of mathematics that studies affine transformations.

❰❰ Affine Velocity

Representative publications: Pollick and Sapiro 1997; Flash and Handzel 2007; Polyakov et al. 2009; Pollick et al. 2009; Maoz et al. 2009

There is some experimental evidence that people perceive the surrounding world in affine geometry. For instance, images of an object on the retina are received via a parallel projection. If a planar object is rotated and translated, the image on the retina undergoes an affine transformation. Despite the image changes, the performer still perceives different object images as representing the same object. The second example is borrowed from driving. Many readers have likely

(continued)

Digression *(continued)*

driven using cruise control. A driver sets a desired speed and enjoys the ride. Then, the car enters a curve and moves along it. The rider immediately has a feeling that the speed is too high even though it is the same as previously (the cruise control is working fine). Hence, our perception of speed depends on the curvature of the trajectory.

A certain function of the movement that possesses some properties of a regular Euclidian speed and does not change its value under affine transformations is called the *equi-affine speed*. For planar movements along a trajectory $r(t) = [x(t), y(t)]$, equi-affine speed is defined as $|\dot{x}\ddot{y}-\ddot{x}\dot{y}|^{1/3}$, where $\dot{x}$, $\dot{y}$ are the first time derivatives and $\ddot{x}$, $\ddot{y}$ the second time derivatives, respectively, of a point position. It has been shown that the power law agrees well with the assumption that people prefer to maintain a constant equi-affine speed. 》

8.1.4 Fitts' Law

Representative publications: Fitts 1954; Fitts and Peterson 1964; Gan and Hoffmann 1988; Buchanan et al. 2006; Duarte and Latash 2007; Jax et al. 2007

Target pointing movements obey another kinematic invariant feature called *Fitts' law*. It describes the relation between the target size (target width along the movement path, W), the distance to the target (D), and the movement time (MT):

$$MT = a + b \cdot \log_2\left(\frac{2D}{W}\right), \qquad [8.9]$$

where a and b are empirical constants. As can be seen from this relation, Fitts' law describes a trade-off between the movement time (or speed) and the task difficulty (or accuracy requirements); that is, the more difficult the pointing task is (the smaller and further away the target), the longer it takes to reach it. This relation was proposed by Paul M. Fitts (1912-1965; an American psychologist). It was named Fitts' law because this empirical relation is thought to have a fundamental significance for our understanding of human motor control. First, it holds for a great number of experimental conditions such as pointing with different effectors: arms, legs, pointers, eyes. It works for pointing movements in different physical environments (e.g., underwater) and for both untrained and trained tasks. The second reason for the success of Fitts' law is that it has a sound theoretical basis and can be derived from simple or well established principles like the isogony principle (see figure 8.5) and information theory.

Derivation of Fitts' Law by Assuming That (1) Target Reaching Is Performed as a Sequence of Path Segments and (2) It Takes the Same Time to Execute Each Path Segment

Source: Card, S.E., T.P Moran, and A.P. Newell. 1983. *The Psychology of Human-Computer Interaction.* Hillsdale, NJ: Erlbaum. (See also http://en.wikipedia.org/wiki/Fitts's_law).

Let us assume, as in figure 8.5, that reaching to the target center at a distance D is executed by a sequence of path segments and each path segment takes constant time t to produce. As the effector (e.g., limb endpoint) passes through each path segment, it moves a constant fraction $(1 - \lambda)$ of the remaining distance to the target center (where $0 < \lambda < 1$). In other words, after movement through the first segment to the target at distance D, the remaining distance is λD. After passing through n path segments, the remaining distance is $\lambda^n D$. Thus the remaining distance to the target center is a function that decreases exponentially with time. Designating N as the number of path segments required to approach the target, we have

$$\lambda^N D = \frac{W}{2}, \qquad [8.10]$$

where the right-hand side of the equation corresponds to the location of target center along the reaching path. After solving for N, one obtains

$$N = \log_\lambda \frac{W}{2D} = \frac{1}{\log_2 \lambda} \log_2 \frac{W}{2D} = \frac{1}{\log_2 1/\lambda} \log_2 \frac{2D}{W}. \quad [8.11]$$

The time to reach the target by passing through all path segments is

$$T = Nt = \frac{t}{\log_2 1/\lambda} \log_2 \frac{2D}{W} = \frac{t}{\log_2 1/\lambda} + \frac{t}{\log_2 1/\lambda} \log_2 \frac{D}{W}. \quad [8.12]$$

By assuming constants a and b equal $\dfrac{t}{\log_2 1/\lambda}$, equation 8.12 can be rewritten in the form similar to Fitts' law:

$$MT = a + b \cdot \log_2 \left(\frac{D}{W} \right) \qquad [8.13]$$

Fitts' law has been also explained in terms of the *information theory* of physical communication systems developed by Claude Elwood Shannon (1916-2001; an American mathematician). According to Fitts, a certain amount of information can be transmitted and processed during motor task performance. Because there is a limit on the amount of information the system can transmit per unit of time, more difficult tasks require longer execution times. The average amount of information required for planning and executing of the task is called the *index of difficulty* and depends on the movement distance and target width:

$$ID = \log_2(2D/W) \qquad [8.14]$$

Thus, the longer the distance to the target and the smaller the target size, the more difficult the target pointing task (greater errors occur during pointing). It has been suggested that Fitts' law might not apply to very simple motor tasks. In more difficult tasks, the movement time increases with the ratio D/W, and the endpoint velocity profiles become asymmetric and less smooth reflecting ongoing trajectory corrections.

8.1.5 Principle of Least Action

Representative publications: Lebedev et al. 2001; Latash et al. 2002; Biess et al. 2007; De Shapiro et al. 2008; Arimoto et al. 2010

The kinematic invariants described previously are acquired during motor practice (see preceding discussion and figure 8.3). To explain why the specific invariant kinematic characteristics of skilled movements are selected during practice, a number of optimization-based models have been suggested. One of them, the minimum jerk model (minimization of jerk, see equation 8.3), predicts straight-line trajectories and bell-shaped velocity profiles. Although the power law for curved trajectories does not exactly follow from the minimum jerk model, it predicts an inverse relation between endpoint velocity and path curvature. The three kinematic invariants—straight line trajectory, bell-shaped velocity profile, and the power law—can be derived from a more general physical principle, the *principle of least action*, according to which the control system chooses the movement trajectory and endpoint velocity such that in a given time period mechanical work done on the endpoint and the arm is minimal. In three-dimensional space, such movements correspond to movement along a *geodesic*—the shortest line connecting two points on a curved surface. However, the minimum work done on the endpoint and the arm does not guarantee the minimum work at the muscle level because many muscle force combinations can produce given kinematics (a consequence on muscle

> ### ▪ ▪ ▪ *FROM THE LITERATURE* ▪ ▪ ▪
>
> ## Minimizing the Variance of Final Arm or Eye Position Predicts Straight-Line Trajectory and Bell-Shaped Velocity Profile of the Effector, the Two-Thirds Power Law, and Fitts' Law
>
> Source: Harris, C.M., and D.M. Wolpert. 1998. Signal-dependent noise determines motor planning. *Nature* 394:780-784
>
> The authors attempted to explain the major kinematic invariances by assuming that the neural control signals are corrupted by noise whose variance increases with the size of the control signal. The authors proposed that in the presence of this signal-dependent noise, the shape of a trajectory is selected to minimize the variance of the final arm or eye position. Simulations conducted using the minimum-variance cost function accurately predicted the trajectories of both arm and saccadic eye movements and the speed-accuracy trade-off described by Fitts' law. Also, the relation between the computed path curvature and hand velocity during drawing movements reproduced the two-thirds power law.

redundancy, discussed later). In addition, the principle of least action does not seem to predict or explain Fitts' law.

All of the four kinematic invariants considered here can be explained by the simple assumption that the control system selects such kinematics of the endpoint that minimize the variance of the final endpoint position. The source of this variance is noise in the control signal. The noise increases with the magnitude of the control signal as observed experimentally; for example, fluctuations of exerted muscle force increase with the magnitude of force. Counteracting noise in the control signals lead apparently to kinematics that are similar to those predicted by the principle of least action.

8.2 KINETIC INVARIANT CHARACTERISTICS OF LIMB MOVEMENTS

Representative publications: Bullock and Grossberg 1988; Kawato 1999; Feldman et al. 2007

The fact that endpoint trajectory and velocity have invariant features in point-to-point arm movements might indicate that these movement characteristics are explicitly specified by the motor control system. If this is the case, then to execute motion with the prescribed endpoint kinematics, the endpoint movement defined in external space (extrinsic) coordinates should then be transformed into joint-space (intrinsic) coordinates to obtain the respective joint angles (transformation corresponding to an inverse kinematics problem solution, see section **2.2** in *Kinematics of Human Motion*). From the computed joint angles and perceived or learned limb inertial properties and external forces applied to the limb, the control system could compute the corresponding joint moments (transformation corresponding to an *inverse dynamics problem* solution (see section **5.4** in *Kinetics of Human Motion*). Finally, given computed joint moments, the control system has to specify patterns of muscle forces or muscle activities (so-called the *distribution problem*, discussed later) in order to execute motion with the specified endpoint kinematics. Whether the motor control system uses this or other approaches to plan and execute motion is still an open question. For instance, smooth endpoint movements satisfying the discussed kinematic invariants can emerge from other motor control schemes that do not require specifications of endpoint and joint kinematics or joint moments and muscle activations (e.g., the *equilibrium point hypothesis*, see section **3.6.2** in *Kinetics of Human Motion*).

8.2.1 Elbow–Shoulder Joint Moment Covariation During Arm Reaching

Representative publications: Gottlieb et al. 1996, 1997

Regardless of how motion kinetics (joint moments or muscle forces) are planned and executed, they change during motor skill acquisition and typically also demonstrate invariant properties after motor learning. The joint moments seem to be rather stereotyped in skilled arm reaching, locomotion, postural reactions, and other tasks. Thus, the motor control system might care not only about kinematics of endpoint motion but also about joint kinetics or muscle forces and muscle activities.

One notable joint kinetic invariant characteristic of reaching movements is a covariation of elbow and shoulder dynamic joint moments (the joint moments with removed static component that depends only on gravity; figure 8.6).

Such a covariation can be explained by the observed similarity of waveforms of the dynamic elbow and shoulder moments. Although the waveforms are movement direction dependent, the waveforms are generally similar between the two joints during movements to different targets. The covariation between the elbow and shoulder dynamic moments seen

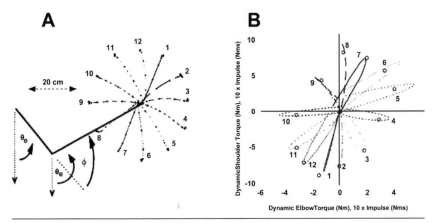

Figure 8.6 Illustration of covariation between the shoulder and elbow dynamic moments (torques) during arm target reaching. *(a)* A schematic of the two-segment arm in the initial position (heavy lines) and 12 targets located in a parasagittal plane 20 cm away from the arm endpoint. Thin lines indicate average finger paths to the targets of one subject (markers are drawn at 50-ms intervals). Angles θ_s and θ_e describe orientation of the upper arm and forearm, and angle $\phi = \theta_e - \theta_s$ is the elbow joint angle. *(b)* Average dynamic shoulder torque plotted as a function of average dynamic elbow torque for reaching movements to 12 targets. Average data of one subject are shown. The dynamic torques were calculated from the resultant joint torques by removing the static component dependent on gravity. Open symbols denote average shoulder and elbow impulse (multiplied by 10) for the 12 directions.

Reprinted, by permission, from G.L. Gottlieb, Q. Song, G.L. Almeida, D.A. Hong, and D. Corcos, 1997. "Directional control of planar human arm movement," *Journal of Neurophysiology* 78(5): 2985-2998.

in figure 8.6*b* is also observed in pointing movements with different speeds and loads. One interpretation of these observations may be that the control system simplifies the control of the multijoint arm by providing a single central command in the form of a time profile of control signals to several joints. This command then could be differentially scaled to activate muscles serving individual joints.

8.2.2 Minimum Joint Moment Change

Representative publications: Nakano et al. 1999; Wada et al. 2001

If one assumes that during reaching point-to-point movements, joint moments are such that the integral of the squared sum of moment time derivatives over the course of movement (equation 8.15) is as small as possible and computes the resulting kinematics accounting for the dynamics of the arm, then the computed kinematics of the arm endpoint have the properties of the kinematic invariants described previously. Correspondingly, the minimally changed joint moments could be considered a kinetic invariant. This kinetic property of reaching movements is called the *minimum moment change* and can be expressed as minimization of the following function:

$$min \ C_M = 1/2 \int_0^{t_f} \sum_{j=1}^{n} (dM_j/dt)^2 \, dt, \qquad [8.15]$$

where $d\tau_j / dt$ is the time derivative of moment at the jth joint, n is the total number of joints, and t_f is the movement time.

■ ■ ■ *FROM THE LITERATURE* ■ ■ ■

Minimum Joint Moment Change Predicts Kinematic Invariants of Reaching Movements

Source: Nakano, E., H. Imamizu, R. Osu, Y. Uno, H. Gomi, T. Yoshioka, and M. Kawato. 1999. Quantitative examinations of internal representations for arm trajectory planning: minimum commanded torque change model. *J Neurophysiol* 81:2140-2155

The authors examined which of the several tested optimization models better predicts experimentally observed kinematic invariant characteristics of arm reaching movements: the nearly straight-line trajectory and the bell-shaped velocity profile. Each of three subjects performed pointing movements to a wide range of targets in the horizontal and sagittal planes. The optimization criteria tested included (1) minimum hand jerk criterion formulated in a Cartesian extrinsic-kinematic space, (2) minimum angle jerk criterion in a joint intrinsic-kinematic space, and (3) minimum moment change criterion in a joint intrinsic-dynamic space. The latter criterion reproduced actual endpoint and joint angle trajectories best for curvature, position, velocity, acceleration, and joint moments. The results suggested that the brain may plan, and learn to plan, the optimal trajectory in the intrinsic coordinates considering arm and muscle dynamics and using representations for motor commands controlling muscle tensions.

8.2.3 Orientation and Shape of the Arm Apparent Stiffness Ellipses

Representative publications: Hogan 1985; Mussa-Ivaldi et al. 1985; Flash and Mussa-Ivaldi 1990; Perreault et al. 2002; Bottasso et al. 2006

In postural tasks, specific combinations of joint stiffnesses at a given arm posture determine a dynamic response of the arm to mechanical perturbations of the endpoint, which can be expressed in terms of orientation, shape, and size of the arm's apparent stiffness ellipse (for mathematical theory and details see **3.2** in *Kinetics of Human Motion*). The *orientation* and *shape of the arm stiffness ellipses* seem to be rather stereotypical during maintenance of a given arm posture at rest and thus can be considered kinetic invariants (figure 8.7). The endpoint stiffness, joint stiffness, and muscle stiffness can be defined unambiguously for a 2-degree-of-freedom (DoF) arm in the horizontal plane, perturbed by a small displacement of the hand (figure 8.7*a*). The following presents an example of such computations.

A simplified relationship between the above representations of stiffness is

$$\mathbf{R} = \mathbf{J}^T \mathbf{K} \mathbf{J} = \mathbf{G}^T \mathbf{M} \mathbf{G}, \qquad [8.16]$$

where

$$\mathbf{R} = \begin{vmatrix} R_{ss} & R_{se} \\ R_{es} & R_{ee} \end{vmatrix}, \quad \mathbf{K} = \begin{vmatrix} K_{xx} & K_{xy} \\ K_{yx} & K_{yy} \end{vmatrix}, \quad \mathbf{M} = diag \left[M_1, M_2, ..., M_6 \right].$$

Matrices **R**, **K**, and **M** represent the apparent joint, endpoint, and muscle stiffness, respectively (*diag* stands for diagonal matrix); matrix **J** is the Jacobian relating the endpoint velocity to the joint angular velocity (see equation 3.12 in **3.1.1.1.3** of *Kinematics of Human Motion* and equation 8.17 below where the joint angles are designated as q_s and q_e, respectively):

$$\mathbf{J} = \begin{vmatrix} -L_1 sin q_s - L_2 sin(q_s + q_e) & -L_2 sin(q_s + q_e) \\ L_1 cos q_s + L_2 cos(q_s + q_e) & L_2 cos(q_s + q_e) \end{vmatrix} \qquad [8.17]$$

Subscripts s and e designate the shoulder and elbow joints, and subscripts x and y designate x and y Cartesian coordinates of the arm endpoint. Matrix **G** is the matrix of moment arms of each of the six muscles of the model with respect to the shoulder (subscript s) and elbow (subscript e) joints (see also equation 5.27 in section **5.3.4** of this book):

$$\mathbf{G} = \begin{vmatrix} d_{1s} & d_{2s} & 0 & 0 & d_{5s} & d_{6s} \\ 0 & 0 & d_{3e} & d_{4e} & d_{5e} & d_{6e} \end{vmatrix}^T \qquad [8.18]$$

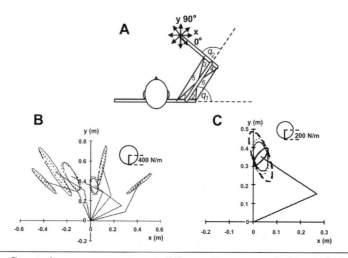

Figure 8.7 Comparison among apparent stiffness ellipses recorded in study by Flash and Mussa-Ivaldi (1990) and predicted using different cost functions. *(a)* Schematic representation of the arm. The arm is a 2-DoF system in the horizontal plane with six muscles acting around the shoulder and elbow joints. The length of the upper arm and forearm is assumed to be 0.31 m. Muscle moment arms at the shoulder and elbow were calculated from the muscle origin and insertion coordinates, the coordinates of joint centers, and the orientation of flexion and extension revolute axes from Wood et al. (1989a, 1989b), assuming that muscles produce a force along the straight line between points of their origin and insertion. The physiological cross-sectional area (PCSA) (in cm^2) of each muscle was estimated by Lemay and Crago (1996) and van der Helm (1994): $PCSA_1 = 20.3$, $PCSA_2 = 16.6$, $PCSA_3 = 10.4$, $PCSA_4 = 12.8$, $PCSA_5 = 2.5$, and $PCSA_6 = 6.3$. Muscles 1 and 2 are one-joint shoulder flexor and extensor, respectively; muscles 3 and 4 are one-joint elbow flexor and extensor, respectively; and muscles 5 and 6 are the two-joint biceps and the long head of triceps, respectively. Arrows at the endpoint correspond to 8 directions of the imposed 1 cm displacement at the endpoint. q_1 and q_2 are shoulder and elbow angles. *(b)* Measured (solid lines; from Flash and Mussa-Ivaldi 1990) and computed (minimizing cost function $\Sigma_i \, [F_i/PCSA_i]^3$; dotted lines) endpoint stiffness ellipses in selected arm postures. The ellipse center coincides with the endpoint of the two-segment arm; the shoulder coordinates are (0, 0). The arm coordinates are measured in meters (horizontal and vertical axes) and stiffness is measured in newtons per meter (scale is shown in the circle). The magnitude, shape, and orientation of the computed ellipses are not significantly ($p > .05$) different from those measured by Flash and Mussa-Ivaldi (1990). The computed ellipse shape and orientation are significantly correlated with the measured ones ($r = 0.565$ and $r = 0.926$, respectively). *(c)* Comparison between stiffness ellipses obtained using different cost functions in one arm posture (shoulder angle 30°, elbow angle 110°). The thin solid line represents the measured stiffness ellipse from Flash and Mussa-Ivaldi (1990); the thin dotted line represents the ellipse computed by minimizing the function $\Sigma_i \, (F_i/PCSA_i)^3$, the thick solid line represents the ellipse computed by minimizing the function $\Sigma_i \, (F_i)^3$, and the thick dashed line represents the ellipse computed by minimizing the function $max_i(F_i/PCSA_i)$. Note that only criterion $min \, \Sigma_i \, (F_i/PCSA_i)^3$ predicted reasonably well the experimentally determined ellipse size, shape, and orientation.

Adapted, by permission, from B.I. Prilutsky, 2000, "Coordination of two- and one-joint muscles: Functional consequences and implications for motor control," *Motor Control* 4(1): 1-44.

Superscript *T* in equations 8.16 and 8.18 denotes the transpose operation. The elements of the joint stiffness matrix **R** derived from equations 8.16 and 8.18 are

$$R_{ss} = d^2{}_{1s} \cdot dF_1/dl_1 + d^2{}_{2s} \cdot dF_2/dl_2 + d^2{}_{5s} \cdot dF_5/dl_5 + d^2{}_{6s} \cdot dF_6/dl_6,$$

$$R_{es} = R_{se} = d_{5s} \cdot d_{5e} \cdot dF_5/dl_5 + d_{6s} \cdot d_{6e} \cdot dF_6/dl_6, \qquad [8.19]$$

$$R_{ee} = d^2{}_{3e} \cdot dF_3/dl_3 + d^2{}_{4e} \cdot dF_4/dl_4 + d^2{}_{5e} \cdot dF_5/dl_5 + d^2{}_{6e} \cdot dF_6/dl_6,$$

where *dF* and *dl* are small changes in muscle force and length caused by the small displacement of the arm endpoint (muscle elongation is defined as negative), and ratios dF_i / dl_i represent diagonal elements of muscle stiffness matrix **M** (i.e., stiffness of the *i*th muscle). Equation 8.19 demonstrates that the elements of apparent joint stiffness and endpoint stiffness of the arm (see equation 8.16) depend on forces of individual muscles. Thus, a specific stereotyped combination of individual muscle forces is evidently required for producing the experimentally observed invariant endpoint stiffness ellipses.

FROM THE LITERATURE

Specific Muscle Coordination Is Required for Producing the Experimentally Observed Invariant Stiffness Ellipses

Source: Prilutsky, B.I. 2010. Coordination of two- and one-joint muscles: Functional consequences and implications for motor control. *Motor Control* 4(1):1-44.

One could try to find a combination of restoring muscle forces resulting from the perturbations of the arm endpoint by solving an optimization problem:

$$\text{minimize objective function } Z = f(\mathbf{dF}) \qquad [8.20]$$

$$\text{subject to constraints } \mathbf{dT} = \mathbf{G}^T \mathbf{dF} \qquad [8.21]$$

$$\mathbf{dF} \geq \mathbf{0}, \qquad [8.22]$$

where *Z* is an objective function whose values depend on the vector of unknown muscle force changes, $\mathbf{dF} = [dF_1, dF_2, \ldots, dF_6]^T$, caused by the imposed displacements of the arm endpoint shown in figure 8.7.

(continued)

Four different objective functions Z were examined: $\Sigma_i (F_i/PCSA_i)^3$, $\Sigma_i F_i^3$, $\Sigma_i F_i/PCSA_i$, and min max$(F_i/PCSA_i)$ (where $PCSA_i$ is the physiological cross-sectional area of the ith muscle; for $PCSA$ values see legend for figure 8.7). All these cost functions are functions of muscle stress or force and have been used to solve the muscle distribution problems (see later discussion). G^T in equation 8.21 is the matrix of known moment arms (equation 8.18); $dT = [dT_s, dT_e]^T$ is the vector of joint moments (joint torques) developed by the muscles at the shoulder and elbow joints. The resultant joint moments were assumed to be known and were obtained from the arm stiffness matrix K (see equation 8.16) as measured in the study of Flash and Mussa-Ivaldi (1990) (see also section **3.2** in *Kinetics of Human Motion*).

The optimization problem (equations 8.20-8.22) was solved for the 15 arm configurations shown in figure 8.7*b* and for 8 directions of imposed hand displacement (figure 8.7*a*). The predicted muscle forces were used to calculate first the elements of the muscle stiffness matrix M (see equation 8.16): $M_i = -dF_i/dL_i$; $i = 1, 2, \ldots, 6$; and then the joint R and endpoint K apparent stiffness matrices (equation 8.16).

The measured and computed apparent endpoint stiffnesses can be compared using graphic representation as an ellipse with three parameters: the area that is proportional to the determinant of matrix K, the shape described by the ratio of the length of the major and minor axes (or by the ratio of the maximum and minimum eigenvalues of K), and the ellipse orientation defined by direction of the major axis with respect to the horizontal axis.

The magnitude, shape, and orientation of the endpoint apparent stiffness ellipses computed using objective function $\Sigma_i (F_i/PCSA_i)^3$ were not significantly different from those obtained from the measured apparent joint stiffness ($p > .05$; figure 8.7*b*). The computed and measured stiffness ellipses resembled each other in shape and orientation. The coefficients of correlation between the computed and measured magnitude, shape, and orientation of the ellipses were 0.492 ($p > .05$), 0.565 ($p < .05$), and 0.926 ($p < .05$), respectively.

Computed ellipses were also sensitive to objective function Z used to calculate muscle forces (see figure 8.7). Ellipses calculated for a selected arm configuration using three other objective functions—$min\Sigma_i F_i^3$, min $\Sigma_i F_i/PCSA_i$, and *min max(F_i/PCSA_i)*—had different area, shape, and orientation (figure 8.7*c*).

Figure 8.7 demonstrates that in order to obtain specific kinetic invariant characteristics such as shape and orientation of the arm stiffness ellipse, a specific combination of muscle forces is required. This is not the case for other kinetic invariants (joint moment covariation and minimum torque change)—many possible force patterns can be used to satisfy the requirements for the above invariant characteristics.

8.3 MUSCLE REDUNDANCY

8.3.1 Sources of Muscle Redundancy

Representative publications: Morecki et al. 1984; Prilutsky 2000b

One important design feature of the human body is a large number of muscles that exceed the estimated total number of kinematic degrees of freedom in the human body, 244 (see section **2.2.3** in *Kinematics of Human Motion*). A conservative estimate of the number of skeletal muscles is 630. Thus, on average, 1 kinematic DoF is controlled by 2.6 muscles (630/244). This feature of the musculoskeletal system is called *muscle redundancy*. As a consequence of muscle redundancy, even in a simplest 1-DoF joint controlled by two anatomical antagonists, there are an unlimited number of muscle force combinations to produce a required moment at the joint. This can be demonstrated by the following joint moment equation

$$M = d_1 \cdot F_1 - d_2 \cdot F_2, \qquad [8.23]$$

where M is the desired (known) joint moment; d_1 and d_2 are known moment arms of two antagonistic muscles; and F_1 and F_2 are unknown muscle forces that need to be specified in order to produce moment M. Equation 8.23 is similar to equation 8.21 in that it has an infinite number of solutions because the number of unknown forces exceeds the number of equations (i.e., kinematic degrees of freedom).

In addition, as discussed in sections **1.8**, **5.3.1.3**, and **6.2.1**, muscles (or their parts, i.e., actons) typically produce moments about several degrees of freedom. In the human arm, for example, 66 actons serve 22 joints and 30 DoFs of the arm. Thus on average three actons serve one joint (66/22) and 2.2 actons serve on average 1 DoF (66/30). The number of functions of an acton equals the number of moments of force that this acton can produce with respect to anatomical joint axes. For instance, the long head of the biceps brachii has five functions. At the shoulder joint it can serve as the flexor, abductor, and arm pronator and at the elbow joint it serves as the flexor and forearm supinator. The total number of functions of actons in the arm is 264, which gives the number of acton functions per joint 12 (264/22). Similarly, there are on average 8.8 acton functions per degree of freedom in the arm (264/30). As a result, the task of activating muscles to produce desired movements seems overwhelmingly complex. Indeed, if a muscle is activated to produce a desired (primary) moment, it might also generate additional undesirable (secondary) moments, which should be neutralized by other actons or muscles, which in turn create undesirable moments about other degrees of freedom (tertiary moments), and so on.

Consider, for example, a forceful arm supination with the elbow flexed at a right angle, as in driving a screw with a screwdriver. During the supination effort, the triceps is also active, even though it is not a supinator. A simple demonstration—and one suitable for class purposes—proves this: perform an attempted forceful supination against a resistance while placing the second hand on the biceps and triceps of the working arm. Both the biceps and the triceps spring into action simultaneously. The explanation is simple: when the biceps acts as a supinator, it also produces an undesirable flexion moment (secondary moment). The flexion moment is counterbalanced by the extension moment exerted by the triceps.

Athletes tend to perform forceful movements in such a way as to minimize secondary moments. For instance, when pull-ups are performed on gymnastic rings, the performers usually supinate the arms while flexing the elbow joints. No one explicitly teaches them this. They select this movement pattern simply because it is more convenient for them.

Despite the described complexities, people activate muscles in a coordinated manner. What principles or rules (if any) does the control system use to select individual muscles for executing a particular movement and to generate required joint moments? The search for the answer to this question constitutes an area of research called the *distribution problem* (also called the *sharing problem*). The fact that many skilled tasks after sufficient practice are characterized by rather stereotypical (invariant) characteristics of muscle activity suggests that these patterns have certain advantages over other possible patterns of muscle activity. This fact encourages researchers to use optimization methods to attempt to find the physiological principles underlying muscle selections.

Before reviewing selected studies on the distribution problem, let us first discuss stereotypic features of muscle activity in skilled tasks.

8.3.2 Invariant Features of Muscle Activity Patterns

Representative publications: Buchanan et al. 1986, 1999; van Bolhuis and Gielen 1997, 1999; Wells and Evans 1987; Flanders and Soechting 1990; Kurtzer et al. 2006; Darainy et al. 2007; Franklin et al. 2007; Nozaki et al. 2005; Nozaki 2009

Several stereotyped features of muscle activity in skilled tasks were identified. They include broad (cosine) tuning of activity, which is sometimes called *synergistic activity*; *reciprocal activation*; and activation of two-joint muscles in accordance with their mechanical advantage.

The level of involvement of individual muscles in the production of endpoint force (or a combination of joint moments) depends to a large extent on the angle between the desired force (or moment) direction and the direction of the muscle moment arm vector (note the word *vector* here, also see figures 5.10 and 6.1). When the joint moment is desired in the direction of the muscle moment arm vector, the muscle commonly demonstrates a high activity, although it is not always maximal. The direction corresponding to maximal muscle activity is called the *preferred direction*. When the desired joint moment direction moves away from the preferred direction, muscle activity decreases, albeit slowly. The muscle can still demonstrate some activity even when it contributes little to the joint moment. This phenomenon is typically referred to as the *cosine tuning* of muscle activity or the *cosine rule* because such muscle behavior can be described by a cosine function fitted to experimental EMG recordings (figure 8.8*a*):

$$EMG = a \cdot F \cdot \cos(\varphi_{PD} - \varphi), \qquad [8.24]$$

where *EMG* is the mean value of recorded muscle electromyographic activity during exerting the endpoint force in direction φ (or the corresponding combination of joint moments); φ_{PD} is the preferred muscle EMG direction, that is, direction corresponding to the maximal muscle activity; F is the exerted force magnitude, and a is a scaling factor.

Figure 8.8*c* exhibits cosine tuning in the EMG recordings. In particular, the preferred direction and rather high activity at force directions far from the preferred direction. The broad tuning of muscle activity across force directions means in particular that for a given force direction several muscles produce force simultaneously. For instance, when force is exerted in the 90° direction, two elbow extensor muscles (triceps lateralis and triceps long head) with different preferred directions are highly active. This is an

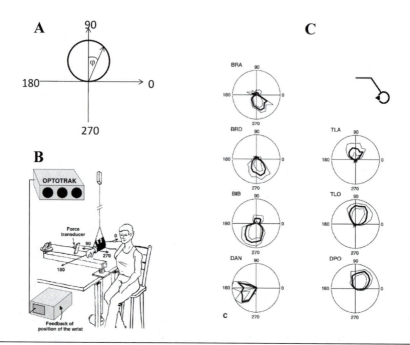

Figure 8.8 Illustration of cosine tuning of muscle activity during exerting an endpoint force in different directions. *(a)* A polar plot of a cosine function illustrating potential changes in muscle activity as a function of direction of the exerted force. The preferred direction in this case is 90°, which corresponds to the maximum amplitude of activity, that is, the distance from the origin to a point on the circle. The direction of the force exerted by a hand (as in the experiment depicted in *b*) is indicated by the line connecting the origin and a point on the circle (or EMG data point) and quantified by angle ϕ. The distance from the origin to the data point corresponds to the EMG magnitude recorded at this particular force direction. EMG activity as a function of force direction can be described by a cosine function (8.24). Parameters of this function are ϕ_{PD}, the preferred muscle EMG direction, that is, direction corresponding to the maximum muscle activity; F, a force magnitude, and a, a scaling factor. *(b)* A typical experimental setup for recording muscle activity for different endpoint force directions. In this particular experiment, the subject's arm was hanging in a long sling attached to the ceiling so that the forearm and upper arm were in a horizontal plane at a shoulder height. The subject exerted an isometric force of 20 N at the wrist in 16 equidistant directions in the horizontal plane. Muscle activity was recorded from several muscles (see panel *c*) and mean amplitude of the rectified EMG activity for each trial was calculated. *(c)* Polar plots of mean EMG activity (mean ± standard deviation) recorded at the elbow angle of 120° from 7 muscles: brachialis (BRA) and brachioradialis (BRD), one-joint elbow flexors; triceps lateralis (TLA), one-joint elbow extensor; biceps brachii (BIB), two-joint elbow flexor and shoulder flexor; triceps long head (TLO), two-joint elbow extensor and shoulder extensor; anterior deltoid (DAN), shoulder flexor; and posterior deltoid (DPO), shoulder extensor. A stick figure in the upper right corner indicates the right arm posture and the head (circle) aligned with the direction of force developed at the wrist.

(b-j) With kind permission from Springer Science+Business Media: *Biological Cybernetics,* "A comparison of models explaining muscle activation patterns for isometric contractions," 81(3), 1999, 249-261, B.M. van Bolhuis and C.C.A.M. Gielen, fig. 3 and fig. 8c.

example of *synergistic activity* when muscles with similar functions at a joint participate in a task.

It can be noticed that biceps brachii, an anatomical antagonist to elbow extensors, is also active at the 90° force direction and thus acts as a joint antagonist counteracting the resultant joint moment. However, antagonistic activity of biceps brachii and other muscles in the considered example is small compared with the activity in the preferred direction. This feature of muscle activity in many skilled tasks in predictable environments is called *reciprocal activation*.

Although on the whole the cosine rule seems valid, reality is usually more complex. Among many reasons for this is the necessity to stabilize the joint: because multiple muscles are usually active and pull in different directions, to prevent dislocation the joint must be stabilized. In experimental research the ideal cosine tuning with the preferred direction coinciding with the muscle moment arm vector direction (its maximal mechanical advantage) is usually not confirmed.

■ ■ ■ *FROM THE LITERATURE* ■ ■ ■

Maximizing a Muscle's Mechanical Advantage Is Not the Only Factor in Determining Muscle Activation

Source: Vasavada, A.N., B.W. Peterson, and S.L. Delp. 2002. Three-dimensional spatial tuning of neck muscle activation in humans. *Exp Brain Res* 147(4):437-448

The authors analyzed the spatial tuning of neck muscle EMG activity while subjects generated moments in three dimensions. EMG tuning curves were characterized by their orientation (mean direction) and focus (spread of activity). For the four muscles that were studied (sternocleidomastoid, splenius capitis, semispinalis capitis, and trapezius), EMG tuning curves exhibited directional preference, with consistent orientation and focus among 12 subjects. However, the directional preference (orientation) of three of the four neck muscles did not correspond to the muscle's moment arm, indicating that maximizing a muscle's mechanical advantage is not the only factor in determining muscle activation. The results indicate that the central nervous system has consistent strategies for selecting neck muscle activations to generate moments in specific directions; however, these strategies depend on three-dimensional mechanics in a complex manner.

Another invariant feature of muscle activation in skilled tasks in predictable environments is specific activation of two-joint muscles. EMG activity of a two-joint muscle or a muscle that controls 2 DoFs is greater if the moments it produces at both joints have the same directions as the resultant joint moments (see, e.g., figure 6.4). If the moments of the two-joint muscle at both joints are opposite to the resultant joint moments, its EMG activity is low. This behavior is consistent with the cosine tuning of muscle activity in accordance with muscle preferred direction. We consider the activity of two-joint muscles as a separate invariant feature because of its more pronounced manifestation, possibly due to a greater mechanical advantage to certain joint moment vector directions (see figure 6.3).

8.4 THE DISTRIBUTION PROBLEM

Representative publications: Nubar and Contini 1961; Tsirakos et al. 1997; Herzog 2000; Prilutsky 2000a, 2000b; Prilutsky and Zatsiorsky 2002; Zatsiorsky et al. 2002; Erdemir et al. 2007; Staudenmann et al. 2010

As mentioned previously, the distribution problem addresses the questions of how and why the central controller (i.e., the part of the brain that is immediately involved in movement control) distributes muscle activity in a specific way despite the enormous musculoskeletal redundancy. As we have seen from the previous sections of this chapter, the solutions to this problem (i.e., the muscle coordination) result in a number of highly robust kinematic, kinetic, and muscle activity features that are called invariant characteristics. These specific invariant movement characteristics suggest that the control system during acquisition of novel tasks selects specific muscle coordination (distributes activity among the muscles) to optimize certain movement performance criteria or physiological functions. This is why many researchers have tried to find solutions to the distribution problem using optimization methods.

There are two basic optimization approaches: static and dynamic.

8.4.1 Static Optimization

8.4.1.1 Problem Formulation

Representative publications: Nubar and Contini 1961; Seireg and Arvikar 1973, 1975; Penrod et al. 1974; Yeo 1976; Pedotti et al. 1978

The use of static optimization is based on several assumptions that are not necessarily mutually exclusive.

  1. The control of joint moments and thus limb dynamics and the control of muscle activity are decoupled. In other words, the central con-

troller determines somehow the desired joint moments and then (or in addition to that) distributes activity among the muscles such that necessary moments are generated. This assumption allows use of experimentally measured joint moments (e.g., by inverse dynamics) in the formulation of optimization problems.

2. Muscle activity is not history dependent; that is, it does not depend on time. This assumption permits formulation of optimization problems using static equations for joint moments independently at each time instance.

3. In tasks with generally stereotypic muscle activity patterns, the control system distributes muscle activity/forces in an optimal way.

A typical static optimization problem is formulated for each instant in time as follows (see also equations 8.20-8.22): minimize an objective function (also called a cost function)

$$Z = f(F_1, \dots, F_m) \qquad [8.25]$$

subject to the equality constraints

$$M_j - \sum_{i=1}^{m} d_{ij} F_i = 0, j = 1, \dots, k; i = 1, \dots, m \qquad [8.26]$$

and the inequality constraints

$$F_i \geq 0, i = 1, \dots, m, \qquad [8.27]$$

where M_j is the known joint moment at joint j, F_i is the unknown force of muscle i, d_{ij} is the moment arm of muscle i with respect to joint j, k is the number of joints, and m is the number of muscles in the system. Equation 8.26 requires that the sum of the moments of force exerted by individual muscles must be equal to the resultant joint moment, and equation 8.27 requires that all muscle forces are positive (a muscle can only pull, but not push, on the bone). In some formulations, an additional equality constraint equation is used, that is, a static equilibrium equation for joint forces.

Many researchers think that cost function Z should be physiologically justified a priori. Thus, static (and dynamic) optimization permits testing hypotheses about functional significance of the observed muscle activation patterns.

8.4.1.2 Cost Functions

Representative publications: Crowninshield and Brand 1981; An et al. 1984; Dul et al. 1984a, 1984b; Herzog and Leonard 1991; Prilutsky and Gregor 1997; Prilutsky et al. 1997; Raikova and Prilutsky 2001; Rasmussen et al. 2001; Schappacher-Tilp et al. 2009

Many cost functions have been tested on their ability to predict muscle coordination in different motor tasks. According to these studies, cost functions such as

$$\sum_{i=1}^{m} F_i^n \rightarrow min, \qquad [8.28]$$

$$\sum_{i=1}^{m} (F_i / PCSA_i)^n \rightarrow min, \qquad [8.29]$$

$$\sum_{i=1}^{m} (F_i / F_{imax})^n \rightarrow min, \qquad [8.30]$$

(where *PCSA* is physiological cross-sectional area and F_{imax} is the maximal isometric muscle force) do not predict synergistic muscle behavior at $n = 1$ but predict quite well an appropriate activation of two-joint muscles (see figures 6.3 and 6.4). Better results are obtained when similar but nonlinear cost functions are used for which $n = 2$ or $n = 3$. They predict synergistic muscle behavior, which is consistent with the cosine tuning of activity. There have been several explanations for why, in some cases, these types of optimization criteria qualitatively predict muscle activity/force distribution. According to one, muscle activation in accordance with quadratic cost functions ($n = 2$) minimizes positional errors in tasks that require accurate control of limb endpoint position or force.

▪ ▪ ▪ *From the Literature* ▪ ▪ ▪

Minimizing Neural Command Noise Leads to Minimum of Exerted Force Error and Predicts Cosine Tuning

Source: Todorov, E. 2002. Cosine tuning minimizes motor errors. *Neural Comput* 14:1233-1260

The author attempted to explain the functional significance of cosine tuning of muscle activity in redundant musculoskeletal systems. He demonstrated using a mathematical model that cosine tuning minimizes the neuromotor noise (e.g., noise of motor commands) whose standard deviation was modeled as a linear function of the mean. Minimization of the neuromotor noise leads in turn to a reduction of endpoint force errors when multiple muscles with similar functions produce force simultaneously. The author concluded that that model predictions are supported by experimental data.

Minimizing cost function (8.29) with $n = 3$ was proposed to predict force distributions that maximize *endurance time* (time during which a person can sustain a given muscle stress magnitude) or minimize fatigue. The muscle stress–endurance time relation, the inverse of function (8.29), has been obtained experimentally, and power $n = -3$ in this relation corresponds to the average value reported in the literature (figure 8.9).

This minimum fatigue criterion has been reported to predict patterns of muscle activation and forces in walking, running, and load lifting (see figure 6.4); cycling; exertion of endpoint forces by the leg; and arm stiffness ellipses (see figure 8.7). However, this fatigue cost function does not seem to work in situations where muscles have extremely different composition of slow and fast muscle fibers, as in the cat triceps surae group.

Mathematically, criterion (8.29) is equivalent to minimizing a norm

$$N = \left[\sum \left(F_i / PCSA_i \right)^n \right]^{1/n} \qquad [8.31]$$

of a vector $F_i/PCSA_i$, where subscript i denotes an individual muscle. Note that in contrast to the introduced previously Euclidian norm (see section 2.3.3 in *Kinetics of Human Motion*, p. 161) exponent n in equation 8.31 is not obliged to be equal to 2; it can be any real number. When the value of

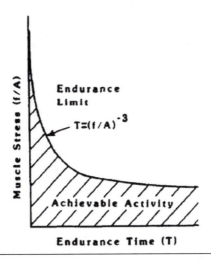

Figure 8.9 The general form of the muscle stress–endurance time relation. According to this relation, high muscle stresses can be sustained for only short durations whereas lower muscle stresses can be sustained longer. The inverse of endurance time T gives the fatigue function in which A corresponds to PCSA.

Reprinted from *Journal of Biomechanics* 14(11), R.D. Crowninshield and R.A. Brand, "A physiologically based criterion of muscle force prediction in locomotion," 793-801, copyright 1981, with permission from Elsevier.

n approaches infinity, the above norm becomes the maximum norm, and its optimal solution approaches the equal distribution of stresses among the involved muscles in a 1-DoF case. This property of criterion 8.29 led to the development of the *min/max* optimization criterion

$$max_i(F_i/PCSA_i) \times min, \qquad [8.32]$$

which predicts the maximum synergist muscle activation possible.

8.4.1.3 Accuracy of the Static Optimization Methods: How Well Do the Methods Work?

Representative publications: Herzog and Leonard 1991; Prilutsky et al. 1997

In many reports, nonlinear optimization criteria (equations 8.29 and 8.30) and other similar criteria (e.g., minimum metabolic cost) with $n > 1$ seem to correctly predict qualitative aspects of muscle activity. In particular, it has been reported that the optimization methods reasonably predict muscle activity or inactivity. However, the quantitative agreement is usually not good: the level of muscle activity is not determined accurately. The inaccuracy is partly caused by the fact that actual muscle forces are not tightly related to EMG measurements. EMG signals are sampled from local muscle areas and may be contaminated by neighboring muscle activity (*cross talk*) and can also be nonlinearly related to muscle forces. In addition, the obtained optimal solutions are very sensitive to model parameter errors (e.g., the muscle moment arms). Even small errors can cause large changes in the predicted activations and muscle force magnitudes.

■ ■ ■ *FROM THE LITERATURE* ■ ■ ■

Errors in Parameters of Optimization Models Substantially Change the Distribution Problem Solution

Source: Raikova, R.T., and B.I. Prilutsky. 2001. Sensitivity of predicted muscle forces to parameters of the optimization-based human leg model revealed by analytical and numerical analyses. *J Biomech* 34:1243-1255

The authors investigated the sensitivity of the optimal solution obtained by minimizing cost function $\sum_{i=1}^{m}(F_i / PCSA_i)^n$ ($n = 2$ and $n = 3$) for a planar 3 DOF model of the leg with nine muscles. The analysis was conducted analytically for $n = 2$ and numerically for $n = 3$. Analytical results revealed that, generally, the nonzero optimal force of each muscle depends in a complex nonlinear way on moments at all three joints as well as moment arms and PCSAs of all muscles. Deviations of the model parameters (moment arms and PCSAs) from their nominal values within a physiologically feasible range affected not only the magnitude of the forces predicted by both criteria but also the number of nonzero forces in the optimal solution and the combination of muscles with nonzero predicted forces. Muscle force magnitudes calculated by both criteria were similar. It was concluded that different opinions in the literature about the behavior of optimization-based models can be potentially explained by differences in model parameters.

In some cases, optimal solutions may be validated against measured muscle forces. However, even if the agreement between predicted and measured forces is reasonable, it is not always easy to understand the physiological meaning of the obtained solution. This point is illustrated in figure 8.10, which shows forces of cat ankle extensor soleus (SO) and ankle extensors–knee flexors gastrocnemius (GA) and plantaris (PL) recorded in 3 cats during locomotion on a treadmill with speeds from 0.4 to 1.8 m/s. The recorded forces shown in the figure were compared with forces of the same muscles computed by solving the static optimization problem (equations 8.25-8.27) for each instant of the normalized cycle time. The equality constraints (see equation 8.26) were

$$M_j - \sum_i d_{ij} F_i = 0, \ j = 1, 2; \ i = 1, 2, 3, \quad\quad [8.33]$$

and the inequality constraints for the three muscles were as equation 8.27. The objective function was

$$Z = \sum_i F_{imax} / F_i \rightarrow min, \quad\quad [8.34]$$

which allocates more force to muscles with relatively small force potential, like soleus, compared with bigger muscles, like gastrocnemius and plantaris. (It is known that during walking SO exerts greater percentage of maximal force than GA or PL, possibly because SO consists almost entirely of slow-twitch fibers, which makes it highly fatigue resistant.) In equations 8.33 and 8.34, F_1, F_2, and F_3 are the unknown forces of SO, GA, and PL, respectively; M_1 is the total moment produced by SO, GA, and PL at the ankle joint; M_2 is the total moment produced by GA and PL at the

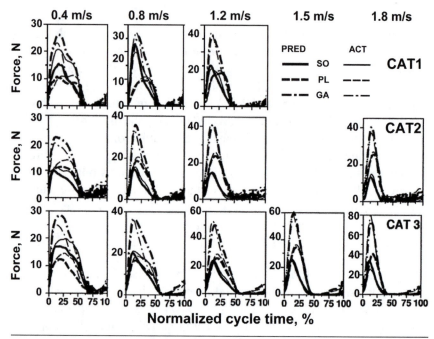

Figure 8.10 Predicted (thick lines) and measured forces (thin lines) of cat soleus (SO, solid lines), plantaris (PL, dashed lines), and gastrocnemius (GA, dashed lines with dots) during locomotion as a function of the normalized cycle time (B.I. Prilutsky, W. Herzog, T. Leonard, and T. Allinger, unpublished data). The measured forces (taken from Herzog et al. 1993) are the averaged forces of 5 to 24 cycles. Data for three cats and different locomotion speeds are presented. Paw contact occurs at time 0. PRED stands for predicted or computed forces; ACT stands for actual or recorded forces. The data were obtained in the laboratory of Dr. W. Herzog at the University of Calgary, Canada. Details of the musculoskeletal model used for calculations can be found in Prilutsky et al. (1996, 1997).

knee joint; and d_{ij} is the moment arm of the ith muscle relative to the jth joint. Moments M_j and moment arms d_{ij} were assumed to be known and the former were determined as the sum of the products of the measured muscle forces during cat locomotion and the corresponding muscle moment arms d_{ij}. The moment arms were calculated from the recorded joint angles and a geometrical model of the cat hindlimb (see, e.g., figure 5.21). F_{maxi} is the maximum force of the ith muscle estimated based on the measured wet mass of SO, GA, and PL; known muscle density; and mean fiber length reported for each muscle.

As seen in figure 8.10, minimization of function 8.34 predicts forces of SO, GA, and PL that match well the respective forces measured in all 3 cats at most of the tested locomotion speeds (at the slowest speed of 0.4 m/s the predictions are not as good, but still reasonable). These results demonstrate that despite muscle redundancy, stereotyped muscle force patterns are used in cat locomotion, and these patterns can be described by a relatively simple cost function. Unfortunately, the physiological meaning of this cost function is unclear. Also, as can be noticed from the cost function (equation 8.34), its minimization would predict unreasonable coactivation of antagonistic muscles (in the example above antagonists were not present).

••• MATHEMATICS REFRESHER •••

Functional

A scalar-valued function of a function: an argument of a functional is a function and its output is a number.

8.4.2 Dynamic Optimization

Representative publications: Hatze 1981; Davy and Audu 1987; Pandy and Zajac 1991; Hatze 2000; Anderson and Pandy 2001; Van Soest et al. 1994; Neptune et al. 2009; Ackermann and van den Bogert 2010

8.4.2.1 Basic Concepts

The optimization models reviewed above seem better suited for estimating activation and forces of individual muscles in static tasks such as depicted in figure 8.8 (e.g., exerting external forces in different directions) rather than in movements. The major limitations of static optimization models applied to dynamic tasks include the following:

1. In many movements, the muscle forces are history dependent. The formulation of the static optimization problem (equations 8.20-8.22) does not account for this fact.

2. Static optimization models cannot predict kinematics and dynamics of movement when only initial or final states of the system are known.

Dynamic optimization-based models do not have these limitations. Dynamic optimization problems are solved only once for the entire movement by minimizing a functional, the value of which depends on the time-dependent system state variables (e.g., joint angles and their time derivatives), the control variables (e.g., muscles activations), and the values of state variables and their time derivatives at the initial or final movement instants. The solution of the dynamic optimization problem, that is, optimal muscle activation or force patterns, must satisfy the equations of musculoskeletal dynamics.

Dynamic optimization models still have limitations. They require knowledge of a large number of model parameters, that is, those required to specify all aspects of a complex musculoskeletal model (e.g., muscle moment arms, slack tendon and muscle fascicle lengths, PCSA, muscle activation and deactivation constants, and the muscle force–length–velocity properties). These parameters can either be measured in a specific subject or estimated from the literature (measurement methods and selected information about these characteristics can be found in the previous chapters of this book).

Dynamic optimization is based on the forward dynamics problem.

❰ 8.4.2.2 Forward Dynamics Problem

Representative publications: Hatze 1976; Zajac 1989; He et al. 1991

A forward dynamics problem formulated for the musculoskeletal system deals with computations of system's motion (i.e., changes in independent degrees of freedom) given known (or assumed) muscle activations and the system's initial conditions. The equations of body limb dynamics can be derived from, for example, Lagrange equations (explained in *Kinetics of Human Motion*, section **5.3.3**):

$$\frac{d}{dt}\left(\frac{\partial K}{\partial \dot{q}_j}\right) - \frac{\partial K}{\partial q_j} = Q_j, \ j = 1,2,...,k, \qquad [8.35]$$

where K is the kinetic energy of the system; q_j and $\dot{q}_j$ are the jth generalized coordinate and velocity, respectively; and Q_i is the jth generalized force. The generalized coordinates are typically either the joint or the segment angles. The dynamics equations can be represented in vector form:

$$\ddot{\mathbf{q}} = \mathbf{I}\,(\mathbf{C}\dot{\mathbf{q}}^2 + \mathbf{R} + \mathbf{S} + \mathbf{M}), \qquad [8.36]$$

where $\ddot{\mathbf{q}}$ and $\dot{\mathbf{q}}$ are vectors of the generalized accelerations and velocities; $\mathbf{I}$ is the system inertia matrix; $\mathbf{C}$ is the Coriolis and centripetal generalized force matrix (Coriolis and centripetal accelerations and forces are explained in sections **3.2.1.2** in *Kinematics of Human Motion* and **5.2.2** in *Kinetics of Human Motion*); $\mathbf{R}$ is the vector of external forces acting on the system (e.g., ground reactions); $\mathbf{S}$ is the vector of generalized viscous and elastic forces restricting joint motion to anatomically feasible ranges; and $\mathbf{M}$ is the vector of generalized muscle forces (cf. with the state-space equation 5.11a in *Kinetics of Human Motion*, p. 383).

The location of each muscle–tendon unit (MTU) with respect to the joints is specified by MTU sites of origin and insertion and by the muscle path (see section **1.6**). The contraction dynamics of the MTU is often based on a Hill-type model (see section **4.4.1**) taking into account muscle mass, angle of muscle fiber pennation, the force–length–velocity relations of the contractile element, and the force–length relation of the tendon and the elastic parallel elastic element (figure 8.11).

Muscle contractile dynamics can be described by the following differential equation:

$$\dot{V}_T = \left[F_T - F_M \cos\alpha\right]/m, \tag{8.37}$$

where

$$F_T = F_T^{Max}[F_T(L_T) + B_T V_T] \text{ and}$$
$$F_M = F_M^{Max}[F_{CE}(L_M)F_{CE}(V_M)k_u^{max}A + F_{PE}(L_M) + B_M V_M]. \tag{8.38}$$

F_T, L_T, and V_T are the tendon force, length and velocity, respectively; F_M, L_M, and V_M are the muscle force, length, and velocity, respectively; α is the pennation angle; B_T and B_M are coefficients of damping (viscosity) for the tendon and muscle; m is muscle mass; $F_T(L_T)$, $F_{PE}(L_M)$, and $F_{CE}(L_M)$ are the normalized force–length relations for the tendon, the muscle elastic parallel element, and the muscle contractile element; $F_{CE}(V_M)$ is the normalized force–velocity relation for the contractile element; F_T^{Max} and F_M^{Max} are the maximal tendon and muscle force, where $F_T^{Max} = F_M^{Max} \cos(\alpha_0)$ (α_0 is pennation angle at the optimal muscle length); k_u^{max} is a parameter of muscle

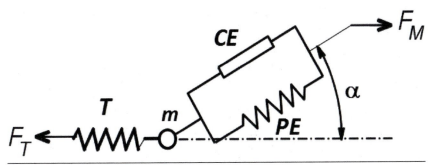

Figure 8.11 A Hill-type muscle model typically used in forward dynamics simulations. *CE*, muscle contractile element; *PE*, parallel elastic element; *T*, tendon; *m*, muscle mass; α, angle of pennation; F_T and F_{CE}, forces developed by tendon and muscle contractile element.

maximal activation during motor task ($0 \le k_u^{max} \le 1$); and A is time-dependent muscle activation which is obtained from the first-order differential equation describing the muscle excitation (EMG)–activation (active state) dynamics:

$$\frac{dA(t)}{dt} + \{\frac{\tau_{act}}{\tau_{deact}} + [1 - \frac{\tau_{act}}{\tau_{deact}}]EMG(t)\}\frac{A(t)}{\tau_{act}} = \frac{EMG(t)}{\tau_{act}}, \quad [8.39]$$

where *EMG(t)* is the rectified, low-pass filtered, and normalized to its peak EMG activity of the muscle recorded during movement of interest or generated by a computer in the process of optimization ($0 \le EMG(t) \le 1$); τ_{ac}, and τ_{deact} are the activation and deactivation time constants.

The set of equations (8.36-8.39) enables calculations of limb motion (i.e., generalized coordinates). The inputs to the calculations are the initial generalized coordinates and velocities and the time-dependent muscle activity. These inputs are typically measured or somehow generated before integrating the equations of muscle and limb dynamics. Forward dynamics models require a large number of parameters describing the geometric, physiological, and mechanical properties of the musculoskeletal system. Unfortunately, only some of them can be measured. The rest should be estimated or the values optimized by requiring that the calculated motion matches the measured motion as closely as possible. 》

■ ■ ■ *FROM THE LITERATURE* ■ ■ ■

Dynamic Minimization of Metabolic Cost and Static Optimization of Muscle Fatigue Predict Realistic Walking Mechanics and Muscle Activity

Source: Anderson, F.C., and M.G. Pandy. 2001. Static and dynamic optimization solutions for gait are practically equivalent. *J Biomech* 34:153-161

Dynamic optimization was used to solve the distribution problem, namely to determine how and why multiple muscles are activated in a specific stereotypical manner during human walking. The problem was solved for one walking cycle of a fixed duration. The equations of limb muscle dynamics were similar to those described in the previous section. Muscle excitation history of each muscle was represented as

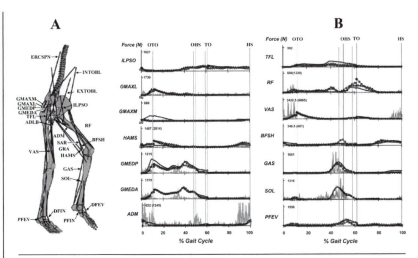

Figure 8.12 *(a)* Musculoskeletal model used for dynamic and static optimization of three selected cost functions during a walking cycle. *(b)* Results of minimization of 3 cost functions (J_1), (J_2), and (J_3); see text for explanations. Forces of selected muscles predicted by the dynamic solution $(J_1$, thick black lines), by the physiological static solution $(J_2$, circles), and by the nonphysiological static solution $(J_3$, triangles) are shown as a percentage of the gait cycle. EMG data from one subject are shown in thin gray lines. The following gait cycle landmarks are demarcated by thin vertical lines: opposite toe-off (OTO), opposite heel-strike (OHS), toe-off (TO), and heel-strike (HS). The plots are scaled to each muscle's maximum isometric strength. Muscles depicted: soleus (SOL), gastrocnemius (GAS), vastii (VAS), rectus femoris (RF), adductor magnus (AD), hamstrings (HAMS); biceps femoris short head (BFSH); anterior gluteus medius/minimus (GMEDA) and posterior gluteus medius/minimus (GMEDP); iliopsoas (ILPSO); medial gluteus maximus (GMAXM); lateral gluteus maximus (GMAXL); tensor fasciae latae (TFL).

Reprinted from *Journal of Biomechanics* 34(2), F.C. Anderson and M.G. Pandy, "Static and dynamic optimization solutions for gait are practically equivalent," 153-161, copyright 2001, with permission from Elsevier.

a function of 15 parameters and was generated in the process of a cost function optimization. The three-dimensional model of two legs and the pelvis had 23 DoFs and was actuated by 54 muscles (figure 8.12*a*).

Three cost functions were minimized by dynamic and static methods. Joint moments computed by dynamic optimization were

(continued)

From the Literature *(continued)*

used as inputs to static optimization. The cost functions were as follows:

1. Metabolic energy expenditure per distance traveled (J_1)—dynamic optimization. Metabolic energy was estimated as a function of the muscles' internal states, that is, as the sum of basal metabolic rate, activation heat rate, the maintenance heat rate, shortening heat rate, and mechanical work.

2. Squared sum of normalized muscle activations with muscle forces constrained by the force–length–velocity properties (J_2)—physiological static optimization.

3. Squared sum of normalized muscle activations without constraining muscle forces by the muscle force–length–velocity properties (J_3)—nonphysiological static optimization.

As can be seen from figure 8.12, muscles for which EMG recordings were available demonstrate a reasonably good qualitative agreement between the computed forces and recorded EMG. The dynamic and static optimization solutions were qualitatively similar. This is consistent with the results of static optimization shown in figure 8.10—forces of three muscles (including the slow-twitch soleus) can be successfully predicted for a wide range of speeds by static optimization without accounting for either the history effects or the muscle force–length–velocity properties.

8.4.3 Inverse Optimization

Representative publications: Kording and Wolpert 2004; Liu et al. 2005; Bottasso et al. 2006; Siemienski 2006; Terekhov et al. 2010; Park et al. 2010; Terekhov and Zatsiorsky 2011

The current use of optimization methods to understand muscle coordination has certain limitations. Essentially, researchers guess what cost functions the central controller might use and then test how good their guesses are. Inverse optimization attempts to find (deduce, compute) an unknown *objective function* (or functions) from experimentally recorded optimal solutions.

In other words, in *direct optimization* the goal is to find a solution that satisfies the task requirements and minimizes a certain known cost function. In *inverse optimization* the goal is to infer an unknown cost function from the experimental observations, that is, to find what exactly is being optimized. The principal pursuit is to "reconstruct" the cost function used by the control system, instead of assuming it a priori. From a mathematical point of view, this problem consists of finding an unknown objective function given the values at which it reaches its optimum. In contrast to direct optimization, which has a long and well-established history, the inverse optimization is comparatively new and its methods are not fully generalizable. Hence, the currently existing studies on inverse optimization mainly propose new mathematical or computational techniques.

To apply the methods of inverse optimization, the experimental data should satisfy some special requirements. In particular, a necessary condition is that the data should be obtained from many trials with different input parameters (constraints). To illustrate this requirement, imagine that a researcher recorded 100 variables in a single step of walking. It seems that the obtained data are sufficient to check for the sought cost function. However, in a 100-dimensional space of variables, one step will be represented by a single point, and because one point can correspond to infinitely many cost functions, the inverse problem in this case cannot be solved in a unique way. The above example illustrates the so-called uniqueness problem.

Non-uniqueness can be illustrated by an example of finding a function defined on the plane $z = f(x,y)$. The experimental data for this problem are the sets of triplets $<x,y,z>$. If the x and y components are distributed over a two-dimensional region of the plane, as shown in figure 8.13*a*, the finding of function f is feasible. However, if the data are such that all x and y values lie on a curve in the plane, then the problem cannot be solved uniquely. Figure 8.13*b* shows three surfaces corresponding to different possible functions $f(x,y)$ that might be estimated from the data shown on the left.

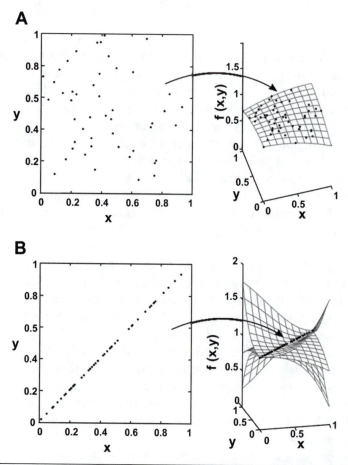

Figure 8.13 Two patterns of the distribution of experimental data on the (x, y) plane (left panels) and functions $z = f(x, y)$ derived from these data (right panels). *(a)* The data cover a two-dimensional region on the (x, y) plane. A function $z = f(x, y)$ can be determined in a unique way. *(b)* The (x, y) values lie on a straight line. Function $z = f(x, y)$ cannot be uniquely determined. Three surfaces representing different $z = f(x, y)$ functions are shown. Three functions satisfy the experimental data.

Courtesy of Dr. A.V. Terekhov.

The Inverse Optimization in Human Movement Research

1. A New Technique for Solving the Problem

Source: Bottasso, C.L., B.I Prilutsky, A. Croce, E. Imberti, and S. Sartirana. 2006. A numerical procedure for inferring from experimental data the optimization cost functions using a multibody model of the neuro-musculoskeletal system. *Multibody System Dynamics* 16:123-154

A method is suggested to overcome the need to hypothesize the cost functions, extracting this non–directly observable information from experimental data. A plausible space of cost functions is initially suggested by a researcher, and then the particular cost function within the above space is determined that yields the best possible match with the experimental data (kinematics, kinetics, and muscle activity). For instance, a researcher assumes that the cost function is a polynomial function with the unknown parameters, that is, a function

$$J = \sum_{k=2}^{N_k} p_{A,k} \sum_{i=1}^{N_m} A_i^k + \sum_{k=2}^{N_k} p_{f,k} \sum_{i=1}^{N_m} F_i^k \rightarrow min, \qquad [8.40]$$

where A_i and F_i are experimental activation and force of the ith muscle; N_k is a maximum exponent value; N_m is the number of muscles in the model; $p_{A,k}$ $(k = 2, \ldots, N_k)$ are the unknown parameters representing the level of participation of the polynomial functions of the muscle *activations* in the cost function; $p_{F,k}$ $(k = 2, \ldots, N_k)$ are the unknown parameters representing the level of participation of the polynomial functions of the muscle *forces* in the cost function; and $\sum_{k=2}^{N_k}(p_{A,k} + p_{F,k}) = 1$. The maximum value of exponent k is assumed by the researcher. The goal is to find unknown parameters $p_{A,k}$ and $p_{F,k}$ and thus the exact form of cost function (equation 8.40). The cost function that best matches the experimental data is identified within the search space by solving a so-called *nested optimization problem*.

(continued)

From the Literature   *(continued)*

2. Solving the Uniqueness Problem

Source: Terekhov, A.V., Y.B. Pesin, X. Niu, M.L. Latash, and V.M. Zatsiorsky. 2010. An analytical approach to the problem of inverse optimization with additive objective functions: an application to human prehension. *J Math Biol* 61423-61453

The authors analyzed inverse optimization problems with *additive cost functions* and linear constraints. Such problems are typical in human movement science. The problem of muscle (or finger) force sharing is an example. For such problems the authors obtained sufficient conditions for uniqueness (the so-called *uniqueness theorem*) and proposed a method for determining the objective functions. The method was applied to the problem of force sharing among the fingers in a grasping task. The objective function has been estimated from experimental data. The function predicted the force-sharing patterns for a vast range of external forces and torques applied to the grasped object. The resulting objective function was quadratic with essentially nonzero linear terms.

8.4.4 On Optimization Methods in Human Biomechanics and Motor Control

Representative publications: Gel'fand and Tsetlin 1971; Latash 2008

Several examples of successful application of optimization methods in human biomechanics and motor control studies have been presented here. The researchers reported that using certain cost functions yielded good correspondence with experimental results. What does this good correspondence signify? How should it be interpreted? Let us consider two (extreme) examples.

1. A ball on an inclined surface. The ball will evidently move downward until it reaches a minimal height, at which its potential energy is minimal. Somebody (a diligent student who studies optimization methods) has decided to apply optimization methods to solve the equations and predict the ball's final position. We know for sure that the agreement between the actual final ball position and the computer-predicted position will be very good, much better than in the above experimental studies. Because the ball follows the predicted path, does it mean that the ball itself minimizes (optimizes) something? Evidently, no. The ball's behavior is governed by the laws of mechanics. The final position of the ball is optimal—it corresponds to the

minimal potential energy—but optimization was used by the student, not the ball.

2. The traveling problem—this is a classic problem of mathematical optimization. A business traveler has to visit many cities. The pairwise distances between the cities (constraints) are known. The traveler needs to visit each city exactly once (this is an additional constraint). She wants to find a shortest possible route. This is a cost function that has to be minimized. Alternatively, the traveler may want to minimize the time of travel or its cost (or another cost function). The traveler's behavior included at least three steps: (a) the cost function was selected (i.e., the traveler decided what should be minimized—the distance, travel time, or cost), (b) the optimization computations were performed and the optimal path was determined prior to the travel, and (c) the traveler followed the precomputed optimal route.

Our question is: does the central controller in our nervous system work as the ball or as the traveler? Does it simply follow some (unknown to us) rules and only we, scientists, apply optimization methods to infer these rules, or does the controller explicitly select a cost function and then seek its optimization?

In case 1, one possible candidate for such a rule is the principle of minimization of neural activity, or minimal final action, a minimization of the total activity of a set of neurons involved in the control of a motor task. As pointed out by Gel'fand and Tsetlin (1971), the ensembles of neurons that are not *pacemakers,* that is, the cells that do not need external input to become active, have a natural tendency to go from activity to rest. For the sake of illustration, imagine a set of active neurons that form a closed system; that is, they do not receive inputs from other parts of the brain and sensory organs. If one cell becomes inactive it ceases to send excitation to other cells. This may be sufficient to deactivate another cell. As a result, a cascade of deactivation occurs and the activity of the entire network decreases or even comes to a complete halt. Such a physiological minimization may be associated with a decrease in many biomechanical variables, such as, for instance, force, power, or energy expenditure. As a result of evolution and learning, the physiological minimization will be associated with minimal values of some biomechanical variables to a larger degree than with minimization of other variables.

We invite the reader to ponder our initial question on whether the central controller in the nervous system works as the ball or as the traveler, or something in between that we do not know, and provide an educated (or uneducated) guess. The authors themselves have two answers to this question, a short one and a long one. The short answer is that we do not know.

The long answer is not much better: (1) we will not tell; (2) we will not tell because we do not know; (3) we do not know because nobody knows; (4) if one tells you that he knows he is a hypocrite.

We think that a proper answer to the preceding question cannot be obtained by using the methods that are presently available and have been used previously. Something new has to be invented. We wish readers good luck in this endeavor.

8.5 SUMMARY

The musculoskeletal system is mechanically redundant. At least two types of mechanical redundancy can be distinguished: kinematic redundancy and muscle redundancy. These two types of mechanical redundancy lead to multiple solutions for performing a given motor task (the so-called degree-of-freedom problem formulated by N.A. Bernstein). In the context of this book the degree-of-freedom problem can be reformulated in terms of combinations of muscles that need to be activated in order to execute given tasks.

After learning the same task people typically demonstrate very similar movement characteristics and muscle activity patterns, which are called invariant characteristics of skilled performance. Four major kinematic invariant characteristics of skilled arm reaching movements are typically distinguished: (1) straight-line limb endpoint trajectory, (2) bell-shaped velocity profile, (3) power law for curved movements, and (4) Fitts' law. Fitts' law describes a trade-off between the movement time (or speed) and the task difficulty (or accuracy requirements); that is, the more difficult the pointing task (the smaller and further away target), the longer it takes to reach it, the poorer the performance, or both. The three kinematic invariants—straight line trajectory, bell-shaped velocity profile, and the power law—can be derived from a general physical principle, the principle of least action, according to which the control system chooses the movement trajectory and endpoint velocity such that, in a given time period, mechanical work done on the endpoint and the arm is minimal. However, the principle of least action does not seem to predict or explain Fitts' law. The four kinematic invariants can be explained by an assumption that the control system selects such kinematics of the endpoint that minimize the variance of the final endpoint position. The source of this variance is noise in the control signal. The noise increases with control signal magnitude as observed experimentally; for example, force fluctuations increase with force magnitude. Counteracting noise in the control signals leads to the kinematics that are similar to those predicted by the minimum action principle.

Besides the kinematic invariants, some kinetic invariants have been also mentioned in the literature. In skilled arm reaching, locomotion, postural reactions, and other tasks, the joint moments seem to be rather stereotyped. Thus, the motor control system might care not only about kinematics of endpoint motion but also about joint kinetics or muscle forces and muscle activities. One notable joint kinetic invariant characteristic of reaching movements is a covariation of elbow and shoulder dynamic joint moments (joint moments with removed static component that depends only on gravity). The minimally changed joint moments could be considered a kinetic invariant. This kinetic property of reaching movements is called minimum change. In postural tasks, specific combinations of apparent joint stiffnesses at a given arm posture determine the dynamic response of the arm to endpoint perturbations, which can be expressed in terms of orientation, shape, and size of the arm stiffness ellipse (for details see **3.2.3** in *Kinetics of Human Motion*). The orientation and shape of the arm stiffness ellipses seem to be rather stereotypic during maintenance of a given arm posture at rest and thus can be considered kinetic invariants.

A number of muscles in the human body (approximately 630) sharply exceed an estimated total number of kinematic degrees of freedom in the human body, around 240. Thus, on average, 1 kinematic degree of freedom is controlled by 2.6 muscles (630/244). This feature of the musculoskeletal system is called muscle redundancy. As a consequence of muscle redundancy, even in a simple one-DoF joint controlled by two anatomical antagonistic muscles, there is an unlimited number of muscle force combinations to produce a required moment at the joint.

Despite these complexities, people activate muscles in a coordinated manner. What principles or rules (if any) does the control system use to select individual muscles for executing a great repertoire of movements? The search for the answer to this question constitutes an area of research called the distribution problem.

Several stereotyped features of muscle activity can be identified in skilled performance. They include broad (cosine) tuning of activity, which is sometimes called synergistic activity; reciprocal activation; and activation of two-joint muscles in accordance with their mechanical advantage. When the joint moment is desired in the direction of the muscle moment arm vector, the muscle commonly exhibits high activity. The direction of movement or exerted force corresponding to maximum muscle activity is called the preferred direction. When the desired movement or force direction moves away from the preferred direction, muscle activity does not decrease sharply, still exhibiting activity in directions in which it can produce little force. This phenomenon is typically referred to as the cosine tuning of muscle activity or the cosine rule because such muscle behavior can generally be described by a cosine function fitted to experimental EMG recordings:

$$EMG = a \cdot F \cdot \cos(\varphi_{PD} - \varphi), \qquad [8.24]$$

where *EMG* is the mean value of recorded muscle electromyographic activity during exerting the endpoint force in direction φ (or the corresponding combination of joint moments); φ_{PD} is the preferred muscle EMG direction, that is, direction corresponding to the maximum muscle activity; F is the exerted force magnitude, and a is a scaling factor. Although on the whole the cosine rule seems valid, reality is usually much more complex. Among many reasons for this is the necessity to stabilize the joint: because multiple muscles are usually active and pull in different directions, the joint must be stabilized to prevent dislocation. In experimental research the ideal cosine tuning with the preferred direction coinciding with muscle mechanical advantage (i.e., the moment arm vector direction), is usually not confirmed.

Another invariant feature of muscle activation in skilled tasks in predictable environments is specific activation of two-joint muscles. EMG activity of a two-joint muscle or a muscle that controls 2 DoFs is greater if the moments it produces at both joints have the same directions as the resultant joint moments (see, e.g., figure 6.4). If the moments of the two-joint muscle at both joints are opposite to the resultant joint moments, then that muscle's EMG activity is low. This behavior is consistent with cosine tuning of muscle activity in accordance with muscle preferred direction.

The fact that specific invariant movement characteristics exist despite kinematic and muscle redundancy suggests that the control system may, during acquisition of novel tasks, select specific muscle coordination patterns to optimize certain movement performance criteria or certain physiological functions. This is why many researchers have tried to find solutions to the distribution problem using methods of optimization. There are two basic optimization approaches: static and dynamic.

Static optimization is based on several assumptions:

1. The control of joint moments and thus limb dynamics and the control of muscle activity are decoupled: the central controller determines the desired joint moments and then (or in addition to that) distributes activity among the muscles such that necessary moments are generated.

2. Muscle activity or forces and their distribution among muscles are not history dependent. This assumption allows for optimization formulation using static equations for joint moments independently at each time instant.

3. In tasks with generally stereotypic muscle activity patterns the control system distributes muscle activity and forces in an optimal way.

A typical static optimization problem is formulated for each instant in time and involves minimizing an objective function that satisfy various equality and inequality constraints. The equality constraints usually require that the sum of the moments of force generated by individual muscle be equal to the measured joint moments. Inequality constraints require that the muscle forces are unidirectional (muscles can only pull but not push on the bones) and do not exceed a certain value.

Cost functions such as

$$\sum_{i=1}^{m} F_i^n \rightarrow min,$$ [8.28]

$$\sum_{i=1}^{m} (F_i / PCSA_i)^n \rightarrow min,$$ [8.29]

$$\sum_{i=1}^{m} (F_i / F_{imax})^n \rightarrow min,$$ [8.30]

(where PCSA is physiological cross-sectional area and F_{imax} is the maximal isometric muscle force) do not predict synergistic muscle behavior at $n = 1$; however, they do predict an appropriate activation of two-joint muscles depending on the direction of the vector of moments at the two joints (see figures 6.3 and 6.4). Better results are obtained when similar but nonlinear cost functions are used for which $n = 2$ or $n = 3$. Minimizing cost function (8.29) with $n = 3$ was proposed to maximize endurance time (the duration with which a person can sustain a given muscle stress magnitude) or to minimize fatigue.

Static optimization models have several limitations:

1. In many tasks, musculoskeletal dynamics are history dependent. Static optimization models do not account for this fact.

2. Static optimization models typically cannot predict kinematics or dynamics when only initial or final states of the system are known.

Dynamic optimization-based models do not have these limitations. In dynamic optimization, the objective function itself is a function of both state and time. Thus dynamic optimization problems are solved only once for the entire movement. Dynamic optimization models require specification of a large number of parameters (e.g., muscle moment arms, slack tendon and muscle fascicle lengths, PCSA, muscle activation and deactivation constants, and muscle force–length–velocity properties). Precise values of these parameters are difficult to measure.

The current use of optimization methods to understand muscle coordination has limitations. Essentially, researchers guess cost functions and then test these guesses. Inverse optimization attempts to find (deduce, compute) an unknown objective function (or functions) from experimentally recorded optimal solutions. From a mathematical point of view, the problem consists of finding an unknown objective function given the values at which it reaches its optimum.

A good correspondence between optimization results and experimental data does not automatically indicate that the central controller actually uses optimization. To those who are interested in this topic we recommend rereading section **8.4.4**.

8.6 QUESTIONS FOR REVIEW

1. Define musculoskeletal redundancy.
2. Define kinematic redundancy. Give an example.
3. What is the degree-of-freedom problem? How, generally, may it be solved?
4. Give a definition of invariant characteristics as it pertains to movement. List the main kinematic invariants.
5. Describe the kinematic invariants of hand trajectories during point-to-point reaching.
6. Describe the two-thirds power law.
7. Formulate the isogony principle.
8. Explain Fitts' law.
9. Derive Fitts' law from the assumptions that (1) target reaching is performed as a sequence of path segments and (2) it takes the same time to execute each path segment.
10. Explain the principle of minimum action.
11. What optimization criteria have been used to explain or predict the main kinematic invariants?
12. List the main kinetic invariants of skilled performance.
13. Formulate the minimum joint moment change principle.
14. Name two kinetic invariants of apparent arm stiffness.
15. Define muscle redundancy. How many degrees of freedom and actons does the human arm have?
16. What are primary, secondary, and tertiary moments?
17. Describe the effects of secondary moments on muscle activity.

18. List the invariant features of muscle activation.
19. What is the cosine rule?
20. Define the terms preferred direction, broad tuning, reciprocal muscle activation, and mechanical advantage.
21. Describe the invariant activation features of two-joint muscles.
22. Give a definition of the distribution problem.
23. Explain static optimization and list its assumptions and limitations.
24. Explain dynamic optimization and list its assumptions and limitations.
25. Give examples of cost functions that have been shown to predict certain invariant movement characteristics.
26. Formulate the inverse optimization problem.

8.7 LITERATURE LIST

Ackermann, M., and A.J. van den Bogert. 2010. Optimality principles for model-based prediction of human gait. *J Biomech* 43(6):1055-1060.

An, K.N., B.M. Kwak, E.Y. Chao, and B.F. Morrey. 1984. Determination of muscle and joint forces: A new technique to solve the indeterminate problem. *J Biomech Eng* 106(4):364-367.

Anderson, F.C., and M.G. Pandy. 2001. Static and dynamic optimization solutions for gait are practically equivalent. *J Biomech* 34(2):153-161.

Arimoto, S., M. Sekimoto, H. Hashiguchi, and R. Ozawa. 2005. Natural resolution of ill-posedness of inverse kinematics for redundant robots: A challenge to Bernstein's degrees-of-freedom problem. *Advanced Robotics* 19(4):401-434.

Arimoto, S., M. Sekimoto, and K. Tahara. 2010. Iterative learning without reinforcement or reward for multijoint movements: A revisit of Bernstein's DOF problem on dexterity. *Journal of Robotics*, Volume 2010, Article ID 217867, 15 pages, doi:10.1155/2010/217867.

Bernstein, N.A. 1947. *On the construction of movement* (in Russian). Moscow: Medgiz.

Bernstein, N.A. 1967. *The co-ordination and regulation of movements*. Oxford, UK: Pergamon Press.

Biess, A., D.G. Liebermann, and T. Flash. 2007. A computational model for redundant human three-dimensional pointing movements: Integration of independent spatial and temporal motor plans simplifies movement dynamics. *J Neurosci* 27(48):13045-13064.

Bottasso, C.L., B.I. Prilutsky, A. Croce, E. Imberti, and S. Sartirana. 2006. A numerical procedure for inferring from experimental data the optimization cost functions using a multibody model of the neuro-musculoskeletal system. *Multibody System Dynamics* 16(2):123-154.

Buchanan, J.J., J.H. Park, and C.H. Shea. 2006. Target width scaling in a repetitive aiming task: Switching between cyclical and discrete units of action. *Exp Brain Res* 175(4):710-725.

Buchanan, T.S., D.P. Almdale, J.L. Lewis, and W.Z. Rymer. 1986. Characteristics of synergic relations during isometric contractions of human elbow muscles. *J Neurophysiol* 56(5):1225-1241.

Buchanan, T.S., G.P. Rovai, and W.Z. Rymer. 1989. Strategies for muscle activation during isometric torque generation at the human elbow. *J Neurophysiol* 62(6):1201-1212.

Bullock, D., and S. Grossberg. 1988. Neural dynamics of planned arm movements: Emergent invariants and speed-accuracy properties during trajectory formation. *Psychol Rev* 95(1):49-90.

Card, S.E., T.P. Moran, and A.P. Newell. 1983. *The Psychology of Human-Computer Interaction.* Hillsdale, NJ: Erlbaum.

Crowninshield, R.D., and R.A. Brand. 1981. A physiologically based criterion of muscle force prediction in locomotion. *J Biomech* 14(11):793-801.

Darainy, M., and D.J. Ostry. 2008. Muscle cocontraction following dynamics learning. *Exp Brain Res* 190(2):153-163.

Darainy, M., F. Towhidkhah, and D.J. Ostry. 2007. Control of hand impedance under static conditions and during reaching movement. *J Neurophysiol* 97(4):2676-2685.

Davy, D.T., and M.L. Audu. 1987. A dynamic optimization technique for predicting muscle forces in the swing phase of gait. *J Biomech* 20(2):187-201.

De Shapiro, V., O. Khatib, and S. Delp. 2008. Least principles and their application to constrained and task-level problems in robotics and biomechanics. *Multibody Systems Dynamics* 19:303-322.

Dounskaia, N. 2007. Kinematic invariants during cyclical arm movements. *Biol Cybern* 96(2):147-163.

Dounskaia, N., D. Wisleder, and T. Johnson. 2005. Influence of biomechanical factors on substructure of pointing movements. *Exp Brain Res* 164(4):505-516.

Duarte, M., and M.L. Latash. 2007. Effects of postural task requirements on the speed-accuracy trade-off. *Exp Brain Res* 180(3):457-467.

Dul, J., G.E. Johnson, R. Shiavi, and M.A. Townsend. 1984a. Muscular synergism: II. A minimum-fatigue criterion for load sharing between synergistic muscles. *J Biomech* 17(9):675-684.

Dul, J., M.A. Townsend, R. Shiavi, and G.E. Johnson. 1984b. Muscular synergism: I. On criteria for load sharing between synergistic muscles. *J Biomech* 17(9):663-673.

Erdemir, A., S. McLean, W. Herzog, and A.J. van den Bogert. 2007. Model-based estimation of muscle forces exerted during movements. *Clin Biomech (Bristol, Avon)* 22(2):131-154.

Feldman, A.G., V. Goussev, A. Sangole, and M.F. Levin. 2007. Threshold position control and the principle of minimal interaction in motor actions. *Prog Brain Res* 165:267-281.

Fitts, P.M. 1954. The information capacity of the human motor system in controlling the amplitude of movement. *J Exp Psychol* 47(6):381-391.

Fitts, P.M., and J.R. Peterson. 1964. Information capacity of discrete motor responses. *J Exp Psychol* 67:103-112.

Flanders, M., and J.F. Soechting. 1990. Arm muscle activation for static forces in three-dimensional space. *J Neurophysiol* 64(6):1818-1837.

Flash, T., and A.A. Handzel. 2007. Affine differential geometry analysis of human arm movements. *Biol Cybern* 96(6):577-601.

Flash, T., and N. Hogan. 1985. The coordination of arm movements: An experimentally confirmed mathematical model. *J Neurosci* 5(7):1688-1703.

Flash, T., and F. Mussa-Ivaldi. 1990. Human arm stiffness characteristics during the maintenance of posture. *Exp Brain Res* 82(2):315-326.

Franklin, D.W., G. Liaw, T.E. Milner, R. Osu, E. Burdet, and M. Kawato. 2007. Endpoint stiffness of the arm is directionally tuned to instability in the environment. *J Neurosci* 27(29):7705-7716.

Gan, K.C., and E.R. Hoffmann. 1988. Geometrical conditions for ballistic and visually controlled movements. *Ergonomics* 31(5):829-839.

Gel'fand, I.M., V.S. Gurfinkel, M.L. Tsetlin, and M.L. Shik. 1971. Some problems in the analysis of movements. In *Models of Structural-Functional Organization of Certain Biological Systems*, edited by I.M. Gelfand, V.S. Gurfinkel, S.V. Fomin, and M.L. Tsetlin. Cambridge, MA: MIT Press.

Gel'fand, I.M., and M.L. Tsetlin. 1971. On mathematical modeling of the mechanisms of the central nervous system. In *Models of the Structural-Functional Organization of Certain Biological Systems*, edited by Gelfand, I.M., V.S. Gurfinkel, S.V. Fomin, and M.L. Tsetlin. Cambridge, MA: MIT Press, 1-22.

Gottlieb, G.L., Q. Song, G.L. Almeida, D.A. Hong, and D. Corcos. 1997. Directional control of planar human arm movement. *J Neurophysiol* 78(6):2985-2998.

Gottlieb, G.L., Q. Song, D.A. Hong, and D.M. Corcos. 1996. Coordinating two degrees of freedom during human arm movement: Load and speed invariance of relative joint torques. *J Neurophysiol* 76(5):3196-3206.

Harris, C.M., and D.M. Wolpert. 1998. Signal-dependent noise determines motor planning. *Nature* 394(6695):780-784.

Haruno, M., and D.M. Wolpert. 2005. Optimal control of redundant muscles in step-tracking wrist movements. *J Neurophysiol* 94(6):4244-4255.

Hatze, H. 1976. The complete optimization of a human motion. *Math Biosci* 28(1-2):99-135.

Hatze, H. 1981. A comprehensive model for human motion simulation and its application to the take-off phase of the long jump. *J Biomech* 14(3):135-142.

Hatze, H. 2000. The inverse dynamics problem of neuromuscular control. *Biol Cybern* 82(2):133-141.

He, J., W.S. Levine, and G.E. Loeb. 1991. Feedback gains for correcting small perturbations to standing posture. *IEEE Transactions on Automative Control* 36(3):322-332.

Herzog, W. 2000. Muscle activation and movement control. In *Biomechanics and Biology of Movement*, edited by B.R. Nigg, B.R. MacIntosh, and J. Mester. Champaign, IL: Human Kinetics.

Herzog, W., and T.R. Leonard. 1991. Validation of optimization models that estimate the forces exerted by synergistic muscles. *J Biomech* 24(Suppl 1):31-39.

Herzog, W., T.R. Leonard, and A.C. Guimaraes. 1993. Forces in gastrocnemius, soleus, and plantaris tendons of the freely moving cat. *J Biomech* 26(8):945-953.

Hicheur, H., Q.C. Pham, G. Arechavaleta, J.P. Laumond, and A. Berthoz. 2007. The formation of trajectories during goal-oriented locomotion in humans: I. A stereotyped behaviour. *Eur J Neurosci* 26(8):2376-2390.

Hogan, N. 1985. The mechanics of multi-joint posture and movement control. *Biol Cybern* 52(5):315-331.

Jax, S.A., D.A. Rosenbaum, and J. Vaughan. 2007. Extending Fitts' law to manual obstacle avoidance. *Exp Brain Res* 180(4):775-779.

Kawato, M. 1999. Internal models for motor control and trajectory planning. *Curr Opin Neurobiol* 9(6):718-727.

Kording, K.P., and D.M. Wolpert. 2004. The loss function of sensorimotor learning. *Proc Natl Acad Sci U S A* 101(26):9839-9842.

Kurtzer, I., J.A. Pruszynski, T.M. Herter, and S.H. Scott. 2006. Primate upper limb muscles exhibit activity patterns that differ from their anatomical action during a postural task. *J Neurophysiol* 95(1):493-504.

Lackner, J.R., and P. Dizio. 1994. Rapid adaptation to Coriolis force perturbations of arm trajectory. *J Neurophysiol* 72(1):299-313.

Lacquaniti, F., C. Terzuolo, and P. Viviani. 1983. The law relating the kinematic and figural aspects of drawing movements. *Acta Psychol (Amst)* 54(1-3):115-130.

Latash, M.L. 2008. *Synergy.* New York: Oxford University Press.

Latash, M.L., J.P. Scholz, and G. Schoner. 2002. Motor control strategies revealed in the structure of motor variability. *Exerc Sport Sci Rev* 30(1):26-31.

Latash, M.L., J.P. Scholz, and G. Schoner. 2007. Toward a new theory of motor synergies. *Motor Control* 11(3):276-308.

Lebedev, S., W.H. Tsui, and P. Van Gelder. 2001. Drawing movements as an outcome of the principle of least action. *J Math Psychol* 45(1):43-52.

Lemay, M.A., and P.E. Crago. 1996. A dynamic model for simulating movements of the elbow, forearm, and wrist. *J Biomech* 29(10):1319-1330.

Liu, C.K., A. Hertzmann, and Z. Popović. 2005. Learning physics-based motion style with nonlinear inverse optimization. *ACM Trans Graph* 24(3):1071-1081.

Maoz, U., A. Berthoz, and T. Flash. 2009. Complex unconstrained three-dimensional hand movement and constant equi-affine speed. *J Neurophysiol* 101(2):1002-1015.

Morasso, P. 1981. Spatial control of arm movements. *Exp Brain Res* 42(2):223-227.

Morecki, A., J. Ekiel, and K. Fidelus. 1984. *Cybernetic Systems of Limb Movements in Man, Animals and Robots.* New York: Wiley.

Mussa-Ivaldi, F.A., N. Hogan, and E. Bizzi. 1985. Neural, mechanical, and geometric factors subserving arm posture in humans. *J Neurosci* 5(10):2732-2743.

Nakano, E., H. Imamizu, R. Osu, Y. Uno, H. Gomi, T. Yoshioka, and M. Kawato. 1999. Quantitative examinations of internal representations for arm trajectory planning: Minimum commanded torque change model. *J Neurophysiol* 81(5):2140-2155.

Neilson, P.D. 1993. The problem of redundancy in movement control: The adaptive model theory approach. *Psychol Res* 55(2):99-106.

Neptune, R.R., C.P. McGowan, and S.A. Kautz. 2009. Forward dynamics simulations provide insight into muscle mechanical work during human locomotion. *Exerc Sport Sci Rev* 37(4):203-210.

Novak, K.E., L.E. Miller, and J.C. Houk. 2003. Features of motor performance that drive adaptation in rapid hand movements. *Exp Brain Res* 148(3):388-400.

Nozaki, D. 2009. Torque interaction among adjacent joints due to the action of biarticular muscles. *Med Sci Sports Exerc* 41(1):205-209.

Nozaki, D., K. Nakazawa, and M. Akai. 2005. Muscle activity determined by cosine tuning with a nontrivial preferred direction during isometric force exertion by lower limb. *J Neurophysiol* 93(5):2614-2624.

Nubar, Y., and R. Contini. 1961. A minimal principle in biomechanics. *Bulletin of Mathematical Biophysics* 23:379-390.

Pandy, M.G., and F.E. Zajac. 1991. Optimal muscular coordination strategies for jumping. *J Biomech* 24(1):1-10.

Park, J., V.M. Zatsiorsky, and M.L. Latash. 2010. Optimality vs. variability: An example of multi-finger redundant tasks. *Exp Brain Res* 207:119-132.

Pedotti, A., V.V. Krishnan, and L. Stark. 1978. Optimization of muscle force sequencing in human locomotion. *Math Biosci* 38:57-76.

Penrod, D.D., D.T. Davy, and D.P. Singh. 1974. An optimization approach to tendon force analysis. *J Biomech* 7(2):123-129.

Perreault, E.J., R.F. Kirsch, and P.E. Crago. 2002. Voluntary control of static endpoint stiffness during force regulation tasks. *J Neurophysiol* 87(6):2808-2816.

Pollick, F.E., U. Maoz, A.A. Handzel, P.J. Giblin, G. Sapiro, and T. Flash. 2009. Three-dimensional arm movements at constant equi-affine speed. *Cortex* 45(3):325-339.

Pollick, F.E., and G. Sapiro. 1997. Constant affine velocity predicts the 1/3 power law of planar motion perception and generation. *Vision Res* 37(3):347-353.

Polyakov, F., E. Stark, R. Drori, M. Abeles, and T. Flash. 2009. Parabolic movement primitives and cortical states: Merging optimality with geometric invariance. *Biol Cybern* 100(2):159-184.

Prilutsky, B., A.N. Klishko, B. Farrell, L. Harley, G. Philips, and C.L. Bottasso. 2009. Movement coordination in skilled tasks: Insights from optimization. In *Advances in Neuromuscular Physiology of Motor Skills and Muscle Fatigue*, edited by M. Shinohara. Kerala, India: Research Signpost, pp. 139-171.

Prilutsky, B.I. 2000a. Coordination of two- and one-joint muscles: Functional consequences and implications for motor control. *Motor Control* 4(1):1-44.

Prilutsky, B.I. 2000b. Muscle coordination: The discussion continues. *Motor Control* 4(1):97-116.

Prilutsky, B.I., and R.J. Gregor. 1997. Strategy of muscle coordination of two- and one-joint leg muscles in controlling an external force. *Motor Control* 1:92-116.

Prilutsky, B.I., W. Herzog, and T.L. Allinger. 1997. Forces of individual cat ankle extensor muscles during locomotion predicted using static optimization. *J Biomech* 30(10):1025-1033.

Prilutsky, B.I., W. Herzog, and T. Leonard. 1996. Transfer of mechanical energy between ankle and knee joints by gastrocnemius and plantaris muscles during cat locomotion. *J Biomech* 29(4):391-403.

Prilutsky, B.I., and V.M. Zatsiorsky. 2002. Optimization-based models of muscle coordination. *Exerc Sport Sci Rev* 30(1):32-38.

Raikova, R.T., and B.I. Prilutsky. 2001. Sensitivity of predicted muscle forces to parameters of the optimization-based human leg model revealed by analytical and numerical analyses. *J Biomech* 34(10):1243-1255.

Rasmussen, J., M. Damsgaard, and M. Voigt. 2001. Muscle recruitment by the min/max criterion: A comparative numerical study. *J Biomech* 34(3):409-415.

Richardson, M.J., and T. Flash. 2002. Comparing smooth arm movements with the two-thirds power law and the related segmented-control hypothesis. *J Neurosci* 22(18):8201-8211.

Sainburg, R.L., C. Ghez, and D. Kalakanis. 1999. Intersegmental dynamics are controlled by sequential anticipatory, error correction, and postural mechanisms. *J Neurophysiol* 81(3):1045-1056.

Schaal, S., and D. Sternad. 2001. Origins and violations of the 2/3 power law in rhythmic three-dimensional arm movements. *Exp Brain Res* 136(1):60-72.

Schappacher-Tilp, G., P. Binding, E. Braverman, and W. Herzog. 2009. Velocity-dependent cost function for the prediction of force sharing among synergistic muscles in a one degree of freedom model. *J Biomech* 42(5):657-660.

Schneider, K., and R.F. Zernicke. 1989. Jerk-cost modulations during the practice of rapid arm movements. *Biol Cybern* 60(3):221-230.

Scott, S.H. 1999. Apparatus for measuring and perturbing shoulder and elbow joint positions and torques during reaching. *J Neurosci Meth* 89(2):119-127.

Seireg, A., and Arvikar. 1975. The prediction of muscular lad sharing and joint forces in the lower extremities during walking. *J Biomech* 8(2):89-102.

Seireg, A., and R.J. Arvikar. 1973. A mathematical model for evaluation of forces in lower extremities of the musculo-skeletal system. *J Biomech* 6(3):313-326.

Shadmehr, R., and F.A. Mussa-Ivaldi. 1994. Adaptive representation of dynamics during learning of a motor task. *J Neurosci* 14(5 Pt 2):3208-3224.

Siemienski, A. 2006. Direct solution of the inverse optimization problem of load sharing between muscles *J Biomech* 39:S45.

Soechting, J.F., C.A. Buneo, U. Herrmann, and M. Flanders. 1995. Moving effortlessly in three dimensions: Does Donders' law apply to arm movement? *J Neurosci* 15(9):6271-6280.

Staudenmann, D., K. Roeleveld, D.F. Stegeman, and J.H. van Dieen. 2010. Methodological aspects of SEMG recordings for force estimation: A tutorial and review. *J Electromyogr Kinesiol* 20(3):375-387.

Terekhov, A.V., Y.B. Pesin, X. Niu, M.L. Latash, and V.M. Zatsiorsky. 2010. An analytical approach to the problem of inverse optimization with additive objective functions: An application to human prehension. *J Math Biol* 61(3):423-453.

Terekhov, A.V., and V.M. Zatsiorsky. 2011. Analytical and numerical analysis of inverse optimization problems: Conditions of uniqueness and computational methods. *Biol Cybern.* 104(1-2):75-93.

Thelen, E., D. Corbetta, and J.P. Spencer. 1996. Development of reaching during the first year: Role of movement speed. *J Exp Psychol Hum Percept Perform* 22(5):1059-1076.

Todorov, E. 2002. Cosine tuning minimizes motor errors. *Neural Comput* 14(6):1233-1260.

Tsirakos, D., V. Baltzopoulos, and R. Bartlett. 1997. Inverse optimization: Functional and physiological considerations related to the force-sharing problem. *Crit Rev Biomed Eng* 25(4-5):371-407.

Turvey, M.T. 1990. Coordination. *Am Psychol* 45(8):938-953.

van Bolhuis, B.M., and C.C. Gielen. 1997. The relative activation of elbow-flexor muscles in isometric flexion and in flexion/extension movements. *J Biomech* 30(8):803-811.

van Bolhuis, B.M., and C.C. Gielen. 1999. A comparison of models explaining muscle activation patterns for isometric contractions. *Biol Cybern* 81(3):249-261.

van der Helm, F.C. 1994. A finite element musculoskeletal model of the shoulder mechanism. *J Biomech* 27(5):551-569.

Van Soest, A.J., M.F. Bobbert, and G.J. Van Ingen Schenau. 1994. A control strategy for the execution of explosive movements from varying starting positions. *J Neurophysiol* 71(4):1390-1402.

Vasavada, A.N., B.W. Peterson, and S.L. Delp 2002. Three-dimensional spatial tuning of neck muscle activation in humans. *Exp Brain Res* 147(4):437-448.

Vieilledent, S., Y. Kerlirzin, S. Dalbera, and A. Berthoz. 2001. Relationship between velocity and curvature of a human locomotor trajectory. *Neurosci Lett* 305(1):65-69.

Viviani, P., and C. Terzuolo. 1982. Trajectory determines movement dynamics. *Neuroscience* 7(2):431-437.

Wada, Y., Y. Kaneko, E. Nakano, R. Osu, and M. Kawato. 2001. Quantitative examinations for multi joint arm trajectory planning: Using a robust calculation algorithm of the minimum commanded torque change trajectory. *Neural Netw* 14(4-5):381-393.

Wells, R., and N. Evans. 1987. Functions and recruitment patterns of one-joint and 2-joint muscles under isometric and walking conditions. *Hum Mov Sci* 6(4):349-372.

Wood, J.E., S.G. Meek, and S.C. Jacobsen. 1989a. Quantitation of human shoulder anatomy for prosthetic arm control: I. Surface modelling. *J Biomech* 22(3):273-292.

Wood, J.E., S.G. Meek, and S.C. Jacobsen. 1989b. Quantitation of human shoulder anatomy for prosthetic arm control: II. Anatomy matrices. *J Biomech* 22(4):309-325.

Yeo, B.P. 1976. Investigations concerning the principle of minimal total muscular force. *J Biomech* 9(6):413-416.

Zajac, F.E. 1989. Muscle and tendon: Properties, models, scaling, and application to biomechanics and motor control. *Crit Rev Biomed Eng* 17(4):359-411.

Zatsiorsky, V.M., R.W. Gregory, and M.L. Latash. 2002. Force and torque production in static multifinger prehension: Biomechanics and control: II. Control. *Biol Cybern* 87(1):40-49.

GLOSSARY

---

abduction—Movement of a body part away from the midsagittal plane.

absolute force—See *specific force*.

Achilles tendon—The tendon connecting the triceps surae and plantaris muscles to the calcaneus.

actin—One of two contractile proteins; the second is *myosin*.

activation heat—The heat corresponding to active state of a muscle.

active flexibility—Maximal range of motion at a joint achieved during self-movement.

active force—Increase in muscle force in excess of the passive force due to activation of the muscle.

active slack length (of sarcomeres)—The length at which no active force can be generated.

active state—State of a muscle that starts during the latent period, from arriving a stimulus to a recordable force increase at the muscle–tendon junction.

acton—A muscle or muscle part with point-to-point attachments that generates a moment of force about a single joint axis.

adduction—Movement of a body part toward the midsagittal plane.

afferent—*adj.* Conveying or transmitting toward the central nervous system. *n.* Sensory cell.

affine transformation—A geometrical transformation, such as shift, rotation, or stretching, retaining the original parallel straight lines.

agonist—A muscle that generates a moment at a joint in the same direction as (a) another muscle (see *anatomical agonists*) or (b) resultant joint moment (see *joint agonists*) or that (c) actively shortens during propulsive phase of movement, or lengthen being active during the yield phase (see *task agonists).*

all-or-nothing law—A nongraded response to a stimulus exceeding a certain threshold.

alpha motor neurons—Large neural cells innervating extrafusal muscle fibers.

analytic inverse optimization (ANIO)—A recently developed method of solving the inverse optimization problem.

anatomical agonists—Muscles that lie on the same side of a bone, act on the same joint, and move it in the same direction (cf. *joint agonists* and *task agonists*).

anatomical antagonists—Muscles that act on the same joint but move it in opposite directions (cf. *joint antagonists* and *task antagonists*).

anatomical cross-sectional area—Area of the muscle slices cut perpendicularly to the longitudinal muscle direction.

anisometric—Nonisometric.

antagonists—Muscles that generate moments at a joint in opposite directions *(anatomical antagonists)*; see also *joint antagonists* and *task antagonists*.

aponeurosis—A flat, broad tendon.

apparent stiffness—The measure of stiffness obtained from active objects, e.g., active muscles. The word *apparent* underscores that these measures differ from the analogous parameters of passive objects.

apparent strain (of a tendon)—Amount of elongation per unit of tendon length.

architectural gear ratio—The ratio of whole muscle velocity to muscle fiber velocity.

ascending limb (of a force–length curve)—A section of the force–length relation where the muscle generates larger active forces at a larger length.

ATP—Adenosine triphosphate, the immediate source of energy for muscle contraction.

basal lamina—The membrane lying next to the basal surface of the adjoining cell layer.

bell-shaped (velocity) profile—A symmetric velocity–time (or velocity–distance) curve with a single maximum.

Bernstein's problem—See *motor redundancy*.

biarticular muscles—See *two-joint muscles*.

binormal vector—Vector perpendicular to the tangent and normal vectors.

biomechanical agonists—See *joint agonists*.

biomechanical antagonists—See *joint antagonists*.

bipennate muscles—Pennate muscles with the fibers converging to both sides of a central tendon.

bowstringing—Tendency of a curved tendon to straighten.

bowstringing force—Lateral force arising when a tendon path is not along a straight line.

capstan equation—see *Eytelwein formula*.

catapult action—The reversible muscle action in which the elastic energy is accumulated over a long time period and then released over a short period of time.

catchlike property (of a muscle)—Increase of muscle tension in response to an extra pulse of high-frequency stimulation applied over a constant frequency stimulation.

Cauchy stress tensor—See *stress tensor*.

Cauchy's stress—Stress measured relative to the instantaneous area of a deformed object.

CC—See *contractile components*.

center of gravity—The point of application of the resultant gravity force. The center of gravity coincides with the center of mass.

center of mass (CoM)—A point with respect to which the mass moment of the first order in any plane equals zero.

central commands (in studying interfinger interaction)—Hypothetical efferent commands to the muscles serving a finger that range from 0 if the finger is not intended to produce force (no voluntary activation) to 1 if the finger is intended to produce maximal force (maximal voluntary activation).

central nervous system—The brain and the spinal cord.

cervical—Pertaining to the neck.

characteristic equation (of muscle dynamics)—Equation 3.26b: $(F + a)(V + b) = b(F_0 + a) = $ Constant.

cineplastic tunnels—Routes in the muscles made surgically to apply muscle pulling forces to a prosthesis.

CNS—See *central nervous system.*

coefficient of shortening heat—See *heat constant.*

collagen—Main protein of connective tissue.

compartment syndrome—Increased long-lasting pressure within a fascial compartment.

complete agonist—A two-joint muscle that is a joint agonist at both joints that it serves.

complete antagonist—A two-joint muscle that is a joint antagonist at both joints that it serves.

compliance—The amount of deformation per unit of force. Compliance is the inverse of stiffness.

components (of a vector)—Elements into which a vector quantity can be resolved.

composite materials (or **composites**)—Materials made from two or more materials.

concentric muscle action—Exerting forces while shortening. Same as *miometric action.*

conservation of energy (in human movement)—Transformation of potential energy into kinetic energy and back.

constitutive equations—Mathematical relations between stresses and strains.

constraint—(a) In mechanics: any restriction to free movement. (b) In mathematics: a restriction on the value of variables.

contractile components (in muscle models)—The force generators in the muscle (actin–myosin crossbridges).

controlled-release method—A technique of muscle research: an active muscle is allowed to shorten and develop force at a new length.

convergent muscles—Muscles that are broad at the origin and converge at one attachment point at the insertion.

cosine rule—See *cosine tuning.*

cosine tuning—Dependence of the muscle activation level on the cosine of the angle between the direction of the exerted endpoint force and direction of the endpoint force at which muscle demonstrates the highest activity (the *preferred direction*).

Cosserat rods—Deformable strands that bend, change their length, and twist.

cost function—A real-valued function whose value is minimized or maximized in an optimization problem.

costameres—The transversal riblike structures, sites of the lateral force transmission in muscles.

couple—Two equal, opposite, and parallel forces acting concurrently at a distance *d* apart. The couple makes the body rotate.

couple vector—A free vector that represents a couple. The couple vector is normal to the plane of the couple.

creep—Increase in length under a constant load; a retarded deformation.

cross-product (of vectors $\mathbf{P}$ and $\mathbf{Q}$)—The vector of magnitude $PQ\sin\alpha$ (α is the angle formed by $\mathbf{P}$ and $\mathbf{Q}$), with direction perpendicular to the plane containing $\mathbf{P}$ and $\mathbf{Q}$, according to the right-hand rule.

cross-talk (of EMG signals)—Contamination of the EMG signals by electric activity of neighboring muscles.

crossbridges—Links between the myosin and actin filaments.

curvature—The change in direction per unit of arc.

curvature vector—See *normal vector.*

damping—Coefficient of proportionality between the load and velocity of deformation.

dashpot (in mechanical models)—An object that resists the deformation in proportion to the loading velocity.

deformable body—A body whose shape changes under external forces.

degree of freedom (DoF)—An independent coordinate used to specify a body, system, or position.

degree of freedom problem—See *motor redundancy.*

delta efficiency—The ratio of change in work to change in metabolic energy expenditure.

density—The amount of mass per unit volume.

descending limb (of a force–length curve)—A section of the force–length relation where the muscle generates smaller active forces at a larger length.

deviatoric stress tensor—The stress tensor that tends to deform the object without changing its volume.

direct forces—Forces exerted by a finger in response to a command sent to this finger; direct forces neglect enslaving effects.

distal—Away from the center of the human body or origin; the opposite of *proximal*.

distribution problem—Selecting specific muscle forces to generate necessary joint moments.

dominant muscle plane—See *muscle–tendon plane*.

Donders' law—The angle of torsion is constant for any given orientation of the line of sight.

dynamic force enhancement—A muscle force increase immediately after muscle stretch.

dynamic optimization—Optimization problems that consider time dependent variables.

dynamic points—Via points of a muscle path which coordinates with respect to the skeleton change during motion.

dynamics—The study of relations between forces and the resulting motion; a branch of kinetics.

eccentric muscle action—Exerting muscle force while lengthening. Same as *plyometric action*.

ectoskeleton—Fascia and tendons with aponeuroses of neighboring muscles.

effective insertion (of a muscle at a joint)—A via point of a muscle at the distal bone forming a joint that determines the muscle moment arm at this joint.

effective moment arm—A moment arm determined with functional methods; such a moment arm fits the equation Moment of Force = Force × Moment Arm.

effective origin (of a muscle at a joint)—A via point of a muscle at the proximal bone forming a joint that determines the muscle moment arm at this joint.

efficiency—See *mechanical efficiency*.

elastic body—A deformable body that returns to its original size and shape when external force is removed.

elastic hysteresis—See *hysteresis*.

elastic modulus—See *Young's modulus*.

elastic resistance (in joints)—The resistance that depends on joint angle and joint displacement.

elastic response (of human muscles to stretch with constant moderate speed)—The first phase of the response when the force rises quickly as a linear function of length change.

elasticity—Capacity of objects to resist external forces and resume their normal shape after force removal.

elastin—Elastic protein in connective tissue.

electromechanical delay—The time period between the first signs of electrical activity of a muscle and the first signs of its mechanical action.

ellipse orientation—The angle between the major axis of the ellipse and the *X* axis of the fixed reference system.

ellipse shape—The ratio of the major to minor axes of an ellipse.

ellipse size—The ellipse area.

endomysium—The connective tissue sheath surrounding a muscle fiber.

energy (in physics)—The capacity for doing work.

energy compensation during time—Transformation of kinetic energy into potential energy and its subsequent use.

energy transfer—Redistribution of kinetic energy among body links; also, redistribution of mechanical energy among muscles: when one muscle shortens extending another muscle the first muscle loses energy while the second absorbs energy. The absorbed energy can be stored as elastic potential energy, dissipated into heat, or both.

enslaving—See *finger enslaving*.

enthesis—Tendon–bone junction.

epimysium—The sheath of connective tissue surrounding a muscle.

epimuscular force transmission—Transmission of muscle force in lateral directions via the adjacent structures.

epitenon—Connective tissue sheath covering a tendon over its entire length.

equi-affine speed—A function of the movement that possesses some properties of a regular Euclidian speed and does not change its value under affine transformations.

equilibrium length (of muscles)—The maximal length at which elastic force is zero; also a length of an isolated passive muscle.

equitonometry—Recording the angular positions of multilink kinematic chains under the instruction of complete relaxation and in the absence of external forces.

Eulerian strain—The ratio of the change in length of a line segment to its final length after the deformation.

excitation-activation coupling—Propagation of excitation inside the fibers from the fiber membranes and linking the excitation with the contractile machinery resulting in force production.

exponential—Raised to the power of *e*, the base of natural logarithms, e.g., e^x.

expressed section (of a force–length curve)—The part of the force–length curve seen in vivo.

external work (in biomechanics)—Work done by the performer on the environment.

extrafusal fibers—Ordinary muscle fibers that make up the main part of the muscle.

extrinsic coordinates—Coordinates of the body parts with respect to the environment.

Eytelwein formula—A relation between the tensions at the proximal and distal ends of a wraparound tendon affected by the friction coefficient.

fanned muscles—See *convergent muscles.*

fascia—Fibrous connective tissue that binds adjacent muscles together.

fascial compartments—Groups of muscles separated by connective tissue.

fascicles—Bundles of muscle fibers.

feedback control—The central controller adjusts command signals based on the movement outcome.

feedforward control—The central controller generates command signals independently of the movement outcome.

Fenn's effect—The amount of energy (Mechanical Work + Heat) liberated during a muscle action increases with the work done.

festoons—Places of the sarcolemma bulging.

fiber-reinforced composites—Composites with the fibers embedded in another material.

filament—See *myofilament.*

finger enslaving—Involuntary force production or movement of a finger when another finger exerts force or moves.

finite element method—A method to analyze stress–strain relations in complex mechanical objects. In the method, the entire structure is divided into smaller, more manageable (finite) elements.

Fitts' law—Law stating that the smaller and farther away is a target, the longer it takes to reach it.

fixed points—Points of a muscle path whose coordinates with respect to the skeleton do not change during motion, e.g., the points of muscle origin and insertion.

flexibility—Maximal range of joint motion.

force deficit—Decrease in peak finger force during multifinger tasks in comparison with the peak force during a single-finger task.

force depression—Force decrease as a result of an immediately preceding muscle action, for instance, the muscle shortening.

force–length curve—Muscle force vs. muscle length relation.

force–velocity curve—Relation between the muscle force and its speed of shortening. Such curves are commonly obtained by interpolating the force and velocity data recorded in several trials while one of the parameters of the task, e.g., the load, was systematically changed.

forward dynamics problem—Computations of system's motion given known (or assumed) muscle activations and the system's initial conditions.

Frenet frame—Orthogonal system of coordinates moving along a continuous, differentiable curve in three dimensions.

functional (in mathematics)—A scalar-valued function of a function: an argument of a functional is a function and its output is a number.

functional agonists—See *task agonists.*

functional antagonists—See *task antagonists.*

functional physiological cross-sectional area—See *projected physiological cross-sectional area* (PCSA).

fusimotor muscles—Muscles that narrow down the end.

gamma motor neurons—Small motor neurons innervating intrafusal muscle fibers of muscle spindles.

geodesic—The shortest line between two points on a curved surface.

give—See *yield effect.*

Golgi tendon organs—Force sensitive receptors located at muscle–tendon junctions.

Golgi tendon organ reflex (GTO reflex)—Reflex accomplishing a force-feedback control: when the tendon force is excessively large, reflex causes the muscle to relax and activates the muscle anatomic antagonists.

gross efficiency—The ratio of mechanical work to metabolic energy expenditure.

Gruebler's formula—The formula used to calculate the number of degrees of freedom of a kinematic chain and a two bone–one -muscle system.

GTO reflex—See *Golgi tendon organ reflex.*

hamstrings—The three posterior thigh muscles: the semitendinosus, semimembranosus, and biceps femoris.

heat constant—Heat per unit of muscle shortening.

helical axis—See *screw axis.*

helix—A smooth curve in 3D for which the tangent line at any point makes a constant angle with an axis.

Hill's equation—Equation describing a parametric force–velocity curve, equation 3.22b.

Hill's force–velocity curve—See *force–velocity curve.*

Hill's model—A lumped-parameter muscle model that includes a contractile component (CC), series elastic component (SEC), and parallel elastic component (PEC).

Hoffmann reflex—See *H-reflex.*

homogeneous muscles—Muscles with fibers of similar length.

Hooke's law—Linear relation between the force and deformation.

H-reflex—A monosynaptic reflex induced by an electrical stimulation of a peripheral muscle nerve.

hydrostatic stress tensor—The stress components are equal in all directions.

hypertonia (muscle hypertonia)—Abnormal increase in the muscle tone.

hypotonia (muscle hypotonia)—Abnormally low muscle tone.

hysteresis—Difference between the loading and unloading curves in the stress–strain cycle.

iliotibial tract—The longitudinal fibrous reinforcement of the fascia lata.

in situ—In the original biological location but with partial isolation.

in vitro—Isolated from a living body and artificially maintained.

in vivo—Within a living body, as it is.

index of architecture—Ratio of mean fascicle length to the muscle belly length at muscle optimum length.

index of difficulty (in arm pointing movements)—The relation $ID = \log_2(2D/W)$, where ID is the index, D is the distance to the target, and W is the target width along the movement direction.

index of softness of landing (ISL)—A ratio: Negative Work of Joint Moments/ Decrease of Total Energy of the Body.

insertion (of a muscle)—The distal attachment of a muscle to a bone or other tissue.

insertional tendinopathy—Tissue inflammation, tear, or rupture at or in the vicinity of an enthesis.

interfinger connection matrix—A matrix of coefficients that relate central commands (mode commands) to individual fingers with actual finger forces.

internal work (in biomechanics)—Work done by the performer to change the mechanical energy of own body or body segments.

intrafusal fibers—The fibers that make up muscle spindles; the intrafusal fibers are innervated by gamma motor units.

intercompensated sources—Sources that compensate for each other. Two-joint muscles are undercompensated sources; they can absorb energy at one joint and generate energy at the second.

intermuscular septum—Tough connective tissue separating fascial compartments.

intramuscular fluid pressure—Interstitial fluid pressure.

intrinsic coordinates—Coordinates of body part with respect to each other.

invariant characteristics (of an arm movement)—Movement characteristics and muscle activity patterns that are similar in skilled performers, e.g., straight-line endpoint trajectory and bell-shaped velocity profile.

inverse kinematic problem—Finding a joint configuration from the known position of the endpoint of a kinematic chain.

inverse myotatic reflex—See *Golgi tendon organ reflex*.

inverse optimization—Finding an unknown objective function (or functions) from experimentally recorded optimal solutions.

isogony principle (of drawing movements)—Equal angular distances are covered in equal times.

isokinetic—Maintaining a constant joint angular velocity.

isometric—Maintaining a constant length.

jerk—The time rate of change of acceleration.

joint agonists—Muscles that generate the moment of force at the same direction as the resultant joint moment (cf. *anatomical agonists*).

joint antagonists—Muscles that generate the moment of force opposite to the resultant joint moment (cf. *anatomical antagonists*).

joint flexibility—Maximal range of motion at a joint or group of joints.

joint moment—See *joint torque.*

joint torque—Two moments of force about the joint rotation axis acting on the adjacent segments. The moments are equal in magnitude and opposite in direction.

Kelvin model—A three-element model that includes a dashpot and two springs, one in series with the dashpot and the second in parallel to the dashpot-spring assemblage.

kinematic redundancy—A number of kinematic degrees of freedom is larger than strictly necessary for motion.

Lagrangian strain—The ratio of the change in length of a line segment to its initial length before the deformation.

Lagrangian stress—Stress measured relative to the original cross-sectional area of a deformed object before the deformation.

Laplace's law—Pressure = Tension/Radius.

lateral force transmission—Force transmission in the lateral direction via shear force.

law of detorsion—Muscles encircling bones when at rest induce untwisting torsional effects when active.

law of hydrostatic pressure—See *Pascal's law.*

linear phase—The first phase of muscle force decay during muscle relaxation.

linear spring (in mechanical modeling)—An object whose length instantaneously changes in proportion to the load.

Lombard's paradox—At certain conditions a two-joint muscle can cause the extension of a joint where it is usually has a flexion action.

longitudinal force transmission—See *myotendinous force transmission.*

lumped parameters (in mechanical modeling)—Mechanical properties of the objects assigned to individual parts of the model, e.g., elasticity (to a linear spring), damping (to a dashpot).

macroscopic strain (of a tendon)—See *apparent strain.*

magnetic resonance elastography—A method to measure the local strain and stress, or stress and strain tensors, in muscular tissue. The method involves low-frequency mechanical vibration of the muscle tissue and visualization of the wave propagation with magnetic resonance technique.

magnetic resonance imaging (MRI)—Method used to visualize body tissues.

master force—See *direct forces.*

material strain (of a tendon)—The tendon fiber length change per unit of fiber length.

matrix (in composite materials)—Material that surrounds the reinforcement materials, e.g., fibers.

maximum maximorum—Highest among maximal values.

Maxwell model—A model that includes a spring and dashpot in series.

mechanical advantage—The ratio of mechanical output (e.g., muscle moment of force) to mechanical input (e.g., muscle force). In this particular example, a mechanical advantage is muscle moment arm.

mechanical efficiency—The ratio of work done and total energy expenditure.

mechanical efficiency of positive work—The amount of positive work done per unit of total available energy, where the total available energy is the sum of metabolic energy spent and the absolute value of muscle negative work.

mechanical energy expenditure—The total amount of mechanical energy expended for a given motion.

mechanobehavior (of tendons)—Mechanics of tendon function.

mechanobiology—Study of changes in response to systematic mechanical loading (adaptation and de-adaptation).

mechanotransduction—Conversion of mechanical stimuli into biochemical signals.

mesh (in the finite element method)—A web-like grid of finite elements.

metabolic—Related to metabolism, the biochemical processes occurring within a living organism.

microscopic strain—See *material strain.*

miometric action—See *concentric muscle action.*

mode (in production of forces by fingers)—A set of forces exerted by all four fingers due to a command sent to one of the fingers.

moment arm (of a force)—A coefficient relating a force and the moment produced by this force about a certain center or axis.

moment arm vector—The moment per unit of force expressed as a vector.

moment axis—The axis along the moment arm vector.

moment of force—A measure of a turning effect of a force.

monosynaptic—A neural circuit involving only one synapse.

motoneuron (also **α-motoneuron**)—A neural cell that directly innervates extrafusal muscle fibers.

motor abundance—See *motor redundancy.*

motor control—Control of movements by the nervous system.

motor redundancy—The availability of more elements that are strictly necessary to solve a motor task. Redundant motor systems are described mathematically by underdetermined sets of equations, i.e., by the equation sets where the number of unknowns exceed the number of equations.

motor unit—A collection of muscle fibers innervated by the same motoneuron.

MRI—See *magnetic resonance imaging.*

MTJ—See *muscle–tendon junction.*

MTU—See *muscle–tendon unit.*

muscle architecture—Internal anatomy of the muscles.

muscle belly—A muscle without the tendons.

muscle centroid—Locus of the geometric centers of the muscle transverse cross-sections.

muscle compartment—A group of muscle fibers with similar mechanical actions within a muscle. Each compartment is surrounded by a sheath of connective tissue called *perimysium.*

muscle coordination—Purposeful activation of many muscles to produce a given motor task.

muscle fiber—A muscle cell.

muscle Jacobian—Matrix of partial derivatives of muscle lengths with respect to joint angles. The elements of the matrix are the muscle moment arms.

muscle matrix—Intramuscular connective tissue that surrounds muscle fibers.

muscle morphometry—Anatomical characteristics of a muscle in the body, e.g., location of its origin and insertion, moment arms, and others.

muscle path—Line of muscle force action from one attachment site to another.

muscle redundancy—A number of muscles exceeds the number of kinematic degrees of freedom that the muscles serve.

muscle spindles—The receptors of muscle length and velocity.

muscle stability—Ability of muscles to resist stretches.

muscle–tendon complex—See *muscle–tendon unit.*

muscle–tendon junction—The connection between a muscle and its tendon.

muscle–tendon plane—The plane that contains all fixed and moving points of a muscle path.

muscle–tendon unit (MTU)—A muscle together with the attached tendon(s).

muscle thickness—Shortest distance between the aponeuroses (internal tendons) of origin and insertion measured in the plane of anatomical cross section.

muscle tone—Residual muscle tension (the term is not precisely defined).

muscle width—Linear size of a muscle measured orthogonally to the direction of muscle thickness measurement.

muscles with uniform architecture—See *homogeneous muscles.*

myofascial force transmission—See *lateral force transmission.*

myofibril—A muscle fibril, a slender striated thread within a muscle fiber, composed of *myofilaments.*

myosin—A contractile protein, the second is *actin*.

myotatic reflex—See *stretch reflex*.

myotendinous force transmission—Force transmission along the fiber axis to the tendon at the myotendinous junction.

myotendinous junction—Site where muscle fibers connect with tendon.

nebulin—The nonextensible protein which makes the actin filaments inelastic.

negative muscle work—Work done by a muscle to resist its elongation. The absolute value of this work equals to the value of work done on the muscle by an external force to extend the muscle.

net efficiency—The ratio of mechanical work to metabolic energy expenditure above the rest level.

neuromuscular transmission—Propagation of excitation from a motor axon to muscle fibers.

nonintercompensated sources of energy—The energy expended by one source is not compensated by the energy absorbed by another source. For instance, joint torques on joints served by single-joint muscles are not intercompensated.

nonlinear summation (of motor unit forces)—The total force from the units is smaller or larger than the sum of the forces from these motor units when they are activated separately.

nonparametric force–velocity relation—Relation between the maximal maximorum force in a muscular strength test (F_{mm}) and the maximal speed V_m against a constant resistance, e.g., body weight. For instance, the relation between the F_{mm} in a leg extension and the takeoff velocity in squat vertical jumps.

nonspanning fibers—Muscle fibers that end in the middle of the fascicle.

normal—Perpendicular.

normal stress—Stress in a plane orthogonal to force direction.

normal vector—Vector in the direction of normal acceleration, also called the *curvature vector.*

normalized maximum shortening velocity—Inverse of *time-scaling parameter.*

normalized moment vector—See *moment arm vector.*

normalized shortening velocity—Velocity of muscle shortening in units of muscle length per second.

number of actions (of a muscle)—Number of independent motions that can be imparted by the muscle on an unconstrained bone.

number of functions of an acton—The number of nonzero moments of force that the acton can produce with respect to anatomical joint axes.

objective function—See *cost function.*

one-third power law—The relation between the limb endpoint tangential velocity v and the curvature C of the movement path segment $v = k_1 \cdot r \cdot C^{1/3}$, where r is the radius of curvature C and k_1 is a coefficient.

optimal length (of a muscle)—The muscle length at which maximal isometric force is produced.

origin (of a muscle)—The proximal attachment of a muscle to a bone or other tissue.

orthogonal—Perpendicular.

osculating plane—The plane spanned by the tangent and normal vectors.

osmolarity—Number of moles of solute per unit of solvent mass.

osmosis—Movement of water or other solvent through a membrane from a more dilute to a more concentrated solution.

osmotic pressure—Pressure required to prevent osmosis.

pacemakers—Neural cells that do not need external neural input to become active.

pascal—Unit of stress or pressure, N/m^2.

Pascal's law—Pressure is conveyed undiminished to every part of the fluid and to the surfaces of its container.

parallel elastic component (in muscle models)—Muscle elements responsible for the passive resistance of an un-stimulated muscle to stretch.

parametric relations—Relations between experimental variables, such as for instance force and velocity, obtained by interpolating data recorded in several trials while one of the task parameters, e.g., the load, systematically varied among the trials.

parallelepipedon—Prism whose bases are parallelograms.

paratenon—Connective tissue between a tendon and its sheath.

passive flexibility—Maximal range of joint motion achieved due to external forces.

passive force—The force of resistance of a passive muscle to stretch.

PCSA—See *physiological cross-sectional area.*

PEC—See *parallel elastic component.*

pennate (or **pinnate**) **muscles**—Muscles with the fascicles attached to the tendon at an angle.

pennation angle—Angle between the direction of muscle fibers and either (a) the line of muscle force action (external portion of the tendon) or (b) the aponeurosis (internal tendon).

pennation plane—Plane formed by the tangent vector of the fascicle and the tangent vector of the aponeurosis at the point of contact; the pennation angle is measured in this plane.

percutaneous—Performed through the skin.

perimysium—The sheath of connective tissue surrounding muscle fascicles.

periosteum—Membrane covering the outer surface of bone.

physiological cross-sectional area (PCSA)—Cross-sectional area orthogonal to the long direction of muscle fibers.

plastic body—A deformable body that does not return to its original size and shape when external force is removed.

plateau region (of a sarcomere's force–length curve)—Length at which the sarcomere generates maximal forces.

plethysmograph—A device used to record volume changes in small objects.

plyometric action—See *eccentric muscle action.*

Poisson's ratio—The ratio of transverse strain (normal to the applied load) divided by the relative extension strain (in the direction of the applied load).

polyarticular muscles—Muscles that serve more than one joint.

pop (popping)—Rapid extension of the weaker sarcomeres in series when the stronger sarcomeres exert large forces and shorten.

positive work (of a force or a moment of force)—Work done over a linear or angular displacement in the direction of the force or moment.

power—The rate of doing work.

muscle power—Power of muscle forces.

power law—Dependence of the endpoint velocity on the trajectory curvature.

preferred direction—See *cosine tuning.*

pressure—The amount of force per unit of area in a direction perpendicular to the surface of an object.

principal stresses—Stresses along the eigenvectors of the stress tensor.

primary moment—An intended moment of force at a joint produced by a given muscle.

principle of least action—A general physical principle according to which an object moves in such a way as to minimize a certain physical variable (e.g., time, distance, work). In accordance with this principle, the control system chooses the movement trajectory and endpoint velocity such that mechanical work done on the endpoint and the arm is minimal.

principle of virtual work—The sum of works of all forces and moments done during virtual displacements in a system with workless constraints is zero.

projected physiological cross-sectional area—Projection of the PCSA of the muscle on the muscle's line of action.

proprioception—Perception of position of and load on body segments.

proximal—Closer to the center of the human body or to the origin, the opposite of *distal.*

pseudo-Hill curve (relation)—Relation between instant values of the force and velocity in a single trial resembling in appearance the Hill's force–velocity curve.

QLV model—See *quasi-linear viscoelasticity model.*

quasi-linear viscoelasticity model—A model of viscoelasticity based on the separation of the stress–strain relation into two parts: an elastic part and a history-dependent part.

quick-release method—A technique of muscle research: a muscle exerts force isometrically and then allowed to shorten against a smaller load.

rate coding—Muscle force control through motor unit firing rate changes.

rate of force development—Time derivative of the force–time function.

reciprocal activation—Decrease of activity of an anatomic antagonist when activity of an agonist increases.

recruitment—Motor unit activation.

recuperation (of energy)—See *energy compensation during time.*

reduced relaxation function—Relaxation stress changes over time.

residual force enhancement—Long-lasting increase in muscle force after stretch.

resilience—See *elasticity.*

rest length (of a muscle)—The natural muscle length in situ.

retinacula—Regional specializations of the deep fascia that envelopes muscles and tendons close to joints.

reversible muscle action—See *stretch–shortening cycle.*

rigid body—A body in which the distance between any two points within that body is constant.

rigidity—A complex of clinical symptoms involving increased velocity-independent resistance to muscle stretch.

sarcolemma—A membrane enclosing a muscle fiber.

sarcomere—A contractile unit of skeletal muscle.

sarcomere nonuniformity—Uneven distribution of muscle strain across sarcomeres.

savasana—A flexibility technique.

scalar—A real number.

screw axis—A line in space. At any given instant, the translation and rotation of a body occur along and around a screw axis.

screw-home mechanism (in the knee joint)—Knee extension combined with the external rotation of the tibia.

SEC—See *series elastic component.*

secondary moment—A moment of force at a joint produced by a given muscle in undesired direction. The moment should be negated by other muscles.

septum—A wall dividing a structure into smaller ones.

serial force transmission—See *myotendinous force transmission.*

series elastic component (SEC)—Muscle elements that behave as spring connected in series to the force generator in the muscle (the contractile component). The SEC is a functional element of muscle mechanics not a precisely defined anatomical structure.

series-fibered muscles—Muscles with nonspanning fibers.

S-gradient—The ratio $F_{0.5m}/T_{0.5m}$, where $F_{0.5m}$ is the half of the maximal force and $T_{0.5m}$, the time of attaining $F_{0.5m}$.

sharing problem—See *distribution problem.*

Sharpey's fibers—Strong collagenous fibers connecting periosteum to bone.

shear stress—Stress in a plane parallel to the force direction.

short-range stiffness—Muscle resistance to stretch during the initial phase of muscle elongation.

shortening factor—Actual fiber length divided by its length at rest.

shortening heat—The heat proportional to the shortening distance of the muscle and independent of load.

shortening-induced force depression—Reduction of muscle force after muscle shortening and redevelopment of isometric force compared to isometric force developed at the same length without preliminary shortening.

single-plane hypothesis (of muscle architecture)—Each muscle fiber lies in only one *osculating plane.*

SISO model—Single input–single output model.

size principle—With increasing muscle force, the motor units are recruited from small motoneurons at low forces to large motoneurons at high forces.

slack length (of a tendon)—The length at which a tendon starts resist extension.

slip—See *yield effect.*

sliding filament theory—Muscle force is generated by the crossbridges causing relative sliding of the actin and myosin filaments.

soft tissue skeleton—See *ectoskeleton.*

sonomicrometry—A method of measuring muscle fiber length based on sound propagation between piezoelectric crystals embedded in the muscle.

sources of mechanical energy—Forces and moments of force.

spanning fibers—Fibers that span from the tendon of origin to that of insertion.

spasmolytics—Centrally acting muscle relaxants.

spasticity—A complex of clinical symptoms involving increased velocity-dependent resistance to muscle stretch.

specific force—Muscle force per unit of the physiological cross-sectional area.

specific tension—See *specific force.*

spindle-shaped muscles—See *fusimotor muscles.*

spline—A smooth polynomial curve composed of pieces that pass through or near the connecting points.

spring constant—See *stiffness.*

standard linear model—See *Kelvin model.*

static optimization—Optimization that neglects time.

stiffness—The amount of force per unit of deformation. Stiffness is the inverse of compliance. Application of this term to active objects, such as muscles or human extremities, is questionable because their resistance to deformation, e.g., to stretch, is time dependent and is under neural control. See also *apparent stiffness*.

stiffness ellipse (at the endpoint)—The ellipse obtained by connecting the tips of the restoring forces in response to a unite deflection in all directions.

strain—A relative elongation $\Delta l/l$, where l is the initial length of the object and Δl is its elongation.

stress—The amount of force per unit of area.

stress relaxation—Decrease of stress over time under constant deformation.

stress tensor—A 3×3 matrix representing normal and shear stresses on an object.

stretch reflex—Muscle activation induced by muscle stretch.

stretch–shortening cycle—The sequence of muscle stretch and shortening.

subadditive summation—The force from two motor units is smaller than the sum of the individual unit forces.

superadditive summation—The force from two motor units is larger than the sum of the individual unit forces.

superficial fascia—The fascia located immediately under the skin.

tangent modulus of elasticity—See *Young's modulus*.

tangent vector—Vector in the direction of the derivative at the point of interest on a curve. If the curve were representing a trajectory of a point in time, the tangent vector would be a unit vector in the direction of the instantaneous velocity of the point.

task agonists—Muscles that assist the ongoing task.

task antagonists—Muscles that resist the ongoing task.

tendon—A band of fibrous tissue that connects a muscle to a bone.

tendon action (of muscles)—Muscles' behavior as nonextensible struts.

tetanic—Pertaining to *tetanus*.

tetanus—A sustained muscle contraction.

thixotropy—Property of a substance to decrease its viscosity when it is shaken or stirred.

three-element muscle model—See *Hill's model*.

time-scaling parameter—Ratio of the optimal muscle length to the maximal shortening velocity expressed in the units of muscle length per second.

titin—Protein responsible for the muscle elasticity.

TNB frame (from *tangent, normal, binormal*)—See *Frenet frame*.

toe region—A region of the stress–deformation curve of a tendon at low levels of the stress.

toe-region modulus—The average slope of the stress–deformation curve in the toe region.

total muscle force—The sum of the passive and active muscle forces.

trabecular bone—Spongy bone.

transverse strain—Strain normal to the applied load.

transient elastography—A method to measure the local strain and stress, or stress and strain tensors, in muscular tissue. The method involves low frequency mechanical vibration of the muscle tissue and visualization of the wave propagation with ultrasound techniques.

triangular muscles—See *convergent muscles.*

true stress—See *Cauchy's stress.*

twitch—A brief muscle contraction in response to a single stimulus.

two-thirds power law—The relation between angular velocity $\dot{\alpha}$ and curvature C of the movement path segment $\dot{\alpha} = k \cdot C^{2/3}$, where k is a constant.

Type I motor unit—Slow motor unit.

Type IIA motor unit—Fast but fatigue-resistant motor unit.

Type IIX motor unit—Fast motor unit with low resistance to fatigue.

two-joint muscles—Muscles that cross and serve two joints.

ultrasonography—A method in which ultrasound waves bounce off tissues and then are converted into an image.

ultrasound—Sound pressure with a frequency exceeding the upper limit of human hearing, approximately 20 kHz.

unipennate muscles—Pennate muscles with the fibers to one side of a tendon.

Valsalva maneuver—Forcible exhaling effort with the glottis closed.

vector—A quantity having magnitude and direction; a unidimensional array of numbers.

vector product—See *cross-product.*

velocity-dependent response (of human muscles to stretch with constant moderate speed)—The second phase of the response during which force continues to increase but with progressively slower rate until the force reaches its peak at the end of stretch.

virtual displacement—A hypothetical small displacement of a body or a system from an equilibrium position.

virtual work—The work done by a force over a virtual displacement.

viscoelastic hypothesis—See *viscosity hypothesis.*

viscoelasticity—Combination of viscous and elastic properties in materials.

viscosity—Internal friction between molecules.

viscosity hypothesis—Under standard stimulation, a muscle generates similar internal forces but these forces are not manifested externally because of the viscous resistance within the muscle. The hypothesis was abandoned after A.V. Hill (1938) showed that the heat production is sharply different in concentric and eccentric muscle actions.

Voigt model—A model that includes a spring and dashpot connected in parallel.

volumetric stress tensor—The stress tensor that tends to change the volume of the stressed body.

weakest-link principle—Force development rate is determined by the slowest muscle fibers.

weight (of an object)—The gravity force exerted on the object.

weight acceptance phase—The first third of the stance during the stair descent when the ankle plantar flexors absorb mechanical energy.

windlass mechanism (in the foot)—Action of the plantar fascia being wound around the heads of the metatarsals. The term comes from the analogy between the function of the plantar fascia and a cable being wound around the drum of a windlass.

Wolff's law—Bones adapt to the loads they are placed under, particularly in the trabecular bone architecture adapts to the direction of transmitted forces.

work—Scalar product of the vectors of force and displacement.

work efficiency—The ratio of mechanical work to metabolic energy expenditure above that in unloaded conditions (e.g., during pedaling without resistance).

wraparound tendons—The tendons that bend around a bone or a pulley.

yield effect—Sudden reduction of muscle force during fast stretches.

Young's modulus—The ratio of the uniaxial stress over the uniaxial strain in the range of the linear stress–strain relation.

Z-disc—See *Z-membrane.*

Z-membrane—A membrane separating two sarcomeres.

INDEX

Page numbers followed by an *f* or a *t* indicate a figure or table, respectively.

ABOUT THE AUTHORS

Vladimir M. Zatsiorsky, PhD, is a world-renowned expert in the biomechanics of human motion. He has been a professor in the department of kinesiology at Pennsylvania State University since 1991 and was a director of the university's biomechanics laboratory.

Before coming to North America in 1990, Dr. Zatsiorsky served for 18 years as professor and chair of the department of biomechanics at the Central Institute of Physical Culture in Moscow. He has received several awards for his achievements, including the Geoffrey Dyson Award from the International Society of Biomechanics in Sports (the society's highest honor), Jim Hay's Memorial Award from the American Society of Biomechanics, and the USSR's National Gold Medal for the Best Scientific Research in Sport in 1976 and 1982. For 26 years he served as consultant to the national Olympic teams of the USSR. He was also the director of the USSR's All-Union Research Institute of Physical Culture for three years.

He has authored and coauthored more than 400 scientific papers and 15 books that are published in English, Russian, German, Italian, Spanish, Portuguese, Chinese, Japanese, Polish, Bulgarian, Romanian, Czech, Hungarian, and Serbo-Croatian. Dr. Zatsiorsky has been conferred doctor honoris causa degrees by the Academy of Physical Education (Poland, 1999) and the Russian State University of Physical Culture and Sport (2003). Among his books are *Kinematics of Human Motion, Biomechanics in Sport: Performance Enhancement and Injury Prevention,* and *Science and Practice of Strength Training* (coauthor).

He and his wife, Rita, live in State College, Pennsylvania.

Boris I. Prilutsky, PhD, is an associate professor in the School of Applied Physiology and director of biomechanics and motor control laboratory at the Georgia Institute of Technology in Atlanta, Georgia. Before that position, he was a senior research scientist in Georgia Tech's Center for Human Movement Studies from 1998 to 2005.

His research interests include muscle biomechanics, neural control of movements, and motor learning. His research contributed to the development of methods for quantifying mechanical energy transfer by two-joint muscles between body segments during locomotion and to the understanding of muscle coordination during human motion. Prilutsky has published more than 50 peer-reviewed research articles and five book chapters, and he is the author of six patents. His research is supported by the National Institutes of Health (NIH) and National Science Foundation (NSF).

While living in the former Soviet Union, Prilutsky received a BS degree in physical education from the Central Institute of Physical Culture in Moscow and a BS degree in applied mathematics and mechanics from the Moscow Institute of Electronic Engineering. He received his PhD in biomechanics from the Latvian Research Institute of Traumatology and Orthopedics in Riga.

From 1978 to 1992, he worked as a research scientist and lecturer in the department of biomechanics for the Central Institute of Physical Culture in Moscow. He was also a postdoctoral fellow in the department of kinesiology at the University of Calgary, Alberta, Canada (1992-1995), and at the department of health and performance sciences at Georgia Tech (1995-1998).

Prilutsky is a member of the American Society of Biomechanics and a 1995 recipient of the organization's Young Scientist Award. He is also a member of the International Society of Biomechanics, Society for Neuroscience, and the Neural Control of Movement Society. He serves as a reviewer for over 30 professional research journals and for the NIH, NSF, South Carolina Space Grant Consortium, Consiglio Nazionale delle Ricerche (CNR), and the Austrian Science Fund.

Prilutsky resides in Duluth, Georgia, and enjoys mountain biking, reading, and traveling in his free time.